COMPUTATIONAL SOLID STATE PHYSICS

THE IBM RESEARCH SYMPOSIA SERIES

1971: Computational Methods in Band Theory
Edited by P. M. Marcus, J. F. Janak, and A. R. Williams

1972: Computational Solid State Physics
Edited by F. Herman, N. W. Dalton, and T. R. Koehler

Sparse Matrices and Their Applications
Edited by D. J. Rose and R. A. Willoughby

COMPUTATIONAL SOLID STATE PHYSICS

Proceedings of an International Symposium Held
October 6-8, 1971, in Wildbad, Germany

Edited by
Frank Herman, Norris W. Dalton, and Thomas R. Koehler

Large-Scale Scientific Computations Department
IBM Research Laboratory
San Jose, California

PLENUM PRESS • NEW YORK-LONDON • 1972

Library of Congress Catalog Card Number 70-188922

ISBN-13: 978-1-4684-1979-5 e-ISBN-13: 978-1-4684-1977-1
DOI: 10.1007/978-1-4684-1977-1

Softcover reprint of the hardcover 1st edition 1972

A Division of Plenum Publishing Corporation
227 West 17th Street, New York, N.Y. 10011

United Kingdom edition published by Plenum Press, London
A Division of Plenum Publishing Corporation
Davis House (4th Floor), 8 Scrubs Lane, Harlesden, NW10, 6SE, London, England

PREFACE

During the past 20 years, solid state physics has become one of the major branches of physics.[1-2] Today over one-third of all scientific articles published in physics deal with solid state topics.[3] During the last two decades, there has also been rapid growth of scientific computation in a wide variety of fields.[4-5] The combination of solid state physics and computation may be termed computational solid state physics. This emerging field is distinguished from theoretical solid state physics only to the extent that electronic computers rather than slide rules or backs of envelopes are used to solve numerical or logical problems, test scientific hypotheses, and discover the essential physical content of formal mathematical theories.

Papers in computational solid state physics are widely scattered in the literature. They can be found in the traditional physics journals and review series, such as The Physical Review and Solid State Physics; in more specialized publications, such as Journal of Computational Physics, Computer Physics Communications, and Methods in Computational Physics; and in the proceedings of a number of recent conferences and seminar courses.[6-9]

Plans for holding an International Symposium on Computational Solid State Physics in early October 1971 were formulated by Dr. Gerhard Hübner, IBM Germany, Sindelfingen, and one of us (F.H.) in Copenhagen in June 1970. The location chosen was Wildbad, Germany, a picturesque health resort in the Black Forest. The objective was to bring together a small number of physicists with a view to exploring future experimental and theoretical opportunities in solid state physics, with particular emphasis on problems that lend themselves to computational investigation. It was felt that such a symposium would also serve as a means for establishing computational solid state physics as a subject of interest in its own right.

In order to encourage in-depth discussions, the scope of the symposium was limited to a few key topics: electronic band structure, exchange and correlation effects, harmonic and anharmonic lattice dynamics, localized imperfections and dislocations, and solid state astrophysics. Experimentalists and theoreticians representing each of these topics were invited, in the expectation that this would lead to a broader exchange of information and ideas, an expectation borne out in practice. Inclusion of the section on solid state astrophysics was inspired by a paper on solid stars presented by Professor Malvin A. Ruderman of Columbia University at a Gordon Conference in the summer of 1970. A popular account of this subject has recently been published.[10]

Most of the papers presented at the symposium are included in this volume, which is being published as the second volume of the IBM RESEARCH SYMPOSIA SERIES. The organizers of the symposium are grateful to the contributors for their excellent lectures and their carefully prepared manuscripts. Informal presentations by Dr. John Hubbard (remarks on the exchange-correlation problem), Prof. Carl Moser (highlights report on the Menton Conference[9]), and Prof. Lars Hedin (highlights report on the Strathclyde Conference[11]) are not included. The organizers are grateful to these three speakers for their valuable off-the-record comments. We also thank Plenum Press for their cooperation and efforts to expedite rapid publication.

This volume is divided into six sections. Each has an introductory essay which sets the stage for the remaining papers in the section. All the papers contained in this volume are intended for graduate students as well as research workers in solid state physics and related disciplines. Most have extensive bibliographies which should facilitate further reading.

The first section is concerned with experimental studies of the electronic structure of solids, particularly spatial dispersion (Yu and Cardona), photoemission (Eastman and Williams); synchrotron radiation (Haensel and Sonntag); and positron annihilation (Berko and Weger). As noted in the introductory lecture (Madelung), current progress in experimental band structure determination represents the culmination of nearly twenty years of fruitful interplay between theory and experiment. With the development of ever more powerful theoretical and experimental techniques, and with the growing importance of numerical studies, this subject has now reached a high level of development. However, much remains to be done on the band structure of important but as yet poorly understood materials such as magnetic semiconductors, and on materials such as the transition metal oxides where one-electron methods appear to be inadequate.

All the papers in this section contain detailed illustrations of the value of strong interaction between theory and experiment. They all emphasize the importance of calculating not only the band structure, but also the physical observables that can be compared directly with experiment. There is still need for improved energy band interpolation schemes, particularly in situations involving extended energy ranges, and for refined zonal integration methods. The analysis of future experiments will undoubtedly stimulate more serious theoretical studies of excited electron transport and surface escape mechanisms, inner shell excitations, multiple electron excitations, and conduction electron momentum distributions.

The second section is devoted to theoretical studies of the electronic structure of solids. Some of the papers emphasize the progress that has been made using empirical methods (Treusch, Schneider and Stoll, Rudge and Ortenburger), while others report progress made with first-principles methods (Dalton, Bross and Stöhr, Christensen, Rössler). These two approaches appear to be complementary, and future progress will undoubtedly depend not only on continuing advances on both sides, but on cross fertilization.

There is also a short contribution on atomic arrangements in amorphous semiconductors (Henderson) and a short research note (Dalton and Schreiber) showing the enormous potential of computer graphics[12-13] in solid state applications.

The third section is concerned with exchange and correlation effects in solids. Following a broad and highly instructive introductory survey of the correlation problem (Löwdin), there are detailed discussions of the effect of electron correlation on charge and momentum distributions in solids based on a density matrix development (March and Stoddart); and of recent progress in the theory and application of local exchange-correlation potentials in realistic solids (Lundqvist and Lundqvist, Hedin). Also included are short contributions on improved statistical exchange approximations (Herman and Schwarz) and on cohesive energy calculations for metallic lithium (Sperber and Calais).

The work reported on local exchange-correlation potentials appears extremely promising, though its merits cannot be judged until such potentials are incorporated into band structure calculations which are free of numerical and physical noise (inadequate convergence, lack of self-consistency, neglect of relativistic effects, low standards of numerical accuracy, etc.). The same can be said for the work on charge and momentum distributions and on improved exchange approximations.

The fourth section on solid state astrophysics (Biermann, Börner, Baym) indicates how such familiar solid state ideas as electron gas degeneracy, equation of state, crustal fracture, plastic flow, superconductivity, and superfluidity are being applied to the extreme states of matter exemplified by neutron stars and white dwarfs, and how these ideas are being used to provide models for these objects and plausible explanations for observations on pulsars.

The fifth section on lattice dynamics includes an introductory survey (Koehler), a highlights report on the Rennes Conference

(Balkanski), discussions of experimental and theoretical studies of lattice vibrational spectra (Cowley, Bilz, Pick), and surveys of recent progress in the theory of anharmonic lattice dynamics (Koehler, Horner). Progress in these areas has been rapid, as all these papers indicate. It is hoped that with continuing progress in electronic structure and lattice dynamical studies these two areas will be brought closer together, and a unified treatment of electrons and lattice vibrations in solids will become possible. There has been some unification at the empirical level, but much remains to be done; at the first-principles level, unification is still remote.

The final section is concerned with selected topics in the defect solid state. This is a field that has already seen considerable computational effort, particularly in studies of radiation damage. Following an introductory survey (Lidiard), there are discussions of infrared and optical studies of localized vibrational modes (Balkanski), the properties of point defects in ionic solids (Lidiard and Norgett), computer simulation of point and line defects in metals (Bullough), and the properties of dislocations in anisotropic media (Lothe). These papers place considerable emphasis on the interplay of microscopic and macroscopic concepts, and on the value of computation in sharpening our understanding of these concepts in complex physical situations.

The symposium was sponsored by the IBM World Trade Corporation, with IBM Germany serving as the regional host. On behalf of all the participants, we wish to express our appreciation to the sponsors for their generous support, particularly Dr. Gerhard Hübner, representing IBM Germany, and Dr. Gary S. Kozak, representing IBM World Trade Headquarters, New York. We are grateful to Dr. Karl Ganzhorn for his Welcoming Remarks on behalf of IBM, and to the staff of IBM Germany for making all the necessary arrangements and for insuring the success of the symposium through their untiring efforts. Finally, the editors of this volume wish to express their appreciation to Mrs. Patricia B. Rodgers for her assistance in the preparation of the final manuscript, and to Miss Gerdi M. Herold for her help with the correspondence.

Frank Herman
IBM San Jose Research Laboratory
Symposium Chairman

Paul Schweitzer
IBM Germany, Sindelfingen
Symposium Manager

REFERENCES

1. "Solid-State Physics and Condensed Matter," National Research Council, National Academy of Sciences, Washington, D.C., 1966, Publication 1295A (reprinted from: Physics: Survey and Outlook Reports on the Subfields of Physics).

2. "Research in Solid-State Sciences," National Academy of Sciences, Washington, D.C., 1968, Publication 1600.

3. L. J. Anthony, H. East and M. J. Slater, Rep. Prog. Phys. 32, 709 (1969).

4. Computers and Their Role in the Physical Sciences, edited by S. Fernbach and A. H. Taub (Gordon and Breach, New York, 1970).

5. Computers and Computation, Readings from Scientific American. (W. H. Freeman and Co., San Francisco, 1971).

6. Computational Methods in Band Theory, Proceedings of a Conference held at IBM Thomas J. Watson Research Center, Yorktown Heights, New York, May 1970, edited by P. M. Marcus, J. F. Janak and A. R. Williams (Plenum Press, New York, 1971).

7. Computational Physics, Proceedings of the Second Annual Conference on Computational Physics held at Imperial College, London, September 7-9, 1970 (The Institute of Physics and The Physical Society, London, 1970).

8. Computing as a Language of Physics, Seminar Course held at the International Center for Theoretical Physics, Trieste, August 2-20, 1971 (International Atomic Energy Agency, Vienna, 1972), in press.

9. Perspectives for Computation of Electronic Structure in Ordered and Disordered Solids, International Colloquium held at Menton, France, September 20-24, 1971, to be published as special issue of Journal de Physique, Spring 1972.

10. M. A. Ruderman, "Solid Stars," in Scientific American, February 1971, p. 24.

11. Band Structure Spectroscopy of Metals and Alloys, Proceedings of an International Conference held at Strathclyde, Glasgow, Scotland, September 27-30, 1971, to be published.

12. I. E. Sutherland, "Computer Displays," in Scientific American, June 1970, p. 56.

13. Advanced Computer Graphics--Economics Techniques and Applications, edited by R. D. Parslow and R. E. Green (Plenum Press, London and New York, 1971).

CONTENTS

1. Experimental Studies of the Electronic Structure of Crystals

INTRODUCTORY REMARKS 3
Otfried Madelung

SPATIAL DISPERSION INDUCED BIREFRINGENCE IN CUBIC SEMICONDUCTORS 7
Peter Y. Yu and Manuel Cardona

PHOTOEMISSION STUDIES OF THE ELECTRONIC STRUCTURE OF SOLIDS 23
Dean E. Eastman and Arthur R. Williams

INVESTIGATION OF INNER SHELL EXCITATIONS IN SOLIDS BY SYNCHROTRON RADIATION 43
Ruprecht Haensel and Bernd Sonntag

POSITRON ANNIHILATION EXPERIMENTS AND THE BAND STRUCTURE OF V_3Si 59
Stephan Berko and Meir Weger

2. Theoretical Studies of the Electronic Structure of Crystals

INTRODUCTORY REMARKS 81
Norris W. Dalton

EMPIRICAL BAND CALCULATIONS 85
Joachim Treusch

ON THE THEORY OF METALLIC LITHIUM 99
Toni Schneider and Eric Stoll

GENERALIZATIONS OF THE RELATIVISTIC OPW METHOD INCLUDING OVERLAPPING AND NON-OVERLAPPING ATOMIC ORBITALS 113
Norris W. Dalton

MAPW-CALCULATIONS WITH SCREENED EXCHANGE APPLICATION TO COPPER 143
Helmut Bross and Herbert Stöhr

SOME REMARKS ON CURRENT RAPW CALCULATIONS FOR SILVER, MOLYBDENUM, AND VANADIUM 155
Niels Egede Christensen

RECENT KKR BAND CALCULATIONS 161
Ulrich Rössler

THE ATOMIC ARRANGEMENTS AND RADIAL DISTRIBUTION FUNCTIONS OF AMORPHOUS SILICON AND GERMANIUM 175
Douglas Henderson

OPTICAL PROPERTIES OF POLYTYPES OF GERMANIUM 179
William E. Rudge and Irene B. Ortenburger

CHARGE DENSITY IN THE GALLIUM ARSENIDE CRYSTAL 183
Norris W. Dalton and Donald E. Schreiber

3. Treatment of Exchange and Correlation Effects in Crystals

INTRODUCTORY REMARKS 191
Per-Olov Löwdin

EFFECTS OF ELECTRON CORRELATION WITH PARTICULAR REFERENCE TO CHARGE AND MOMENTUM DENSITIES IN CRYSTALS 205
Norman H. March and J. Colin Stoddart

LOCAL EXCHANGE-CORRELATION POTENTIALS 219
Bengt I. Lundqvist and Stig Lundqvist

EXCHANGE-CORRELATION POTENTIALS 233
Lars Hedin

AN IMPROVED STATISTICAL EXCHANGE APPROXIMATION 245
Frank Herman and Karlheinz Schwarz

COHESIVE ENERGY OF THE LITHIUM METAL BY THE AMO METHOD 253
Gunnar Sperber and Jean-Louis Calais

4. Solid State Astrophysics

INTRODUCTORY REMARKS 257
Ludwig Biermann

OBSERVATIONAL EVIDENCE FOR SOLID STATE PHENOMENA IN PULSARS 261
Gerhard Börner

NEUTRON STARS AND WHITE DWARFS 267
Gordon Baym

5. Lattice Dynamics

INTRODUCTORY REMARKS 289
Thomas R. Koehler

NOTE ON THE RENNES CONFERENCE ON PHONONS 293
Minko Balkanski

NEUTRON AND X-RAY INELASTIC SCATTERING BY LATTICE VIBRATIONS 299
Roger A. Cowley

PHONON DISPERSION RELATIONS 309
Heinz Bilz

MICROSCOPIC THEORY OF PHONONS IN SOLIDS 325
Robert M. Pick

COMPUTATIONAL ASPECTS OF ANHARMONIC LATTICE DYNAMICS . . 339
Thomas R. Koehler

ANHARMONIC LATTICE DYNAMICS: RENORMALIZED THEORY . . . 351
Heinz Horner

6. Localized Imperfections and Dislocations

INTRODUCTORY REMARKS 363
Alan B. Lidiard

LOCALIZED VIBRATIONAL MODES IN CRYSTALS 367
Minko Balkanski

POINT DEFECTS IN IONIC CRYSTALS 385
Alan B. Lidiard and Michael J. Norgett

COMPUTER SIMULATION OF POINT AND LINE DEFECTS IN IRON AND COPPER 413
Ronald Bullough

DISLOCATIONS IN ANISOTROPIC MEDIA 425
Jens Lothe

List of Contributors 441

Subject Index 445

I

Experimental Studies of the Electronic Structure of Crystals

INTRODUCTORY REMARKS

Otfried Madelung

Institut für Theoretische Physik (II)

der Universität Marburg/Lahn, W.Germany

The use of electronic computers has increased our knowledge of the electronic structure of solids essentially. The significance of large scale computation can be shown by the development of the semiconductor model within the last twenty years. When - after the discovery of the transistor - an intensive investigation of the properties of Ge and Si began the structure of the conduction and valence bands of these semiconductors was thought to be isotropic and parabolic. This means that the only difference in the properties of the electrons and holes to free charge carriers should be a constant scalar effective mass. The only possible improvement of this simple phenomenological model was a directional dependence of the effective mass. In the frame of this concept a transport theory could be developed in a rough approximation, but a theory of the optical spectra going beyond the explanation of the absorption edge was impossible.

Frank Herman's first calculation of the band structure of Ge and Si in 1954 and the improvements of this model in the following years changed the situation. The interpretation of optical spectra became possible. But the main progress connected with the possibility of the computation of detailed band structures was the possibility of a systematic exploration of the electronic structure of semiconductors, of the relationship between the properties of different semiconductors. Examples are the interrelation of the band structure of the cubic semiconductors and its mixed crystals or between the cubic and the hexagonal modifications of certain semiconductors. The overall picture of the electronic properties of semiconductors that we possess to-day is based on the successful use of large electronic computers.

A similar situation prevails in metal theory. Often it is possible to gain a rough picture of the shape of a Fermi-surface by the free electron approximation. But detailed results can only be obtained by extensive numerical calculations.

But the task of a physicist is not a mere comparison of numerical results with experimental data. We need concepts for our imagination, for our ability to perceive the connections between physical phenomena, to discriminate the different physical processes involved in an experiment. Numerical calculations are no substitute for theoretical concepts and models. Only the joint use of experimental means, theoretical concepts and numerical calculations lead to a deeper understanding of the solid state. In this sense the phrase "computational solid state physics" should not be misunderstood. There are different ways in solid state physics, but no way alone leads to the goal. This aspect is shown in the program of the to-day's sessions. The morning session is dedicated to the experimental point of view that is based mostly on intuitive concepts. The afternoon session is reserved to the theoretical aspect connected with large scale numerical calculations.

Let me close with a sketch of the present status of our effort of understanding the electronic structure of solids. The time of inventing new (isolated) methods with advantages for the computational technique is over. All methods for the determination of band structures can be connected by an unifying point of view. This is emphasized in a recent review article by Ziman in the last volume of Solid State Physics. The question of relationships between different band structures of semiconductors, between Fermi-surfaces of metals is still up-to-date. But the main interest of the theorists is shifted to questions like the exact form of the crystal potential entering the one-electron-Schrödinger-equation. Exchange, correlation, many body corrections are important topics.

The reason for the importance of the band model is that the function $E_n(k)$ contains all information about the quasi-particle "crystal electron", about its properties and its behavior in interactions with other elementary excitations in solids. But the necessity increases to calculate directly physical quantities like densities of states or dielectric constants. Here it is often difficult to get the information with the necessary degree of accuracy out of a given band structure. Or the concept of a band structure itself has only a limited value as in the case of amorphous semiconductors.

There are several experimental methods that give direct information about parameters of the band model. These topics are discussed in the following lectures. The dielectric constant is the key to the understanding of all optical effects. Its frequency dependence is well known. Its spatial dispersion, i.e. the q-dependence of the dielectric constant is a new and exciting problem. Measurements of the photoemission of electrons supply information about the density of states of the bands the photoelectrons are coming from. Beyond it it seems possible to measure the k-dependence of the photoelectrons, that is the band structure function itself. A similar situation exists for the excitation of electrons from core states to the conduction band by synchrotron radiation. Here the density of the final states can be determined. The shape of Fermi-surfaces has mostly been determined by measurements of magnetic phenomena like the de Haas-van Alphen effect. Here positron annihilation measurements seem to yield more detailed information. Thus the following lectures give us a survey of the most modern methods that are in close connection with the efforts of the theoretical physicists to understand the electronic structure of solids in all its aspects.

SPATIAL DISPERSION INDUCED BIREFRINGENCE IN CUBIC SEMICONDUCTORS [+]

P. Y. Yu[++] and Manuel Cardona[*]

Brown University, Providence, Rhode Island

ABSTRACT

A procedure to calculate the lowest order spatial dispersion effects in the dielectric constant of a germanium-type semiconductor is presented. The contribution of the lowest direct edge (E_o, $E_o + \Delta_o$), the $E_1 - E_1 + \Delta_1$ edge, and the Penn gap is given. These results are used to estimate the birefringence for light propagation along [110] . Good agreement with recent experimental results for Ge, GaAs, and Si is found.

INTRODUCTION

Numerical calculations of the optical constants of semiconductors within the framework of the one-electron band theory present, in principle, no major difficulties. After the pioneering work of Brust et al[1] calculations have been performed by a number of other authors.[2] Good agreement with the experimental results is usually obtained, as shown in Fig. 1 for InAs. However these calculations require considerable amounts of computer time and also have the disadvantage of not yielding a simple physical picture of where the main contributions to the oscillator strength arise from. Every time band structure parameters have to be changed in order to obtain a better fit to the experimental data, one has to make use of the computer. The same thing happens if one wants to calculate the effect of

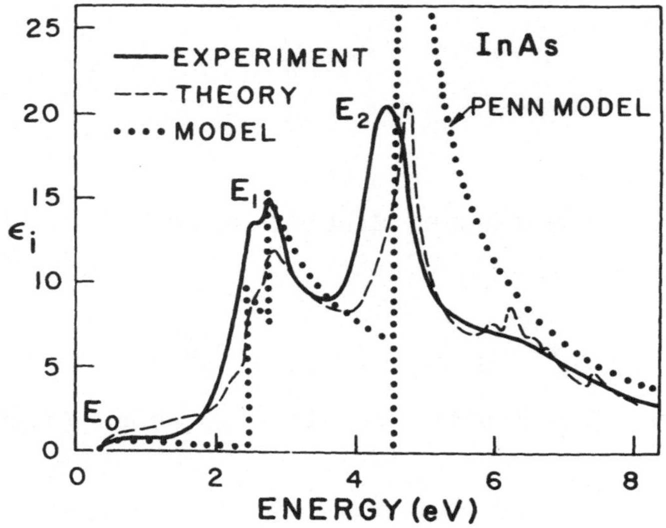

Fig. 1. Imaginary part of the dielectric constant of InAs. Solid curve from experiment H. R. Philipp and H. Ehrenreich, Phys. Rev. 129, 1550 (1963). Dotted curve, simplified model used for the calculations described in the text.

perturbations on the optical constants, a problem of considerable current interest.

An analysis of the ϵ_i spectra calculated numerically has revealed that their main features can be represented by direct transitions between a few pairs of parabolic bands osculating the true bands at critical points.[3] These pairs of bands correspond to the E_o critical points, split into E_o and $E_o + \Delta_o$ by spin-orbit interaction, the $E_1 - E_1 + \Delta_1$ critical points, also spin-orbit split, and the strong, nearly isotropic E_2 critical point. The ϵ_i spectrum obtained with this simplified model density of state in the manner described below is represented by the dotted line of Fig. 1: it contains the main features and approximately the correct strengths of both the experimental and the numerically calculated spectra. The discrepancies which exist are a reflection of the oversimplified nature of the model: they should not, however, affect the dispersion of ϵ_r below the lowest absorption edge and the effects of perturbations on ϵ_r in the same region. This effect can usually be measured rather accurately in the region of transparency. The model band structure described above has therefore been quite successful in explaining the magnitude and dispersion of the piezobirefringence, the pressure and temperature coefficient of ϵ_r, the Raman tensor[3] and the electro-optic tensor.[4,5] In this paper we shall describe its application to the calculation of spatial dispersion effects in ϵ_r.

The $E_o - E_o + \Delta_o$ edge is, in most semiconductors with zincblende structure, the lowest absorption edge. The only exceptions are materials with an indirect edge such as GaP and AlSb. The E_o edge is produced by an M_o critical point. The contribution of this edge to the real and imaginary parts of the dielectric constant are:

$$\Delta\epsilon_r = \begin{cases} C''_o \left[2 - (1+x)^{1/2} - (1-x)^{1/2} \right] & \text{for } x < 1 \\ C''_o \left[2 - (1+x)^{1/2} \right] & \text{for } x > 1 \end{cases}$$

$$\Delta\epsilon_i = \begin{cases} 0 & \text{for } x < 1 \\ C''_o \; (x-1)^{1/2} & \text{for } x > 1 \end{cases} \qquad (1)$$

In Eq. (1) $x = \omega/\omega_o$ is the reduced frequency, with ω the frequency of the radiation and ω_o that of the critical point. The dimensionless constant C''_o is related to the corresponding electron and hole masses and to ω_o. For a material with small spin-orbit splitting ($\Delta_o \ll \omega_o$) it can be shown that $C''_o \approx P^{-1}$, where P is the corresponding dipole matrix element. A similar expression holds for the $E_o + \Delta_o$ edge with C''_o replaced by $C''_o / 2$ and ω_o by $\omega_o + \Delta_o$.

The E_1, $E_1 + \Delta_1$ critical points occur along the {111} directions of k-space. The corresponding valence and conduction bands are nearly parallel over most of the Brillouin zone along [111]. Therefore the E_1, $E_1 + \Delta_1$ critical points can be regarded as two-dimensional minima. The contribution of E_1 to ϵ_r and ϵ_i are, for $\Delta_1 \ll \omega_1$ (atomic units are used):

$$\Delta\epsilon_r = - \frac{16\sqrt{3}\,\omega_1}{9\,a_o\,\omega^2} \ln \left| \frac{\omega_1^2 - \omega^2}{\omega_1^2} \right|$$

$$\Delta\epsilon_i = \frac{16\sqrt{3}\,\omega_1 \pi}{9\,a_o\,\omega^2} \; H\,(\omega - \omega_1) \qquad (2)$$

where ω_1 is the frequency of the E_1 critical point, H(x) the step function and a_o the lattice constant. A similar expression, with ω_1^o replaced by $\omega_1+\Delta_1$, applies to the $E_1+\Delta_1$ edges. Equations (1) and (2) have been used to calculate the dotted curve of Fig. 1 below E_2.

The E_2 peak corresponds approximately to a one-dimensional critical point and, because of the strong singularity associated with such critical points, contains most of the oscillator strength. Its strength can be estimated with the Penn model for the energy bands of a semiconductor: isotropic, free-electron-like bands with an isotropic gap at the edge of the nearly spherical Brillouin zone. The corresponding contribution to the dielectric constant is:[3]

$$\Delta\epsilon_r \begin{cases} = \dfrac{\pi}{2\sqrt{2}} \dfrac{\omega_p^2}{\omega_g^2} \left[(1-x)^{-1/2} + (1+x)^{-1/2} - 2\right] & \text{for } x<1 \\ = \dfrac{\pi}{2\sqrt{2}} \dfrac{\omega_p^2}{\omega_g^2} \left[(1+x)^{-1/2} - 2\right] & \text{for } x>1 \end{cases} \quad (3)$$

$$\Delta\epsilon_i = \begin{cases} 0 & \text{for } x<1 \\ \dfrac{\pi}{2\sqrt{2}} \dfrac{\omega_p^2}{\omega_g^2} (1-x)^{-1/2} & \text{for } x>1 \end{cases}$$

In Eq. (3) ω_p is the plasma frequency of the valence electrons and ω_g that of the E_2 critical point. As shown in Fig. 1, Eq. (3) overestimates ϵ_i for $x>1$ because it neglects the fact that some of the oscillator strength occurs for $x<1$.

SPATIAL DISPERSION: PHENOMENOLOGY

It is useful to expand the dielectric constant tensor $\epsilon(\vec{q},\omega)$ in power series of the components of the light wave vector $\vec{q}$:[6,7]

$$\epsilon_{ij}(\vec{q},\omega) = \epsilon_{ij}(\omega) + \epsilon_{ijk}(\omega)q_k + \epsilon_{ijkl}(\omega)\, q_k q_l \dots \quad (4)$$

where summation convention over repeated indices has been used. For cubic materials with reflection planes, such as the germanium and zincblende-type materials, $\epsilon_{ijk} = 0$. The fourth-rank tensor ϵ_{ijkl} has, like the elastic tensor, three independent components ϵ_{1111}, ϵ_{1122}, and ϵ_{1212}. ϵ_{1111} determines the effect of spatial dispersion on the "longitudinal" dielectric constant for $\vec{q} \parallel [001]$, while ϵ_{1122} determines the corresponding effect on the transverse dielectric constant. The linear combination $\xi_{001} = \epsilon_{1111} - \epsilon_{1122}$ determines the difference between longitudinal and transverse dielectric constant for $\vec{q} \parallel [001]$, to second order in the components of $\vec{q}$:

$$[\epsilon_{\parallel} - \epsilon_{\perp}]_{\vec{q} \parallel [001]} = \xi_{001}\, q^2 . \tag{5}$$

Similarly, for $\vec{q} \parallel [111]$ the difference between longitudinal and transverse dielectric constant is determined by $2\epsilon_{1212} = \xi_{111}$:

$$[\epsilon_{\parallel} - \epsilon_{\perp}]_{\vec{q} \parallel [111]} = 2\epsilon_{1212} q^2 = \xi_{111} q^2 . \tag{6}$$

It is unfortunately not possible to measure ξ_{001} and ξ_{111} directly. It is, however, possible, to measure directly the difference between ξ_{001} and ξ_{111} rather accurately in the region of transparency. This difference is related to the birefringence observed for propagation along the $[110]$ direction. This birefringence is given by:

$$\epsilon(\vec{E} \parallel [1\bar{1}0]) - \epsilon(\vec{E} \parallel [001]) = 1/2(\xi_{001} - \xi_{111}) q^2 \tag{7}$$

This birefringence vanishes whenever the Cauchy isotropy condition $\xi_{001} - \xi_{111} = 0$ is fulfilled.

MICROSCOPIC THEORY

The one-electron expression for ϵ with $\vec{q} \neq 0$ [8] differs from that for $\vec{q} = 0$ in two respects. For $\vec{q} = 0$ transitions do not occur between states of the same k but from a valence state of crystal momentum $\vec{k}$ to a conduction state of crystal momentum $\vec{k} + \vec{q}$. This has the effect of increasing the energy gap at a <u>minimum</u> critical point slightly changing the matrix elements for the transitions. The two effects just mentioned, changes in energy gaps and in matrix elements, are typical of practically all effects of perturbations (the finite $\vec{q}$ can be regarded as a perturbation) on the dielectric constant. The shifts in energy gaps give line shapes proportional to the derivative of the unperturbed dielectric constant. These line shapes have strong dispersion near the critical point and become dominant over the effect of changes in matrix elements. The matrix elements for finite $\vec{q}$ are:

$$\int u_v\,(\vec{k},\vec{r})\left[\,\vec{p}+\vec{k}+(1/2)\vec{q}\,\right] u_c(\vec{k}+\vec{q},\vec{r})\; dV, \tag{8}$$

where $u_v(\vec{k},\vec{r})$ and $u_c(\vec{k},\vec{r})$ are the periodic components of the valence and conduction Bloch functions, respectively.

E_o Gap

The transitions take place for a finite wave vector $\vec{q}$ from a valence state of crystal momentum $\vec{k}$ to a conduction state of crystal momentum $\vec{k}+\vec{q}$. The corresponding frequency is:

$$(\omega_c - \omega_v)_q = \omega_c + (1/2m_c + 1/2m_v)\, k^2 + 1/2m_c q^2 + 1/m_c \vec{k}.\vec{q} \tag{9}$$

Minimizing Eq. (9) with respect to $\vec{k}$ we find for the "effective" energy gap $(\omega_c - \omega_v)_{min}$ in the presence of spatial dispersion:

$$(\omega_c - \omega_v)_{min} = \omega_o + \delta\omega_o = \omega_o + q^2/2M, \tag{10}$$

where $M=m_c+m_v$. Equation (10) can be used to obtain the energy shift contribution to the second-order spatial dispersion tensor ϵ_{ijkl}. While the notation used assumes implicitely isotropic bands (a good approximation for the E_o and $E_o+\Delta_o$ gaps), it is also valid for parabolic, non-isotropic bands, provided one uses for m_c and m_v the effective masses along the direction of $\vec{q}$. The $E_o+\Delta_o$ gap is non-degenerate and isotropic. The corresponding valence band wave functions have the form:

$$(1/2;1/2) = \frac{1}{\sqrt{3}}\,(X + iY)\downarrow + \frac{1}{\sqrt{3}}\, Z\uparrow \quad , \tag{11}$$

and its time reversed. In Eq. (11) X, Y, Z are the orbital functions of x, y, and z symmetry, respectively and the arrows indicate spin functions. The standard angular momentum notation (1/2;1/2) has been used. When calculating matrix elements between Eq. (11) and the s-like conduction band we find that they are isotropic, independent of the direction of $\vec{p}$. As a result the $E_o + \Delta_o$ gap produces no energy shift contribution to $\xi_{001} = \epsilon_{1111} - \epsilon_{1122}$ and $\xi_{111} = 2\epsilon_{1212}$. There is, however, an energy shift contribution of the same amount for both ϵ_{1111} and ϵ_{1122}. From Eqs. (1) and (10) we find this contribution to be , for $\omega < \omega_o + \Delta_o$:

$$[\epsilon_{1111}]_{E_o+\Delta_o} = [\epsilon_{1122}]_{E_o+\Delta_o} = -(1/8)C_o'' \frac{3f(x)+g(x)}{M[\omega_o+\Delta_o]}, \tag{12}$$

where M is the sum of the corresponding electron and hole mass, $x = \omega/\omega_o+\Delta_o$, and the function $g(x)=2-(1+x)^{-1/2} -(1-x)^{-1/2}$.

The above conclusion $\xi_{111} = \xi_{001} = 0$, does not hold for the energy shift contribution of the E_o gap. This fact arises from the degeneracy of the corresponding valence bands. The wave functions of these bands at the "effective" gap are:

$$\begin{aligned} (3/2\,;\,3/2) &= 1/\sqrt{2}\;(X+iY)\uparrow \\ (3/2\,;\,1/2) &= 1/\sqrt{6}\;(X+iY)\downarrow - \sqrt{2/3}\;Z\uparrow , \end{aligned} \tag{13}$$

and their time reversed. In Eq. (13) the direction of quantization (z) has been chosen as that of $\vec{q}$. Modifying Eq. (1) so as to take into account Eq. (10) and the selection rules implied by Eq. (13) we find:

$$\begin{aligned} \epsilon_{1111} &= -C_o''\, g(x)/4M_{lh}^{(001)}\omega_o \\ \epsilon_{1122} &= -3C_o'' \left[1/6M_{lh}^{(001)} + 1/2M_{hh}^{(001)}\right] g(x)/8\omega_o \\ 2\epsilon_{1212} &= -3C_o'' \left[1/M_{hh}^{(111)} - 1/M_{lh}^{(111)}\right] g(x)/16\omega_o \end{aligned} \tag{14}$$

In Eq. (14) $M_{hh}=m_e+m_{hh}$ and $M_{lh}=m_e+m_{lh}$, where m_{hh} and m_{lh} are the masses of the heavy and light valence bands, respectively. The superscripts (111) and (001) express the fact that these masses depend on the direction of $\vec{k}$ because of the well known warping of the valence bands. In terms of the band parameters A, B, and C[9] the hole bands are:

$$(1/m_{hh})_{001} = -A + B \; ; \; (1/m_{lh})_{001} = -A - B \qquad (15)$$

$$(1/m_{hh})_{111} = -A - (B^2 + C^2/3)^{1/2} ; \; (1/m_{lh})_{111} = -A + (B^2 + C^2/3)^{1/2}$$

We note that the corresponding contribution to the spatial-dispersion-induced birefringence represented by Eq. (7) is a consequence of the valence band warping, and thus a direct measure of C.

It is also possible to estimate the less dispersive contribution of $\vec{q}$-induced changes in the matrix elements. To second order in $\vec{q}$ and including only the interaction between one valence and one conduction band, the conduction band, wave functions are:

$$\langle \vec{k}+\vec{q}, c| = \langle \vec{k}, c| \left(1 - \frac{|\langle \vec{k}, c| \vec{q}\cdot\vec{p}| \vec{k}, v\rangle|^2}{2\omega_o^2} \right) + \frac{\langle \vec{k}, c| \vec{q}\cdot\vec{p} | \vec{k}, v\rangle}{\omega_o} \langle \vec{k}, v| \qquad (16)$$

Two types of contributions of order q^2 to the matrix element of $\vec{p} + \vec{k} + (1/2)\, \vec{q}$ arise from Eq. (16), one due to the quadratic term in q^2 in Eq. (16) and the other to the linear term. The former contribution which can be shown to be dominant, gives:

$$\Delta\epsilon_{1111} = -(2C_o''/3) f(x)\, P^2/\omega_o^2$$

$$\Delta\epsilon_{1122} = -(C_o''/6)\, f(x)\, P^2/\omega_o^2 \qquad (17)$$

$$\Delta 2\epsilon_{1212} = -(C_o''/2)\, f(x)\, P^2/\omega_o^2 ,$$

where $P = \langle c| p_x |X\rangle$. P^2 is typically of the order of 10 eV. Equation (17) yields no contribution to Eq. (7). A contribution is nevertheless obtained if the mixing with the second lowest (Γ_{15}) conduction band is included. This contribution can be shown to be small compared with that of Eqs. (14).

We should point out, however, that the derivation of Eqs. (17) assumes truly parabolic bands and thus, effective masses which are independent of $\vec{q}$. Since this is obviously not exact, the effective shift in the bands produced by $\vec{q}$ results in an increase in the reduced mass which enters into Eq. (17). This effect, which can also be easily evaluated, results in a reduction of Eq. (17) by approximately a factor of two. We believe, however, that while the strongly dispersive terms of Eqs. (14) should be accurate near ω_o, Eqs. (17), only weakly dispersive, are subject to a large inaccuracy because of the crude assumptions of the calculation.

E_1 Gap

For $\vec{q} \parallel [001]$ all E_1 valleys are equivalent and the same spatial dispersion effect is obtained regardless of whether $\vec{E}$ is parallel or perpendicular to $\vec{q}$; Thus the corresponding contribution to $\xi_{001} = \epsilon_{1111} - \epsilon_{1122}$ vanishes. There is, however, an equal contribution to both ϵ_{1111} and ϵ_{1122}. Let us estimate the long-wavelength value of this contribution from the long-wavelength limit of Eq. (2):

$$\Delta\epsilon_r\ (\omega=o) = \frac{16\sqrt{3}}{9\, a_o\, \omega_1} \tag{18}$$

Some error will be incurred in using Eq. (18) for the evaluation of spatial dispersion effects because the matrix element of $\vec{p}$ which determines effective masses is not the same as that appropriate to the transition probability in the presence of spatial dispersion. Both matrix elements are assumed equal in Eq. (18). The shift in the ω_1 gap induced by spatial dispersion is, for $\vec{q} \parallel [001]$:

$$\delta\omega_1 = \frac{q^2}{9\, m_\perp} = \frac{2P^2 q^2}{9\,\omega_1}\ , \tag{19}$$

where $m_\perp$ is the transverse electron mass for the E_1 transitions. By replacing Eq. (19) into Eq. (18) we find:

$$\Delta\epsilon_{1111} = \Delta\epsilon_{1122} = -\frac{32P^2\sqrt{3}}{81\, a_o\, \omega_1^2}\ . \tag{20}$$

For $\vec{q} \parallel [111]$ the $[111]$ gap becomes inequivalent to the $[\bar{1}11]$, $[1\bar{1}1]$, $[11\bar{1}]$ set. As a result it is easy to see that the contribution to ϵ_{1212} does not vanish. The $[111]$ gap does not experience any gap shift since the longitudinal mass is practically infinite. The $\{\bar{1}11\}$ gaps, however, have a transverse component of $\vec{q}$ and thus the corresponding gap shift:

$$\delta\omega_1 = -\frac{4}{9} \frac{q^2}{m_{e\perp} + m_{h\perp}} . \tag{21}$$

In the spirit of Eq. (19) this gap shift gives rise to the following contribution to ϵ_{1212}:

$$\Delta\epsilon_{1212} = \frac{C_1''}{9\omega_1} \cdot \frac{q^2}{m_{e\perp}} , \tag{22}$$

with $C_1'' = \dfrac{4\sqrt{3}}{a_o\omega_1}$.

We should point out again that Eqs. (18) and (22) are only expected to give an order of magnitude estimate of the effects under consideration.

Penn Model

The long-wavelength dielectric constant[3] can be obtained within the Penn Model with Eq. (3):

$$\epsilon_r(\omega = o) = 1 + \frac{3\pi}{8\sqrt{2}} \frac{\omega_p^2}{\omega_g^2} . \tag{23}$$

We shall, as usual, neglect in Eq. (4) the factor $3\pi/8\sqrt{2}$, which is close to unity.

Because of the assumed isotropy of the Penn gap, the corresponding spatial dispersion effect should be isotropic, i.e., $\Delta(\epsilon_{1111} - \epsilon_{1122}) = 2\epsilon_{1212}$, and hence no contribution to the birefringence for $\vec{q} \parallel [110]$ should result. Let us now evaluate ϵ_{1111} and ϵ_{1122} within the Penn model. The gap shift is:

$$\delta\omega_g = \frac{1}{4m} q^2 \cos^2\theta , \tag{24}$$

where θ is the angle between $\vec{q}$ and the $\vec{k}$ vector of the electrons under consideration, and m is the electron mass at the edge of the Penn Brillouin zone:[3]

$$\frac{1}{m} \simeq 2 \left(\frac{3\pi}{4}\right)^{4/3} \frac{\omega_p^{8/3}}{\omega_g} \quad . \tag{25}$$

The contribution of the energy shift of Eq. (24) to the longitudinal and transverse dielectric constant can be easily evaluated. We thus find:

$$\begin{aligned} \epsilon_{1111} &= -3\left(\frac{3\pi}{4}\right)^{4/3} \langle \cos^4\theta \rangle \frac{\omega_p^{14/3}}{\omega_g^4} \\ \epsilon_{1122} &= -3\left(\frac{3\pi}{4}\right)^{4/3} \langle \cos^2\theta \ \sin^2\theta \rangle \frac{\omega_p^{14/3}}{\omega_g^4} \end{aligned} \tag{26}$$

The angular brackets denote averages over the "spherical" Brillouin zone. They are given by:

$$\begin{aligned} \langle \cos^4\theta \rangle &= \frac{1}{5} \\ \langle \cos^2\theta \ \sin^2\theta \rangle &= \frac{2}{15} \quad . \end{aligned} \tag{27}$$

We should point out that Eqs. (26) give qualitatively the correct $\vec{q}$ dependence of $\epsilon_r(0)$ for small $\vec{q}$. Numerical calculations based on the complete band structure give sometimes spurious increases in $\epsilon_r(0)$ with increasing q due to difficulties in the numerical limiting process for $q \rightarrow 0$[10].

EXPERIMENT AND RESULTS

The experiments described here consist of measuring the birefringence for propagation along the [110] direction of light of frequency below that of the fundamental absorption edge. The measurements can be performed by compensating the observed birefringence with a soleil-Babinet compensator.[6] It is also possible to measure the phase difference for the two principal directions of polarization ($\vec{E}$ || [001] and $\vec{E}$ || [$1\bar{1}0$]) by placing the sample between either parallel of crossed polarizers and rotating it about the direction of propagation ([110]).[11] For parallel polarizers the signal at the minimum of the transmited intensity is:

$$I_{\parallel} = Ae^{-\alpha d}\left[1 + \cos\Delta\right]^2 \tag{28}$$

while for crossed polarizers the maximum is:

$$I_{\perp} = Ae^{-\alpha d}\left[1 - \cos\Delta\right]^2 , \tag{29}$$

where α is the absorption coefficient and d the sample thickness. The phase difference Δ can be easily obtained from Eqs. (28) and (29). It is related to the birefringence of Eq. (7) through:

$$\Delta = \frac{2\pi d}{\lambda}\left[\epsilon(\vec{E} \parallel [1\bar{1}0]) - \epsilon(\vec{E} \parallel [001])\right] \tag{30}$$

where α is the wavelength of light in the material under consideration. Equation (30) shows that an enormous sensitivity to the birefringence can be achieved by choosing a very thick sample. This is only possible in the region of transparency and if birefringence due to internal strains is absent. This last requirement is easy to check by observing whether the measured birefringence has the symmetry expected for the effect of spatial dispersion (maximum for [110] propagation). Internal stresses should either be at random or along the direction of growth.[11] The effect of internal stresses has been shown to be small and separable from that of spatial dispersion for GaAs and Ge; these stresses are usually along the direction of growth[11] and the associated birefringence can be easily corrected for.

Figure 2 shows the spatial-dispersion-induced birefringence found by Pastrnak and Vedam for Si. It can be represented by the equation:

$$\epsilon[1\bar{1}0] - \epsilon[001] = D\, q^2 \tag{31}$$

We thus infer that the tensor coefficients ϵ_{ijkl} are non-dispersive in the region of transparency of Si. This is a consequence of the fact that Si has an indirect edge at 1.1eV, with the lowest direct edge at much higher energy ($\sim$3 eV). Indirect edges are too weak and smooth to produce significant spatial dispersion. Considerable dispersion in ϵ_{ijkl} is expected for GaAs when approaching the direct gap at 1.4eV. This is shown in Fig 3. The triangles are the experimental points corrected for the effect of internal stresses. The solid curve was calculated with Eq. (14) plus a non-dispersive term to include higher energy gaps:[11]

$$\epsilon(\vec{E} \parallel [1\bar{1}0]) - \epsilon(\vec{E} \parallel [001])$$

$$= \left\{ \frac{3}{32} \frac{C''_o}{\omega_o} \left[\frac{1}{M_{hh}^{(111)}} - \frac{1}{M_{lh}^{(111)}} - \frac{1}{M_{hh}^{(100)}} + \frac{1}{M_{lh}^{(100)}} \right] g(x) + D \right\} q^2$$

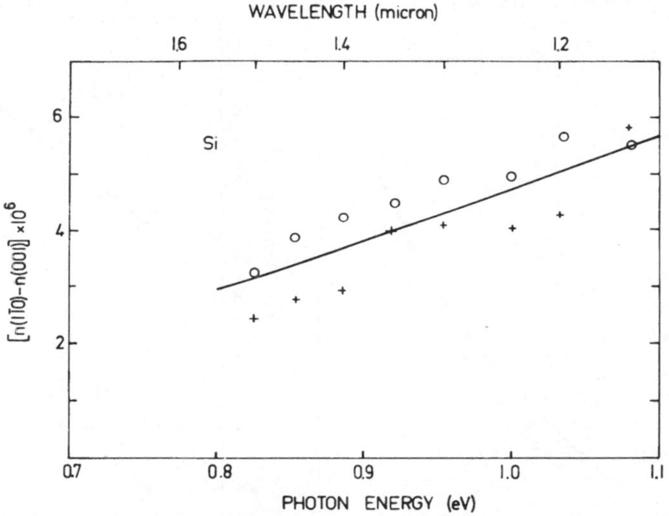

Fig. 2. Spatial-dispersion-induced birefringence found by Pastrnak and Vedam[6] for Si at room temperature. The solid curve corresponds to Eq. (31).

The value of the constant D required for the fit is +60 (Bohr radii)^2, which has the sign and the order of magnitude of the predictions of Eq. (22). The constant D for Silicon has also a similar value [Eq. (31), D=+34 (Bohr radii)^2].

Since germanium has at room temperature a direct E_o gap at 0.8eV, only slightly above the indirect gap, some dispersion of the ϵ_{ijkl} coefficients is expected even in the region of transparency. This can be seen in Fig. 4. In order to show the dispersion in the ϵ_{ijkl} coefficients, λ^2 [n(1$\bar{1}$0) - n(001)] is also plotted. The measurements were performed at liquid nitrogen temperature so as to bring the gap closer to the visible and to reduce the phonon-absorption component of the indirect edge. Below the indirect gap E_{IG} the effect predicted with Eq. (32) is seen. Above E_{IG} a decrease in the effect is seen. We cannot, at the present time, give an explanation of this phenomenon. It is interesting to note, however, that superimposed on this decrease, a peak at the energy of the phonon-absorption-aided indirect exciton is seen.

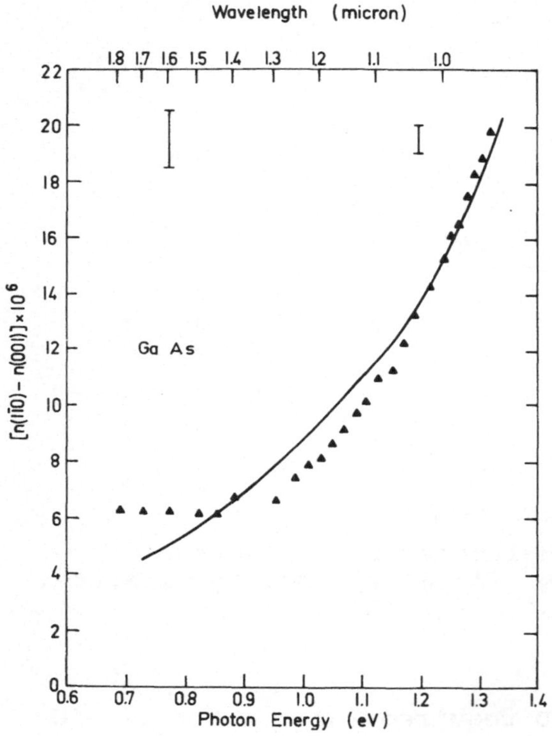

Fig. 3

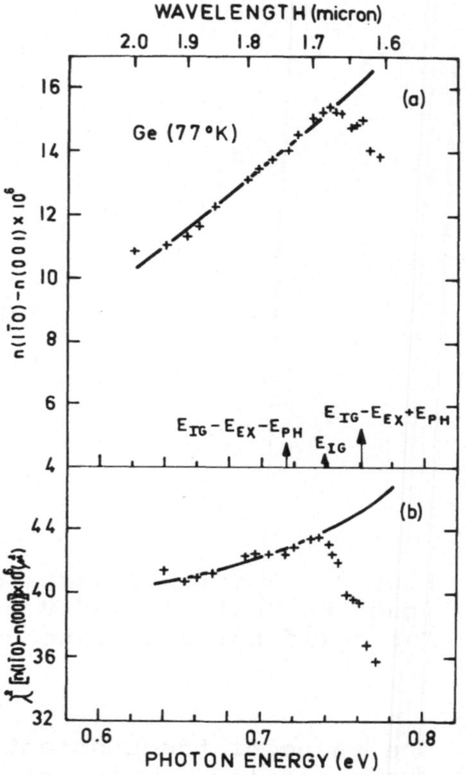

Fig. 4

Fig. 3. Spatial-dispersion-induced birefringence observed by Yu for GaAs at room temperature.[11] The solid curve was obtained with Eq. (32) using only D as an adjustable parameter.

Fig. 4. Spatial-dispersion-induced birefringence observed by Yu in Ge at 77°K.[12] The solid curve was obtained with Eq. (32) using only D as an adjustable parameter.

REFERENCES

+ Supported in part by the National Science Foundation and the Army Research Office, Durham.

++ Present Address: Physics Department, University of California, Berkeley.

* Present Address: Max Planck Institut für Festkörperforschung, Stuttgart.

1 D. Brust, J. C. Phillips, and F. Bassani, Phys. Rev. Letters 9, 94(1962).

2 See for a review M. Cardona, J. of Research of the Nat. Bureau of Standards 74A, 253(1970).

3 For a detailed discussion see M. Cardona, Proceedings of the 1971 Enrico Fermi Summer School, to be published.

4 M. I. Bell, Proceedings of the symposium on Electronic Density of States, Nat. Bureau of Standards, 1969, to be published.

5 J. A. Van Vechten, M. Cardona, D. E. Aspnes, and R. M. Martin, Proceedings of the 10th International Conference in the Physics of Semiconductors, Cambridge, Mass. 1970, p. 83.

6 T. Pastrnak and K. Vedam, Phys. Rev. 3, 2567(1971)

7 V. M. Agranovich and V. L. Ginzburg, "Spatial Dispersion in Crystal Optics and the Theory of the Exciton", (J. Wiley, New York, 1966).

8 H. Ehrenreich and M. H. Cohen, Phys. Rev. 115, 786(1959).

9 G. Dresselhaus, A. Kip, and C. Kittel, Phys. Rev. 98, 368(1955).

10 J. P. Walter and M. L. Cohen, Phys. Rev. 2, 1821 (1971).

11 P. Y. Yu and M. Cardona, Solid State Commum., in press.

12 P. Y. Yu, Ph.D. Thesis, Brown University, 1971.

PHOTOEMISSION STUDIES OF THE ELECTRONIC STRUCTURE OF SOLIDS*

D. E. Eastman and A. R. Williams

IBM Thomas J. Watson Research Center

Yorktown Heights, New York 10598

ABSTRACT

A discussion of ultraviolet photoemission spectroscopy is given. A central theme which is applied to photoemission is that both theory and experiment profit from mutual interaction. Examples which are discussed include experimental and theoretical studies of d-bands in Cu, studies of Au including liquid Au, studies of the Pd/H system, the magnetic semiconductor of EuS, and chemisorbed CO on Ni.

I. INTRODUCTION

Ultraviolet photoemission spectroscopy is a useful method for probing the electronic structure of solids over a wide range of energies away from the Fermi level E_F. This is a relatively new method, and rapid advances in both experimental techniques and theoretical understanding are presently occurring. The primary quantity which is experimentally measured is the kinetic energy distribution, i.e. energy distribution curve (EDC), of electrons which are excited by monochromatic radiation and ejected into vacuum. Thus the one-electron energies of occupied electron states which are excited and emitted are directly determined.

Photoemission spectroscopy has been applied to a variety of solids, liquids and gases, and more recently, to energy level measurements of adsorbed atoms on clean surfaces. In this survey paper, we describe applications of photoemission spectroscopy to

*Based in part on work sponsored by the Air Force Office of Scientific Research.

several types of systems. These include studies of the electronic structure of d-bands in Cu and Au including liquid Au, studies of the palladium-hydride system, the magnetic semiconductor EuS, and chemisorbed CO on Ni. These applications are intended as illustrative systems of physical interest for which theory and experiment either have yielded or hopefully will yield fruitful interactions.

II. RELATION OF THEORY AND EXPERIMENT

A central theme of this paper is that both theory and experiment profit from mutual interaction and that for many experiments the state of the theoretical art is sufficient to allow this. This interaction is beneficial in both directions; theory can assist in the interpretation of the data, and data can be used to judge theoretical models of comparable plausibility. The block diagram in Fig. 1 schematically illustrates that a microscopic model and an experimental measurement are separated/connected by a body of analysis which in all but the simplest cases involves substantial computation. This body of analysis is one of the subjects of this conference and we propose to describe its relationship to optical and photoemission measurements.

We begin by describing the analysis relating theory to measurement in greater detail. Fig. 1 shows that the first step in the analysis is the reduction of the many body problem to an effective one-electron problem. A good deal of progress has been made in this area particularly with respect to the ground state, due to the work of Kohn and Sham[1] and others, and more recently with respect to excited states by Hedin and Lundqvist[2] and others.

The next step is to determine the states of the effective one-electron problem--for ordered materials this is a band structure calculation. For many particle systems there are, of course, far too many states to determine directly. Unfortunately, however, experimental measureables are averages over all these states and we are thus forced to interpolate among the states we can compute directly in order to obtain a sufficient number to accurately estimate the average. The interpolation step is followed by the integration over the states required for the average. With the average computed, there is often a final phase, referred to in Fig. 1 as auxiliary theory, required to relate the quantity of fundamental significance to the experimental observable. We give a concrete example of this auxiliary theory for optical absorption and photoemission below, but before going on we wish to point out that this final step is just as

I.   MODEL: MANY-BODY PROBLEM
                        ↓
        EFFECTIVE   1-electron PROBLEM

EXAMPLE: Cu

CHODOROW POTENTIAL

II. ANALYSIS: (a) BAND CALC: (1) ENERGIES,

                    (2) MATRIX ELEMENTS

            (b)   INTERPOLATION TO SUFFICIENT

                  NUMBER OF E, $\vec{k}$ STATES

            (c) $\vec{k}$-SPACE INTEGRATION

                  (INCLUDING AUX. THEORY)

(a) KKR, 240 $\vec{k}$-PTS IN

    1/48 BZ, $E(\vec{k})$, $\vec{P}_{fi}(\vec{k})$

(b) $\vec{k} \cdot \vec{P}$ EXTRAPOLATION TO

     10^4 PTS IN 1/48 BZ

(c) GR METHOD, 10^4 CUBES,

     INCLUDE LIFETIMES,

     TRANSPORT, ESCAPE.

III. CALCULATED OBSERVABLES: (a) $\varepsilon_2(\omega)$.

        (b) ENERGY DISTRIBUTIONS $N(E,\omega)$

COMPARE

EXPERIMENTAL OBSERVABLES

$\varepsilon_2(\omega)$, $N(E,\omega)$

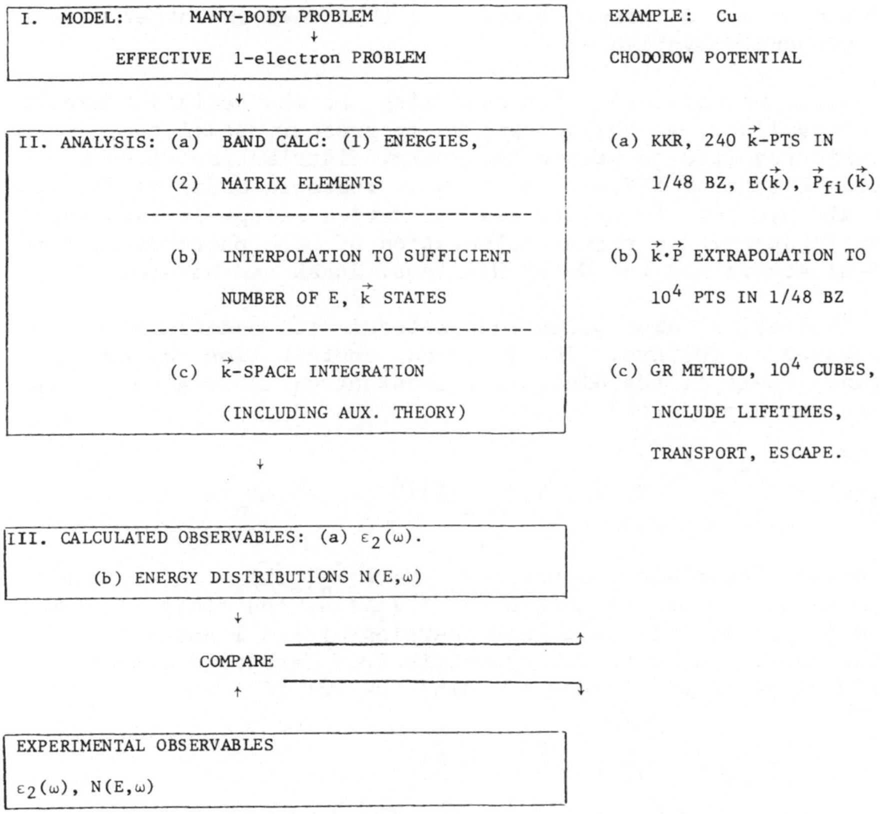

Fig. 1 Theory and Experiment--Optical absorption and
        Photoemission.

essential as the other links in the chain and is often neglected
in this highly specialized field.

The averaging over the states of the system required to
compute an observable means, of course, that there is seldom a
direct relation between an experimental measurement and individual
states. Some experimental quantities involve more of this averaging than others and herein lies the special significance of
photoemission. Photoemission involves less averaging and
therefore provides a more direct probe of the individual electronic

states over a wide range of energies than does any other experiment presently available.

With the motivation for analyzing the photoemission experiment established we shall describe in greater detail the analysis required to relate the energy distribution of photoemitted electrons $N(E,\omega)$ back to a microscopic model of the many particle system. To illustrate the various steps in this analysis we shall describe a recent calculation of both photoemission and optical absorption for Cu by Williams, Janak and Moruzzi.[3]

Theoretical absorption and photoemission spectra have been calculated as follows. The interband contribution to the imaginary part of the dielectric constant $\varepsilon_2(\omega)$ is given by the well-known expression[4]

$$\varepsilon_2(\omega) = \frac{e^2\hbar^2}{3\pi m^2\omega^2} \sum_{n'n} \int d^3k\,|\vec{P}_{n'n}(\vec{k})|^2\,\delta(E_{n'}(\vec{k})-E_n(\vec{k})-\hbar\omega) \tag{1}$$

for direct interband transitions. Here $\vec{P}_{n'n}(\vec{k}) = \langle n'\vec{k}|\vec{p}|n,\vec{k}\rangle$ is the momentum matrix element between initial and final bands n and n'; and $E_{n'}(\vec{k})$, $E_n(\vec{k})$ are band energies of final and initial states above and below E_F, respectively. Interband momentum matrix elements were computed using the relation[3]

$$\vec{P}_{n'n} = -\frac{i\hbar\langle n',\vec{k}|\nabla V|n,\vec{k}\rangle}{E_n - E_{n'}} \tag{2}$$

where $V(r)$ is the muffin-tin potential. This relation follows directly from the commutator $[H,\vec{p}] = i\hbar\nabla V$, where H is the one-electron Hamiltonian. Eq. (2) is computationally convenient since $\nabla V(r)$ is non-zero only inside the muffin tin where KKR-based wavefunctions $|n,\vec{k}\rangle$ are easily specified.

The energy distribution of primary (non-scattered) photoemitted electrons can be written as[5-7]

$$N(E_i,\omega) = C\sum_{n'n}\int d^3k\,|\vec{P}_{n'n}|^2\,\delta(E_{n'}(\vec{k})-E_n(\vec{k})-\hbar\omega)\;\delta(E_i-E_n(\vec{k}))\,T(E_{n'},\vec{k}) \tag{3}$$

where C is a normalization factor, E_i is the energy of the initial state which is optically excited, and $T(E_{n'},\vec{k})$ is an escape function which describes the probability for an excited electron in state $E_{n'}(\vec{k})$ to reach the surface and escape.[5-7] An important feature of the photoemission experiment is its ability to measure one-electron energies relative to the Fermi level. In reality, the kinetic energy E* of a photoemitted electron is measured. However, E_i is directly related to E* according to

$$E_i = E^* + \phi - h\nu \qquad (4)$$

where the work function ϕ and photon energy $h\nu$ are known quantities.

Comparing Eqs. (3) and (1), $N(E_i,\omega)$ contains an extra delta function $\delta(E_i-E_n(\vec{k}))$ which picks out all transitions of initial energy E_i. Thus $N'(E_i,\omega)$ gives a surface of information while $\varepsilon_2(\omega)$ gives a curve of information. The escape function $T(E_{n'},\vec{k})$ in Eq. (3) smooths out conduction band structure but plays a lesser role than the delta function at high energies $h\nu >> \phi$. Thus, at high energies $N(E_i,\omega)$ is approximately proportional to the fraction of the total absorption $\varepsilon_2(\omega)$ due to transitions from states at E_i, i.e. $\int N(E_i,\omega)dE_i \propto \varepsilon_2(\omega)$. As shown by Eq. (3), structure in the EDC, $N(E_i,\omega)$, at each $h\omega$ corresponds to structure in the joint interband density of states rather than the occupied band density of states.

We now turn to the calculation of $\varepsilon_2(\omega)$ and $N(\varepsilon,\omega)$ as outlined in Fig. 1. The reduction of the many body problem to an effective one electron problem was done in this case by Chodorow in 1939.[8] His result is embodied in a one electron potential which provided input to a KKR band calculation. This calculation resulted in the energies and interband momentum matrix elements at 240 points in 1/48 of the Brillouin zone. These quantities in turn provided us with 240 independent $\vec{k}\cdot\vec{p}$ eigenvalue equations, each one accurate in the vicinity of one of the original $\vec{k}$-points. These $\vec{k}\cdot\vec{p}$ matrices were then solved at 10,000 $\vec{k}$'s in the 1/48 of the zone to obtain energies and momentum matrix elements of sufficient density to perform state averages accurately. The latter were performed using the Gilat-Raubenheimer method for optical absorption and a generalization of the Gilat-Raubenheimer method to the photoemission problem due to Janak.[9] Finally, what we referred to earlier as auxiliary theory (in the case of photoemission, the description of the transport of the optically excited electrons through the crystal to its surface and escape through the surface into vacuum) were combined with the Gilat-Raubenheimer state average to obtain the final results.

III. PHOTOEMISSION AND OPTICAL ABSORPTION FOR COPPER

Figure 2 shows a portion of the photoemission surface

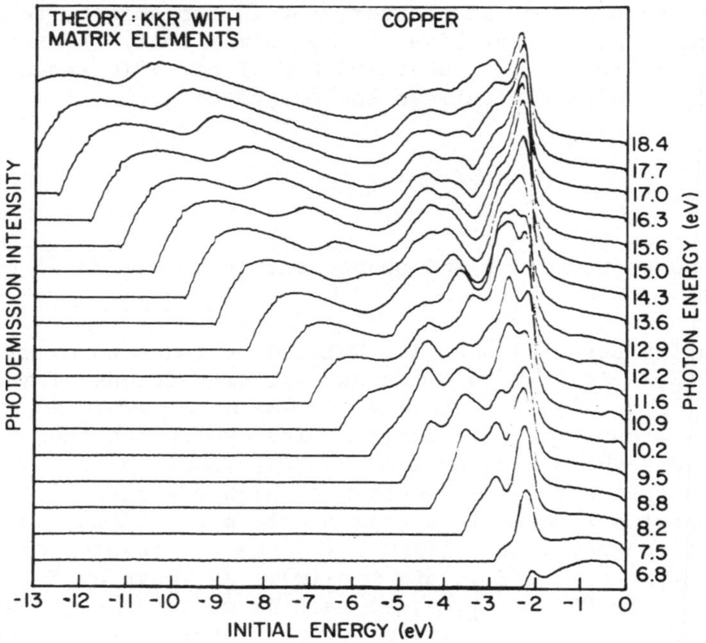

Fig. 2 Theoretical energy distributions $N(E_i, \omega)$ for Cu. All curves are plotted versus the initial energy E_i measured relative to the Fermi level E_F which is placed at zero.

$N(E_i,\omega)$ for copper which is at present experimentally accessible without cesiation of the crystal surface.[10] The surface results from the calculation described above and represents an expenditure of approximately 1 1/2 hours of computation on an IBM 360/91. In Fig. 2, emission from the d-bands of Cu extends over initial energies from 2 eV to 5 eV below E_F. The surface shown in Fig. 2 demonstrates two points quite clearly. First, the surface exhibits considerable variation with photon frequency, meaning that we have indeed a surface of information to work with. Second, the surface possesses several regions of very little ω-dependence demonstrating that such regions do not imply the existence of non-direct transitions (the surface results entirely from direct transitions).

Fig. 3 compares specific slices through the surface shown in Fig. 2 with the corresponding experimental data.[11] It is seen that there is general agreement over the entire 20 volt range of photon energies. Even at high photon energies where the agreement is worst, the main trends agree. For example, the emission intensity from d-band states near the middle of the d-bands ($\sim$3.5 eV below E_F) increases from a minimum at $h\nu$ = 16.8 eV to a maximum at $h\nu$ = 26.9 eV. For the most part the agreement is much better. At $h\nu$ = 16.8 eV, for example, not only do we find the large peak, the shoulder, the valley and the two smaller peaks all at the right initial energies, but the relative

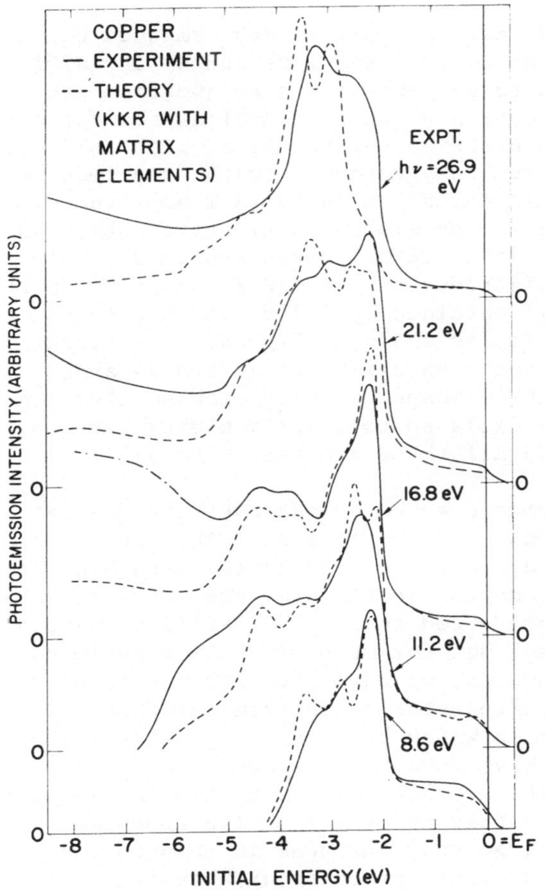

Fig. 3 Experimental and theoretical energy distributions for Cu for $8.6 \leq h\nu \leq 26.9$ eV.

amplitudes are in agreement as well. The portion of the calculation for initial energies less than -5 eV should not be considered since it is dominated by secondary electrons which have undergone electron-electron scattering. This portion of the emission spectra is not our principal interest here.

Optical absorption, while providing somewhat less experimental information than photoemission, is an important probe of the system particularly at low photon energies where photoemission is either prohibited or greatly complicated by the work function. In the case of copper it has provided the first experimental information away from E_F which we have used to improve the one-electron potential.

Mueller and Phillips[4] previously computed the interband absorption of copper in 1967 and were led by the difference between their results and experiment to question the entire effective one-electron picture of optical absorption. We have resolved the two principal difficulties with their calculation and obtain the excellent agreement with experiment shown in Fig. 4. The small deviation in overall amplitude is comparable to uncertainties in the experimental curve, which was determined from reflectance data via a Kramers-Kronig analysis. A measure of the uncertainty introduced by the K.K. analysis is the difference between the $\varepsilon_2(\omega)$ obtained by Pells and Shiga[12] and that obtained by Nilsson[13], which is shown in Fig. 4. The overall difference in amplitude in these experimental curves is about 30% while the differences in curve shape are very much smaller--in fact, our calculated $\varepsilon_2(\omega)$ falls approximately midway between the two experimental curves and all three are nearly parallel.

The improvements we have made over the Mueller and Phillips calculation demonstrate two facets of the present state of the art. First, we are able to compute the interband transition probabilities (momentum matrix elements) directly whereas Mueller and Phillips were forced to rely on Phillips' partial sum rules[14] and the latter account for much of their experimental--theoretical disagreement. Second, we have found that most of the remaining error in their calculation stems from the Chodorow effective one-electron problem on which both our original calculation and theirs were based. We have found that widening the L-gap only 0.3 eV (by increasing the d-phase shift slightly for energies above E_f) results in much better agreement of the computed $\varepsilon_2(\omega)$ with experiment. A large spurious peak is removed and the s-p bands near L are brought into agreement with experimental studies by

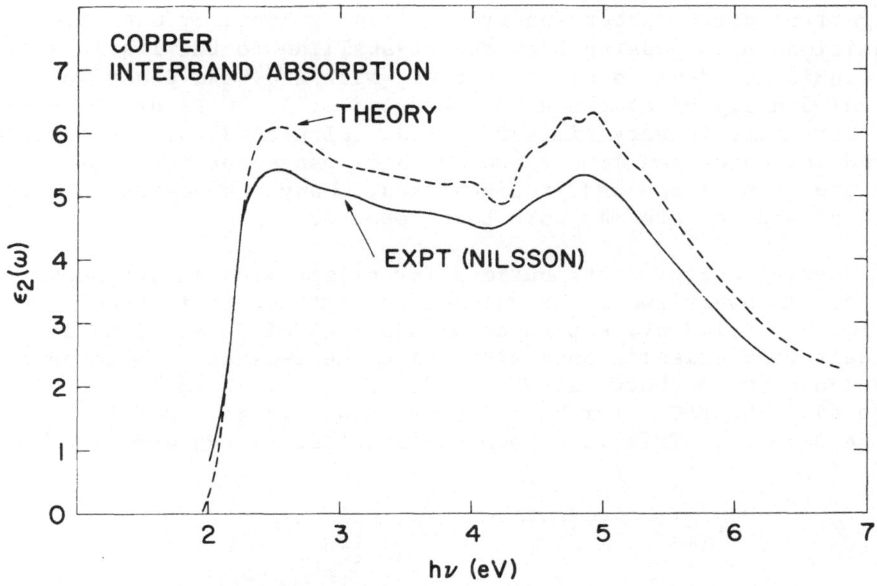

Fig. 4 Calculated and measured interband dielectric constant ϵ_2 (ω) for Cu.

Gerhardt[15] and by Lindau and Walden.[16] The success we have had in using the experimental $\epsilon_2(\omega)$ to refine our potential makes us hopeful that the even greater quantity of photoemission data can be used to advance this process even further.

IV. GOLD

In this section we describe experimental studies of liquid and solid gold.[17,18] Gold is one of the prototype metals for many band structure studies and has wide d-bands which permit structure to be easily observed. Photoemission measurements for crystalline Au in the range $7.7 \leq h\nu \leq 26.9$ eV give perhaps the most dramatic and clear-cut evidence for direct interband transitions. Measurements for liquid Au show that optical excitations change from direct interband transitions to local or nondirect transitions upon passing from the crystalline to the liquid state. For liquid Au, d-bands of width 6 eV, with 2 d-band peaks in the optical density of states at 3.2 and 5.8 eV below E_F are observed. This structure is very similar to that calculated for crystalline Au and indicates the importance of short range interactions in both the crystalline and liquid states. Many photoemission studies on clean and cesiated Au have been reported.

Several energy distributions for crystalline Au are shown in Fig. 5. An overview of the valence band structure is seen in the EDC for hν = 21.2 eV; s-p bands within 2 eV of E_F are identified by their weak emission intensity while the d-bands show intense emission between about -2 eV and -7.7 eV (i.e. about 5.7 eV wide d-bands). The EDC's for hν ≤11.6 eV do not probe the full range of the d-bands. This is a common limitation of UPS when limited

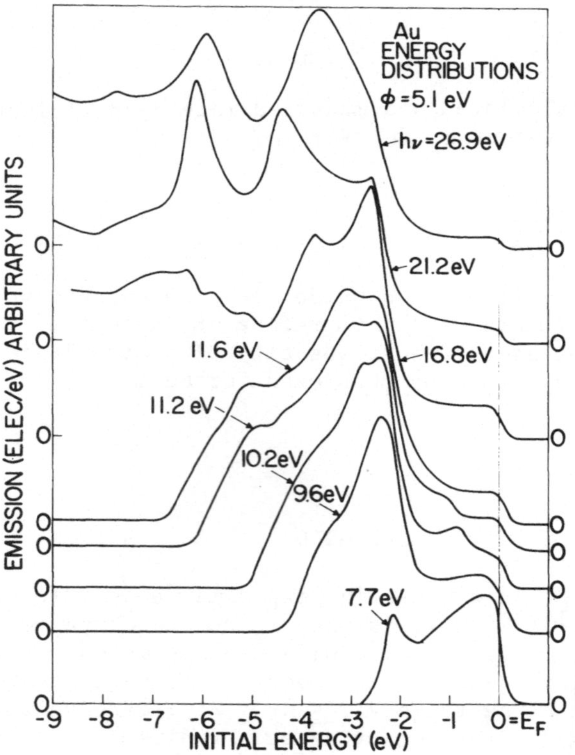

Fig. 5 Experimental energy distributions for crystalline Au.

to hν ≤11.6 eV by LiF windows.

The d-band emission spectra in Fig. 5 show much structure which is indicative of direct transitions (i.e. peak positions and amplitudes change markedly with photon energy). For example, the strongest d-band peaks are at -2.6 and -3.7 eV for hν = 16.8 eV, at -2.6 eV, -4.4 eV and -6.0 eV for hν = 21.2 eV and at -3.6 eV and -5.8 eV for hν = 26.9 eV. For Au, calculated energy distributions for direct transitions have been reported for photon energies hν ≤11.6 eV and in general show agreement with experiment.

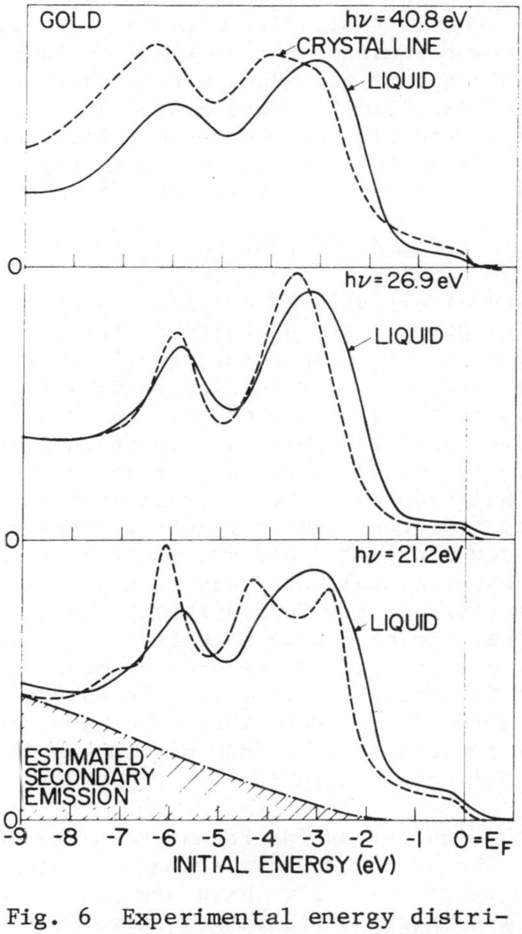

Fig. 6 Experimental energy distributions for liquid and crystalline Au.

Photoemission energy distributions for liquid Au in Fig. 6 do not show the above-mentioned peak shifts, etc. For liquid Au, structure in the EDC's does not change with hν as seen for crystalline Au in Fig. 5. That is, energy distributions for liquid Au can be described by an optical density of states (i.e. non-direct transitions). This comparison of solid and liquid Au indicates that momentum conservation plays an important role in the crystalline state and becomes relatively unimportant in the liquid state, i.e. when long range order is destroyed. Thus the photoemission spectra for liquid Au give a fairly direct picture of the d-band density of states. In Fig. 6, d-bands are seen to extend from about -1.6 eV to -7.6 eV below E_F (about 6 eV wide) and show peaks at about -3.2 eV and -5.7 eV and a dip near -5 eV. This structure shows an overall similarity both in width and shape to the density of states calculated for crystalline Au by Connolly. This similarity would appear to indicate that short range interactions are the most important in determining the overall d-band structure. The existence of such short-range interactions common to liquid and crystalline Au is consistent with reported structural data, i.e. both have close-packed, 12-coordination structures.

V. PALLADIUM HYDRIDE

Pd/H is an extensively studied metal/gas system, and much has been reported concerning its hydrogen-diffusion characteristics, pressure-composition-temperature characteristics, and use as a hydrogenation catalyst. Several interesting theoretical models for the electronic structure of metal hydrides and chemisorbed hydrogen have recently appeared. These include the one-electron augmented-plane-wave (APW) energy-band calculations for metal hydrides by Switendick,[19] the Anderson resonant bound-state model as applied to hydrogen chemisorption by Newns,[20] and the induced-covalent-band model of Gomer and Schrieffer.[21] However, little definitive experimental data have been published concerning the electronic structure of the Pd/H system. The simple "proton model" of Pd/H appears to be widely accepted. In this model the hydrogen electrons simply fill up the d-band holes (and possibly s states) of Pd at the Fermi level E_F and the nature of electron screening of the proton is not described. We have reported photoemission measurements for the Pd/H system and find that the name "proton model" is misleading.[22]

Energy-distribution curves for Pd and Pd/H are shown in Fig. 7 for hν = 21.2 eV. The Pd curve shows strong emission from d-band states within ∿4.4 eV of E_F. As we have previously described for the noble metals, the energy-distribution curve for Pd reflects structure in the energy distribution of the joint density of states and is not equal to the band density of states. Structure is also seen in the secondary electron emission of Pd at about 7.5 eV below

E_F which corresponds to inelastic scattering via plasmon excitation. The Pd/H curve, which is for a two-phase mixture of Pd and β-phase PdH, shows new energy levels (shaded region) centered at 5.4 eV below E_F as well as the Pd d-bands. Emission from these states near -5.4 eV is also seen for hν = 16.8 and 26.9 eV, and is very weak for hν = 40.8 eV. We associate these levels with hydrogen-palladium bonding states in β-phase PdH.

Switendick has analyzed our photoemission results for non-stoichiometric β-phase PdH via APW band calculations for Pd and hypothetical PdH, Pd_4H_3 and Pd_4H.[22] These results show a new set of states for PdH about 5-6 eV below E_F, i.e., centered about 1.5 eV below the bottom of the Pd d-bands. The appearance of these

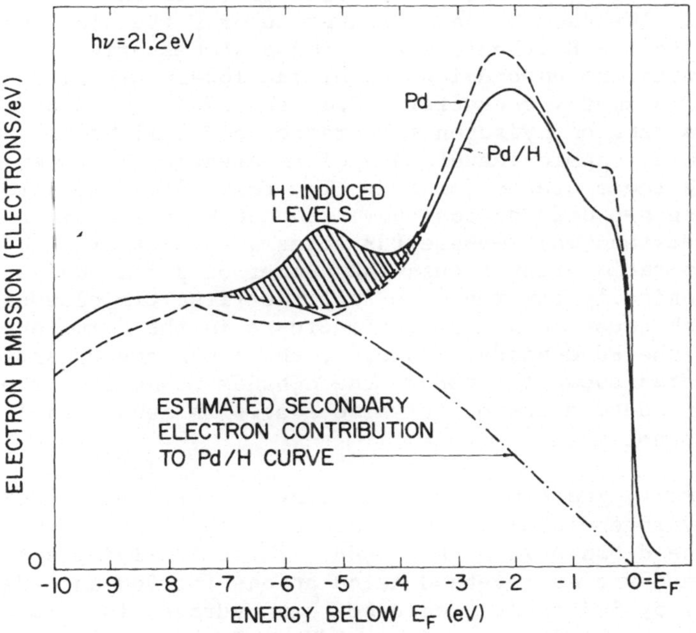

Fig. 7 Experimental energy distribution curves for Pd (dashed line) and a two-phase mixture of β-PdH and Pd (solid curve). The estimated secondary emission contribution is shown by the dashed-dotted line in the PdH curve.

new states agrees with our measurements for β-phase PdH in Fig. 7. These states are formed by the strong interaction of the hydrogen 1s electron states and the low-lying Pd states. Consequently, a modified band with strong hydrogen-palladium bonding character is developed below the Pd d-bands. Charge-density calculations show that these states are hybridized bonding states with greater than 0.6 electron of 1s character inside the hydrogen APW sphere (radius = 0.704 Å). This charge is larger than the 0.5 electron inside the same size sphere for the hydrogen atom. Thus the proton is well screened in PdH and has about the same screening charge as it does in neutral hydrogen.

Since these "new" low-lying states in PdH are modified states which already existed (and were occupied) in Pd metal, the additional electrons associated with the hydrogens must fill previously unoccupied states. Switendick's calculations show that these states are of three classes. The first class is the ∿0.36 hole in the d-bands. The second class is a group of states located around the L point in the Brillouin zone. These states originally were associated with the unoccupied top of the lowest s-p band in Pd and have L_2' symmetry character. Upon the addition of hydrogen, these states take on hydrogen s character and fall below the top of the d-bands. Approximately 0.5 of an electron is taken up upon filling these states for PdH. The third class of states is the remaining s-p band states above the top of the d-bands. We can now understand why β-phase PdH forms with an average hydrogen-to-palladium ratio greater than the number of d-band holes (0.36). It is energetically favorable to fill the first two classes of states, which cause only a slight increase in the Fermi energy relative to the Pd d-bands. However, the low state density of the s-p band states above the top of the d-bands means that the energy increase per added electron to these states is substantial, and thus they do not fill.

In summary, photoemission measurements for β-phase PdH have shown new hydrogen-induced energy states about 1 eV below the bottom of the d-bands in β-phase PdH. These low-lying states have been explained via energy-band calculations for PdH and Pd_4H. The calculations by Switendick indicate that hydrogen is more electronegative than palladium in β-phase PdH, and also explain the limiting composition in terms of filling s-p hydrogen-hydrogen bonding states which are pulled down below the top of the d-bands as well as filling the Pd d-band holes.

VI. EuS

The Eu chalcogenides EuO, EuS, EuSe and EuTe are magnetic semiconductors which possess many novel features that are due to localized 4f electrons. These 4f electrons strongly influence the transport and optical properties of these compounds. For example, the resistivity and photoconductivity of doped samples change by many orders-of-magnitude with modest changes in magnetic field or temperature.[23] Very large magneto-optical effects are observed for $h\nu \lesssim 3$ eV. For example, a Faraday rotation of about 520,000 deg/cm at λ = 7000 A is observed for EuO.[23]

We have made photoemission measurements for EuS up to 41 eV which show that the $4f^7$ level lies in the gap above the top of the valence "p" bands.[24] In addition to determining the position of the $4f^7$ level and position and width of the valence "p" bands, these measurements show striking energy-dependent emission strengths for the f and p states. The emission strength of the

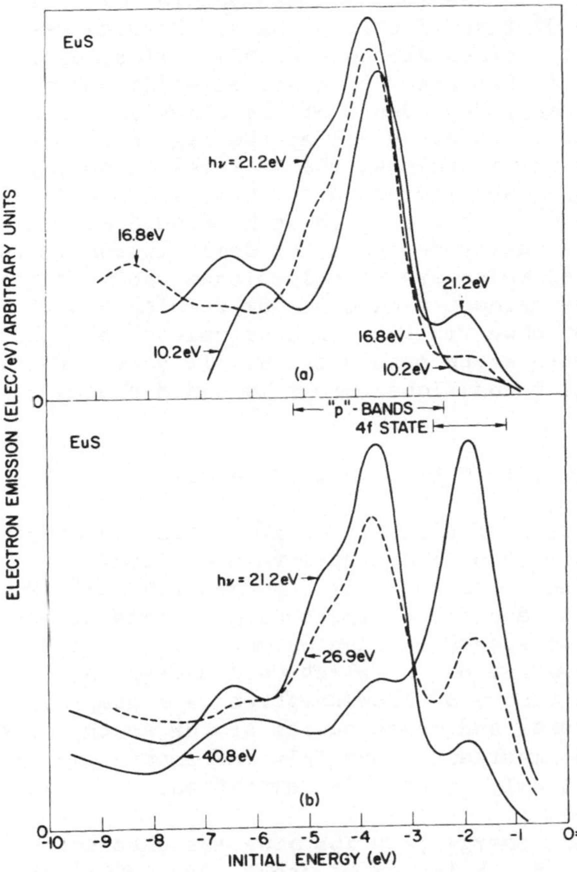

Fig. 8 Energy distribution curves for EuS.

$4f^7$ level in EuS increases about 100-fold relative to the 3 eV wide valence "p" band which lies just below the $4f^7$ level. Such energy-dependent relative emission intensities can be used to determine the character of occupied valence band energy levels in solids. Spin-polarized photoemission measurements have been reported for the Eu-chalcogenides which also give interesting information.[25]

Photoemission energy distribution curves for EuS are shown in Fig. 8. The $4f^7$-multiplet binding energy is centered at about 1.8 eV below E_F and the filled valence band extends from ~2.3 to 5.2 eV below E_F. This conclusion is confirmed by the rapid increase in 4f-level emission relative to the "p"-band emission with increasing photon energy. For $h\nu = 10.2$ eV in Fig. 8(a), the emission strength (or absorption cross section σ) of the 4f level is only $\sigma_f/\sigma_p \simeq 0.03$ that of the "p" band. Here we assume for each EDC that absorption cross sections of the various electron states are proportional to the areas of their emission peaks. That is, we neglect the energy dependence of the electron-transport and-escape processes, which are small at the high energies involved. As the photon energy increases, the 4f-level intensity increases very rapidly; σ_f/σ_p has increased to $\simeq 0.14$ at $\hbar\omega = 21.2$ eV, to $\simeq 0.32$ at $\hbar\omega = 26.9$ eV, and to $\simeq 3.0$ at $\hbar\omega = 40.8$ eV. As we have discussed,[24] this energy-dependent emission intensity is characteristic of f-states, which are more localized and have higher angular momentum than the valence p-bands. Our results clarify why 4f electrons are not observed in optical studies[26] and photoemission studies of the rare earth metals for $h\nu < 11.6$ eV. For such low energies, optical transitions from s-,p-and d-states dominate the f-states.

VII. CHEMISORBED CO ON Ni

The application of ultraviolet photoemission spectroscopy to measurements of electronic energy levels of adsorbed gases on clean metal surfaces has recently been described.[27] While many studies have been reported on the configurations of adsorbed gases on surfaces and on adsorbate binding energies, relatively little work has been reported on the electronic energy levels of such adsorbates. The theory of chemisorption is a complicated problem of current interest, and measurements of the energy levels and level widths are important since this information is closely related to theoretically accessible quantities.

Photoemission energy distributions for chemisorbed CO on Ni are shown in Fig. 9. Films of Ni about 1000-2000 Å thick were prepared by evaporation (~3Å/sec) at pressures of $\sim 10^{-8}$ Torr using an electron beam gun. Curve (1) in Fig. 9(a) shows the energy distribution for a polycrystalline Ni film prepared on a smooth quartz substrate at room temperature. Emission from the d-bands is observed in the energy range 0 to about 3.3 eV below E_F and a smooth background of secondary emission due to inelastically scattered electrons is observed at lower energies. Upon one Langmuir (1L) exposure of this film to CO (1L ≡ 1 x 10^{-6} Torr-sec), the work function was observed to increase by +0.31 eV and curve (2) was obtained. We now observe two additional levels centered at about 7.5 and 10.7 eV below E_F (cross-hatched areas). Upon further exposure, these CO-derived levels increase in amplitude and the work function increases until maximum coverage is reached at about 7L (curve 3). Widths of about 1 eV for the -10.7 eV level and about 2 eV for the -7.5 eV level are observed. The -7.5 eV level has about 1/3 the emission intensity of the Ni d-bands.

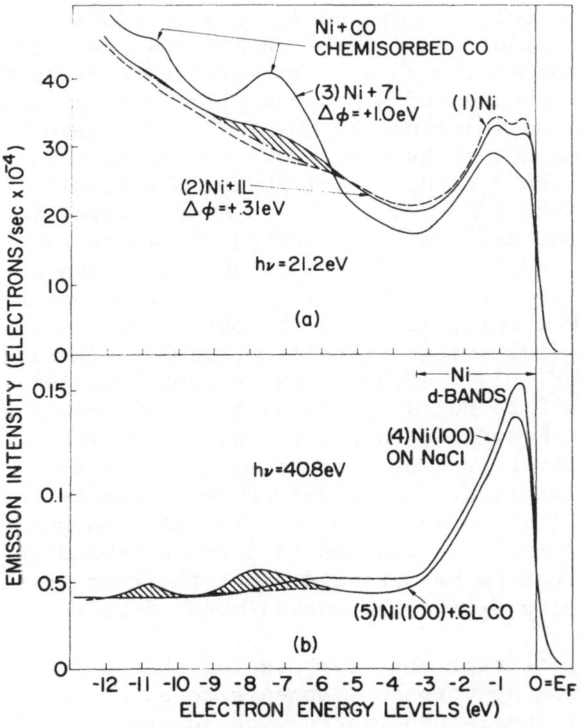

Fig. 9 (a) Energy distributions at hν = 21.2 eV for (1) Ni (2) Ni + 10^{-6} Torr-sec exposure of CO and (3) Ni + 7 x 10^{-6} Torr-sec CO. (b) Energy distributions at hν = 40.8 eV for (4) Ni and (5) Ni + 0.6 x 10^{-6} Torr-sec of CO.

Similar results were obtained using a (100) oriented Ni film prepared by evaporation onto a cleaved NaCl substrate at about 250°C. Spectra for Ni (100) and Ni (100) + 0.6L exposure of CO are shown in Fig. 9(b) for $h\nu$ = 40.8 eV. Compared with curve (1) for $h\nu$ = 21.2 eV, Ni d-band emission within about 3.3 eV of E_F exhibits a different shape (due to momentum selection rules), and the secondary emission background below -3.3 eV is much lower. In curve (5), CO-derived energy levels (cross-hatched) are observed at about 7.5 and 10.7 eV below E_F with widths of about 1 and 1.8 eV, respectively and are about 20% as intense as the background. The observation of these energy levels at two different photon energies also indicates that they correspond to occupied levels and are not due to structure in the final state or to Auger transitions.

Considerable work has been reported on annealing effects, binding energies, and surface configurations of chemisorbed CO on Ni[28] and Pd.[29] Within the framework of a molecular orbital (MO) picture, it is commonly believed[29] that CO is bound to the surface through the carbon end, with the carbon lone pair $\sigma 2p$ MO donating electrons to the d-bands of Ni; the d electrons in turn being back donated into the antibonding $\Pi 2p$ MO of CO. For gaseous CO, photoemission measurements at $h\nu$ = 21.2 eV give ionization potentials of 14.01 eV for the $\sigma_g 2p$ MO, 16.91 eV for the bonding $\Pi_u 2p$ MO, and 19.7 eV for the $\sigma_u 2s$ MO, which has a much lower intensity.[30] Comparing with our data for chemisorbed CO, we tentatively associate the level at -7.5 eV (∿12.5 eV below the vacuum level for low coverages, since $\phi \simeq 5$ eV for Ni) with the carbon lone pair $\sigma 2p$ MO, which is shifted upwards in energy by about 1.5 eV and broadened to about 2.0 eV (full width at half maximum). Likewise, the level at -10.7 eV (∿15.7 eV below the vacuum level) is associated with the bonding $\Pi_u 2p$ MO, which is shifted upwards in energy by about 1.2 eV and broadened to about 1 eV. The greater broadening of the -7.5 eV level is expected since it interacts more strongly with the Ni d electrons. The observed level widths for polycrystalline and (100) oriented Ni are roughly independent of CO coverage. Electrons back donated into the antibonding $\Pi 2p$ MO of CO are not expected to be observed since their energy levels are probably just below E_F and are masked by Ni d-band emission.

In summary, initial measurements for chemisorbed CO on Ni show that photoemission spectroscopy appears to be a promising tool for measuring energy levels associated with adsorption and other surface reactions. The sensitivity of the photoemission technique for observing adsorbate energy levels is due to the short hot electron scattering length, which limits the escape depth of photoemitted electrons to typically 10-20 Å.

REFERENCES

1. W. Kohn and L. J. Sham, Phys. Rev. 140, A1133 (1965).
2. L. Hedin and S. Lundqvist, in Solid State Physics 23, F. Seitz, D. Turnbull and H. Ehrenreich, editors, Academic Press N.Y., 1969, pp. 2-180. Also, B. Lundqvist and S. Lundqvist, in this conference proceedings.
3. A. R. Williams, J. F. Janak and V. L. Moruzzi, to be published.
4. F. M. Mueller and J. C. Phillips, Phys. Rev. 157, 600 (1967).
5. D. Brust, Phys. Rev. 139, A489 (1965).
6. J. F. Janak, D. E. Eastman and A. R. Williams, in Proc. of the Electronic Density of States Symposium, National Bureau of Standards, Washington, D. C. Nov. 1969 (unpublished).
7. N. V. Smith, Phys. Rev. B3, 1869 (1971); Phys. Rev. Letters 23, 1232 (1969).
8. M. Chodorow, Phys. Rev. 55, 675 (1939).
9. J. F. Janak, in Computational Methods in Band Theory, edited by P. M. Marcus, J. F. Janak and A. R. Williams, (Plenum Press, New York, 1971), p. 323.
10. The use of cesiated surfaces to lower the work function has been very successful in studies of the noble metals (C. N. Berglund and W. E. Spicer, Phys. Rev. 136, A1030 (1964); ibid, Phys. Rev. 136, A1044 (1964); Ref. 7) where it has permitted new structure to be observed, and has stimulated theoretical calculations (see Ref. 7).
11. Currently, measurements above $h\nu = 11.6$ eV are generally limited to the strong spectral lines of He and Ne gas discharge radiation ($h\nu = 16.8$, 21.2, 26.9 and 40.8 eV). Experiments underway by several groups using synchrotron radiation or spark source radiation are expected to extend measurements to all photon energies in the range 5-50 eV.
12. G. P. Pells and M. Shiga, J. Phys. C. 2, 1835 (1969).
13. P. O. Nilsson, Phys. Kond. Mat. 11, 1 (1970).
14. J. C. Phillips, Phys. Rev. 153, 669 (1967).
15. U. Gerhardt, Phys. Rev. 172, 651 (1968).
16. I. Lindau and L. Walldén, Solid State Commun. 9, 1147 (1971).
17. D. E. Eastman and J. K. Cashion, Phys. Rev. Letters 24, 310 (1970).
18. D. E. Eastman, Phys. Rev. Letters 26, 1108 (1971).
19. A. C. Switendick, Solid State Commun. 8, 1463 (1970).
20. D. M. Newns, Phys. Rev. 178, 1123 (1969).
21. J. R. Schrieffer and R. Gomer, Surface Sci. 25, 315 (1971).
22. D. E. Eastman, J. K. Cashion and A. C. Switendick, Phys. Rev. Letters 27, 35 (1971).
23. S. Methfessel and D. C. Mattis, Encylopedia of Physics XVIII, Part 1 (Springer, Berlin, 1968) pp. 389-562.
24. D. E. Eastman and Moshe Kuznietz, Phys. Rev. Letters 26, 846 (1971).

25. G. Busch, M. Campagna and H. C. Siegmann, Solid State Commun. 7, 775 (1969); G. Busch, M. Campagna and H. C. Siegmann, J. Appl. Phys. 41, 1044 (1970).
26. J. G. Endriz and W. E. Spicer, Phys. Rev. B2, 1466 (1970).
27. D. E. Eastman and J. K. Cashion, Phys. Rev. Letters, Nov., 1971.
28. M. Onchi and H. E. Farnsworth, Surface Sci. 11, 203 (1968).
29. J. C. Tracy and P. W. Palmberg, J. Chem. Phys. 51, 4852 (1969).
30. D. W. Turner, et al, Molecular Photoelectron Spectroscopy, Wiley-Interscience 1970, p. 34.

INVESTIGATION OF INNER SHELL EXCITATIONS IN SOLIDS BY SYNCHROTRON RADIATION

Ruprecht Haensel and Bernd Sonntag

II. Institut für Experimentalphysik der Universität Hamburg, Hamburg, Germany and
Deutsches Elektronen-Synchrotron DESY, Hamburg, Germany

I. INTRODUCTION

The optical properties of many gases and solids are well known in both the photon energy range below 10 eV (infrared, visible and near ultraviolet) and in the X-ray region. However, twenty years ago only a few data on optical properties were available in the intermediate range, in the vacuum ultraviolet or soft X-ray region because of great experimental difficulties. Since then this situation has been improved thanks to the development of new vacuum spectrometers, radiation detectors and light sources. One of the most important steps was the introduction of synchrotron radiation as a light source for this spectral range.

Synchrotron radiation is emitted by electrons radially accelerated in the magnets of high energy electron synchrotrons or storage rings. The general features of synchrotron radiation may be found in several textbooks on classical electrodynamics[1], more detailed information may be obtained from the paper of Schwinger[2] and the book by Sokolov and Ternov[3].

The spectrum of synchrotron radiation emitted by running high energy synchrotrons and storage rings (electron energy 10^8 eV to 10^{10} eV, circulating current 1 mA to 100 mA) spans the whole range from the soft X-ray region to the infrared. The intensity in the vacuum ultraviolet is higher than that of any conventional light source. The spectral distribution is continuous and smooth. The intensity is concentrated in a small cone tangential to the electron orbit. Further advantages are the high degree of polarization and the fact that the source works in a very good vacuum ($<10^{-6}$ Torr).

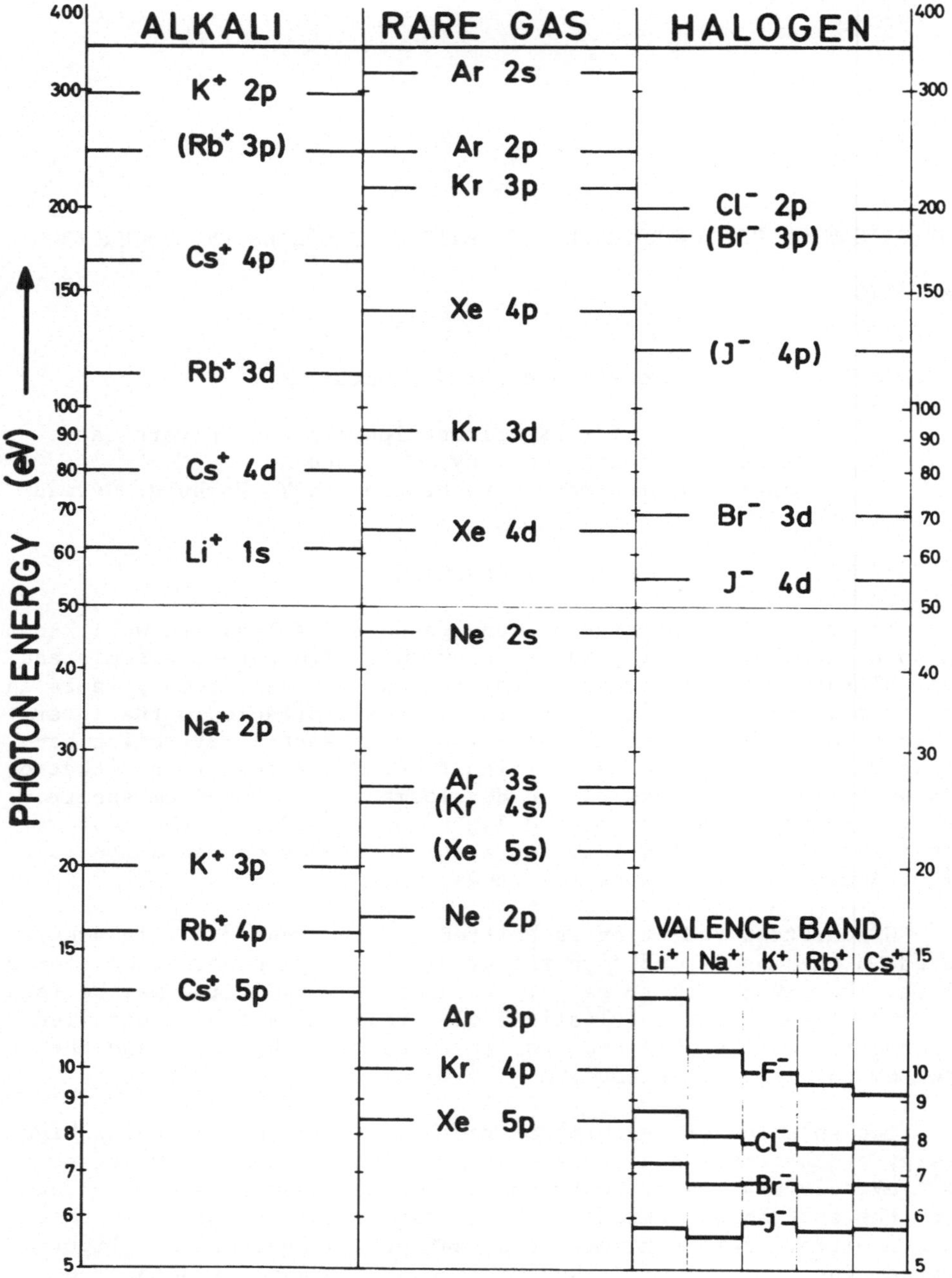

Fig. 1 Threshold energies for transitions from the different occupied shells in the alkali halides and rare gases

The first attempts to use synchrotron radiation for optical experiments in the extreme ultraviolet were made by Hartman and Tomboulian[4] as early as 1953, but systematic use for spectroscopic purpose only began after 1960. Several papers[5-7] may be consulted for more detailed information on the state of spectroscopic work with synchrotron radiation at different accelerators. In addition to the synchrotrons and the storage rings which have been used so far (Cornell, NBS, Frascati, Tokyo, Glasgow, Bonn, Moscow and DESY synchrotrons and the University of Wisconsin storage ring) new facilities will be opened in the near future (Daresbury synchrotron and storage rings in Tokyo, Orsay, Stanford and Hamburg). Nothing will be said here on experimental techniques since they are, essentially, all reviewed in Samson's book[8].

The aim of spectroscopic measurements is to obtain information on the electronic structure of solids. With conventional sources ranging up to ∿15 eV mostly electronic excitations from the valence band can be studied. For metals and low band gap semiconductors transitions to states several eV above the bottom of the conduction band can be reached. For wide band gap materials such as alkali halides and solid rare gases only the first excitations can be studied; in solid Ne and He they even lie above that limit. In the energy range above 15 eV transitions from tightly bound core levels to unoccupied states above the Fermi level give the main contribution to the optical absorption. These core states are very well localized and the wave functions of neighbouring atoms do not, therefore, overlap. As a result, no k-dependence of the energy is observed.

The width of core levels is determined by lifetime broadening (Auger effect) but for levels with excitation energies in the extreme ultraviolet this broadening is much smaller than for those in the X-ray region.

The optical spectra in the extreme ultraviolet are different from those found in the fundamental absorption or X-ray region. On the following pages we are going to review some of their main characteristics.

II. ONE ELECTRON MODEL

A. Energy Levels

The electronic structure of atoms and solids can be described in a first order approximation by the one electron model or by the energy band scheme resp. Optical measurements yield information on the position of the tightly bound initial states and the final states (in solids the positions of the energy bands) with respect to the Fermi energy. Fig. 1 shows the onset of absorption due to the excitation of electrons from the different subshells for the alkali halides and (solid) rare gases. Most of the positions have been established by optical experiments; the positions of energy levels in brackets are taken from the tables of Bearden and Burr[9], where the atomic energy levels of the other elements are also listed.

For an experimentalist the first step is to compare his experimental curves with theoretical energy level diagrams, taking into account the optical selection rules. Because of the small width of the initial state the spectra should reflect the structure of the conduction bands. Many attempts have been made to interprete the experimental results by assigning the absorption maxima to electron transitions from the flat core band to the different conduction bands at points of high symmetry or at distinct regions of the Brillouin-zone. However, in most cases the knowledge of the energy band scheme alone cannot lead to an unambiguous interpretation and thereby to an understanding of the spectra above threshold.

B. Density of States

The optical absorption of solids is proportional to ε_2 as given in the following relation:

$$\varepsilon_2 = F \int \frac{M(E,k)^2}{|\nabla_K(E_f - E_i)|}\, dS$$

$M(E,k)$ is the transition matrix element between initial state E_i and final state E_f. For transitions from core states $\nabla_K E_i$ is equal to zero. Furthermore, we assume that $M(E,k)$ = const. In this case ε_2 is proportional to the density of states of the conduction band.

We are now going to compare some experimental data with theoretical calculations. Generally, this can only be done for a range of several eV above threshold which is covered by the energy band calculations. Our comparison will be made for some transitions in the alkali halides and solid rare gases. In the core transitions of the solid rare gases detailed fine structure has been found for Xe 4d[10,11], Kr 3d[11] and Ar 2p[12] excitations. Rössler has calculated the energy bands and density of states curves of the solid rare gases (Ne, Ar, Kr and Xe) and has shown that the major part of these fine structures can be explained as being caused by density of states effects[13,14]. Fig. 2 shows the experimental curve of the 4d transitions in solid Xe[10,11] together with the density of states curve, taking into account the spin orbit splitting energy of the 4d shell of 2 eV and the degree of degeneracy of the two subshells[14]. The first peak in the density of states curve has been energetically adjusted to the first broad and pronounced peak D in the experimental curve. After doing so we see that the energy position of most of the maxima of both curves coincide. Furthermore, the shapes of both curves are similar. Deviations can be seen at peak F' and around H

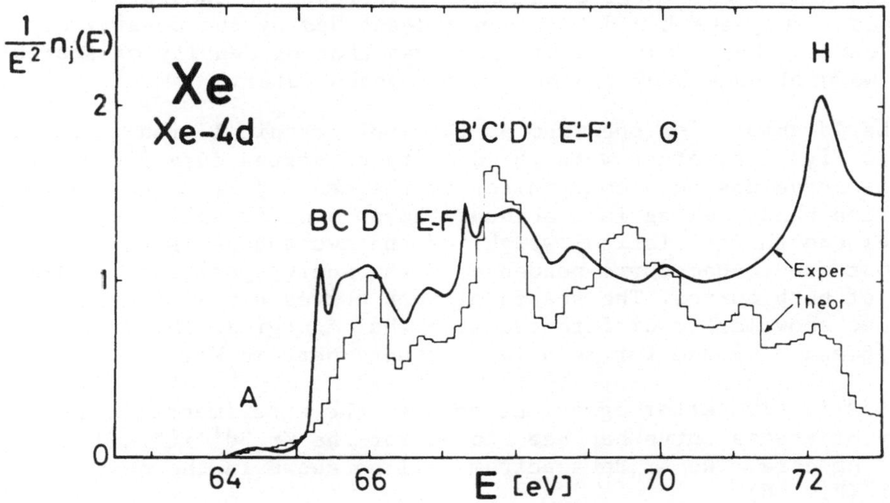

Fig. 2 Absorption spectrum from 4d states and joint density of states in solid Xe (fig. 4 of ref. 14 improved by Rössler)

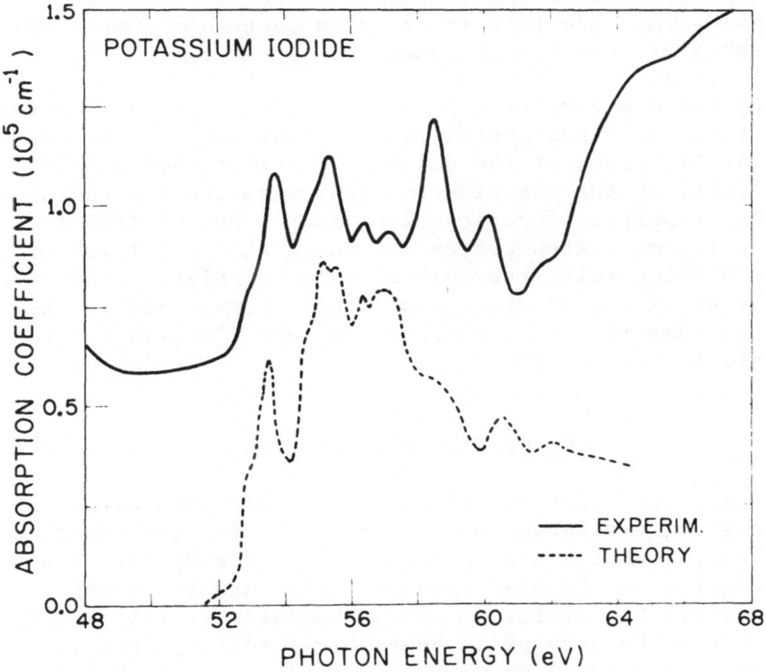

Fig. 3 Absorption spectrum from I^- 4d states and joint density of states in KI (fig. 13 of ref. 15)

where the experimental curve increases, whereas the theoretical curve decreases. The peaks B and B' are not described by the density of states curve. Their width is smaller than that of density of states peaks. We will come back to these extra-peaks later.

The 4d absorption spectrum of the isolectronic I^- ion in KI is shown in Fig. 3 together with the density of states curve[15]. As for Xe, this curve has been constructed by the density of states of the conduction band, taking into account the spin orbit splitting of the 4d level and the statistical weight of the two subshells. We find an almost one-to-one correspondence of the energy positions of the maxima of both curves. The shapes of both curves are similar at the onset but show marked differences at higher energies. The "extra lines" B and B' found for Xe have no counterpart in KI.

A slightly better agreement between the experimental and the density of states curve has been found for the Kr 3d[14,16] and Ar 2p[14] spectra. The Ar 2p spectrum will be shown in the next chapter (Fig. 6).

These examples show that many of the features of the spectra can be explained by the density of states of the conduction band. There are, however, still many discrepancies between theory and experiment which indicate that the spectra cannot be completely understood in terms of this simple model.

In optical experiments one can derive the optical constants ε_1 and ε_2 from the experimental results, but they do not directly yield the distribution of the density of states. For a better test of the validity of the one electron approximation for solids it is necessary to construct theoretical ε_1 and ε_2 curves from the theoretical density of states curves including the transition matrix elements. Unfortunately, the bulk of band calculations was made without the extension to density of states curves and to the authors' knowledge no case is known, where a complete ε_2 curve for core transitions has been calculated.

C. Continuum Absorption

For transition into states far above threshold almost no band calculations are available for a comparison with the experiments. For this case, however, atomic models, in the simplest case one electron models, can be used which describe atomic absorption far above the ionization limit. In the vacuum ultraviolet many cases are known where the absorption behaviour strongly deviates from the hydrogen case. For a pure Coulomb potential the absorption coefficient shows a saw-tooth-like behaviour typical in the X-ray region. The absorption increases step-like at threshold and decreases above treshold with increasing photon energy E according to $E^{-\alpha}$ (α being constant).

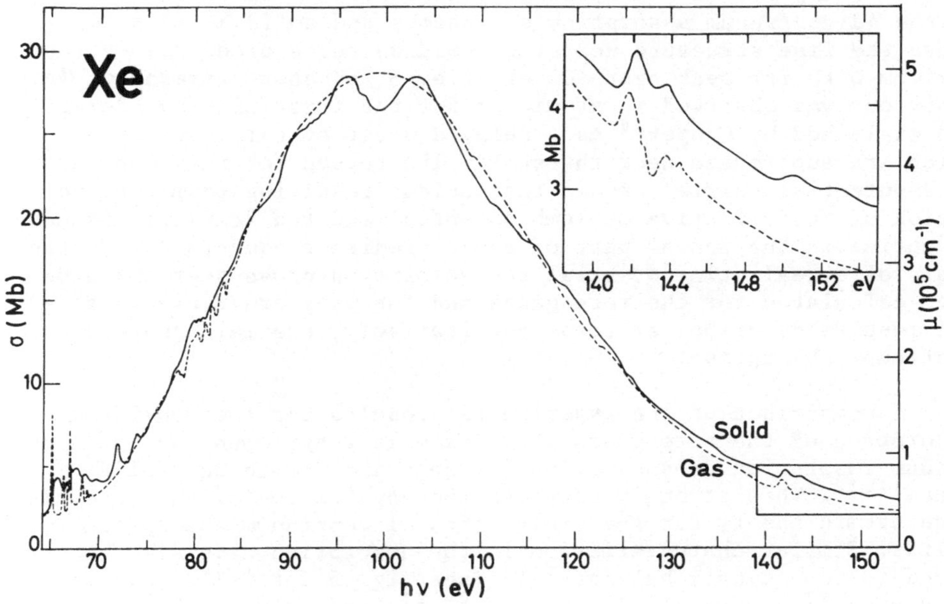

Fig. 4 Absorption spectrum of solid (solid curve) and gaseous (dotted curve) Xe in the range of 4d transitions (ref. 11)

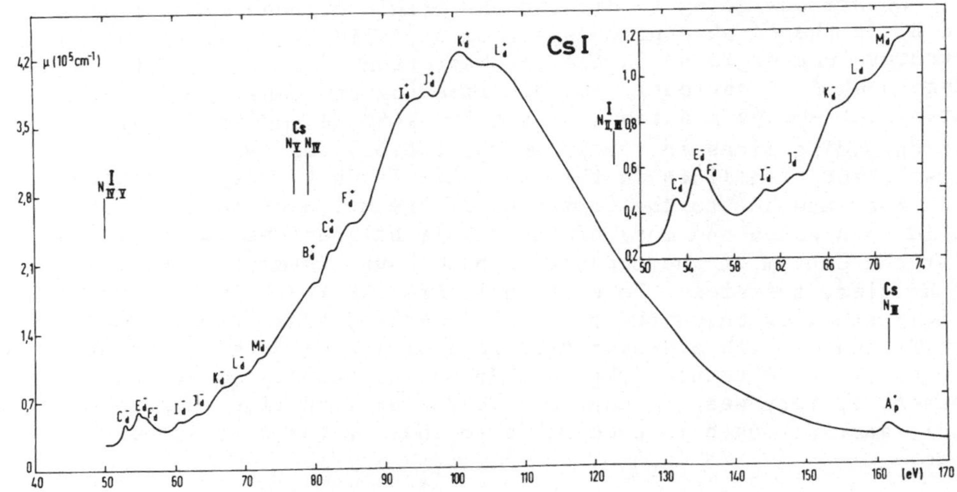

Fig. 5 Absorption spectrum of CsI from 50 to 170 eV (ref. 22)

An example of a non-hydrogenic absorption behaviour can be seen in the 4d continuum absorption of gaseous and solid Xe (Fig. 4). Above the fine structure near threshold we see a broad absorption maximum with its peak near 100 eV, i.e. 35 eV above threshold. This behaviour was observed in atomic Xe for the first time by Ederer[17] and explained by Cooper[18] as a delayed onset of d→f transitions which are suppressed near threshold. The reason for this suppression is a potential barrier around the nucleus resulting from superposition of the effective Coulomb potential and the $\ell(\ell+1)/r^2$ pseudopotential in the radial part of the Schrödinger equation[19]. On the basis of a realistic potential the absorption cross-sections have been calculated for the rare gases and for many other elements[20,21]. The results describe, at least qualitatively, the main features of continuum absorption.

A comparison of the experimental results for the continuum absorption of the rare gases shows excellent agreement of the continuum absorption cross section for both the atomic and solid state[11,12]. This strongly supports the application of this one electron atomic theory for the explanation of continuum absorption of solids. Similar characteristics in the absorption have also been found in some alkali halides[15,22] (see Fig. 5 for CsI), some semiconductors[23], almost all lanthanides[24,25], and many metals[26,27].

III. BEYOND THE ONE ELECTRON MODEL

A. Coulomb interaction between hole and electron

We are now going to discuss the origin of the "extra lines" B and B' in the 4d absorption spectrum of solid Xe (Fig. 2). A similar structure can be found in the 2p absorption of solid Ar (Fig. 6) where peak A is obviously not described by the density of states curve. The energy positions of the lines in Xe and Ar (as well as corresponding lines in the Kr 3d absorption) are very close to the first absorption lines in the gas. This leads to the assumption that they are due to the formation of Frenkel excitons[28]. In all solid rare gases and most of the alkali halides the wavefunctions near the bottom of the conduction band have s-symmetry. According to Rössler, therefore the exciton in Ar with its hole in a p-symmetric state lies below the onset of interband transitions, whereas for Xe and Kr with a d-symmetric initial state the excitons are above the onset. As a result, they can decay into continuum states and are, therefore, very weak in contrast to the exciton line in Ar whose oscillator strength is comparable to that of the corresponding gas line.

Whereas the exciton line in Xe is very weak it completely disappears in KI[15]. For Ar 2p-absorption a corresponding case in the

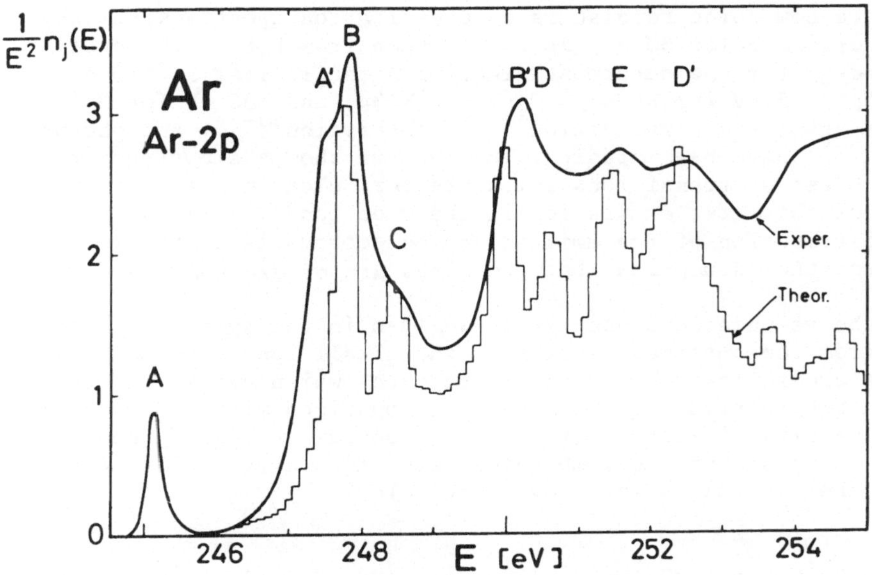

Fig. 6 Absorption spectrum from 2p states and joint density of states in solid Ar (fig. 6 of ref. 14)

alkali halides is the Cl^- 2p absorption which has been discussed for NaCl by Lipari and Kunz[29]. The authors believe that this spectrum can also be entirely explained by density of states effects. It should, however, be noted that the width of the first peaks in the experimental curve is obviously due to the instrumental energy resolution which becomes worse for higher energies. The same is true for the Ar 2p curve (Fig. 6) where the exciton line A has the intrumental width, and could possibly be otherwise smaller than the density of states peaks of the theoretical curve. As for NaCl only three peaks, two of which are spin orbit mates, have been compared, a final statement cannot be made at present. The difference in the oscillator strength of the exciton lines in Xe and Ar makes one believe that the absence of exciton lines in the I^- 4d spectrum must not necessarily also lead to the absence of exciton lines in the Cl^- 2p spectrum.

So far, exciton effects have mainly given rise to additional absorption lines. The rest of the absorption features can be explained as being due to density of states effects; however, the agreement between theory and experiment is not so good that an additional influence of Coulomb interaction onto the shape of the interband absorption lines can be excluded.

We are now going to discuss core excitation spectra with excitation energies below 50 eV. As can be seen from Fig. 1 the excitation energy for the outermost alkali ion shells lies at ∿13 eV for Cs^+ 5p, ∿15 eV for Rb^+4p, ∿20 eV for K^+3p, and ∿30 eV for Na^+2p. For these transitions reflection[30-32], absorption[33-36], and photoemission[37-40] have been studied. The spectra show sharp structures near the onset of transitions which are very close to the energy position of the first excitation in the free ion[41]. This and the energy distribution of the emitted photoelectrons is a strong support for the assumption that the lines are of excitonic origin.

In the valence band spectra as well as in the spectra of transitions from the outermost shells of the alkali ions the sharp exciton lines are followed by broader structures which can be interpreted as interband transitions. Here the comparison with the pure one electron density of states curves shows considerable disagreement. It turns out that here Coulomb interaction also drastically influences the spectral shape above threshold[16].

B. Exchange interaction, multiplet splitting, configuration interaction and autoionization

We have already mentioned the 4d continuum absorption maximum in the lanthanides. Below this d→f maximum we find an area of fine structure near threshold[25,42] (at ∿110 eV) which extends over an energy range of ∿20 eV. These fine structure absorption lines have been ascribed to transitions from the 4d shell into the unoccupied states of the 4f shell[42]. This assumption is supported by the fact that the absorption spectra of the metals and their oxides are very similar[25]. In an energy band scheme these unoccupied 4f levels should be close together so that no explanation for the spread of fine structure over 20 eV can be given from this picture. Dehmer et al.[43] have calculated the splitting of the $4d^9 4f^2$ configuration for Ce. The levels of this configuration are spread over a range of $\geq$ 20 eV, primarily by the effect of exchange interaction between the 4f electrons and the 4d vacancy. Dehmer et al.[43] also calculated the line intensities and obtained surprisingly good agreement with the experimental results. Both experimental and theoretical results for Ce are given in Fig. 7.

The calculations show that most of the oscillator strength is concentrated in transitions to the higher levels of the configuration. These levels, lying far above threshold and, therefore, strongly broadened by autoionization, give rise to the broad maximum ∿20 eV above threshold. Similar results have been obtained for other lanthanides[44]. The total oscillator strength of the lines is proportional to the number of 4f vacancies. This explains the disappearance of the absorption peaks upon filling the 4f shell.

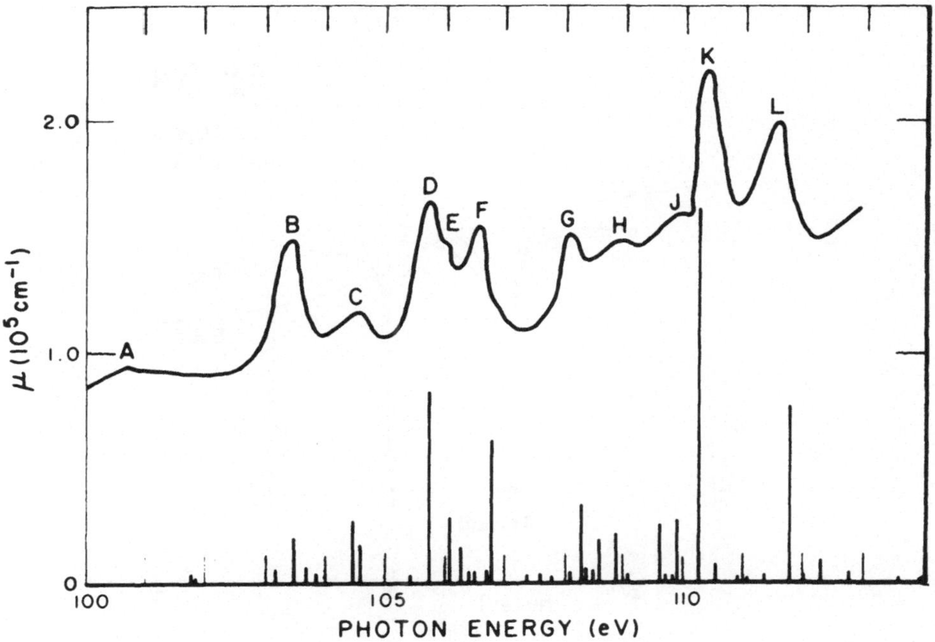

Fig. 7 Comparison of μ, the measured absorption coefficient for Ce in the region of the $N_{IV,V}$ edge with calculated relative positions and line strengths of $4d^{10}4f^1 \rightarrow 4d^9 4f^2$ transitions (the broad d→f absorption maximum about 125 eV is not included)

Dehmer et al.[43] pointed out that this interpretation of rare earth spectra may extend to any optical transition whose final state critically depends on a centrifugal barrier. This explanation should hold especially for the 3p absorption of transition metals in the third series[26] and the 5p absorption of the transition metals in the fifth series[27]. These spectra show, analogous to the 4d rare earth spectra, broad absorption maxima extending over an energy range of ∿20 eV above threshold. In contrast to the rare earth metals no marked fine structure can be seen at the onset. The characteristic features of the transition metals spectra result from 3p→3d or 5p→5d transitions resp. The width of the broad maximum above threshold is much larger than the width of the empty d-band given by band calculations.

Besides the broadening of absorption lines autoionization processes may also give rise to characteristic asymmetric line shapes. Asymmetric lines have been observed in the 2s absorption of Ne[45] and so-called window lines in the 3s absorption of Ar[46]. In the solid rare gas spectra[47] these lines are broadened and slightly

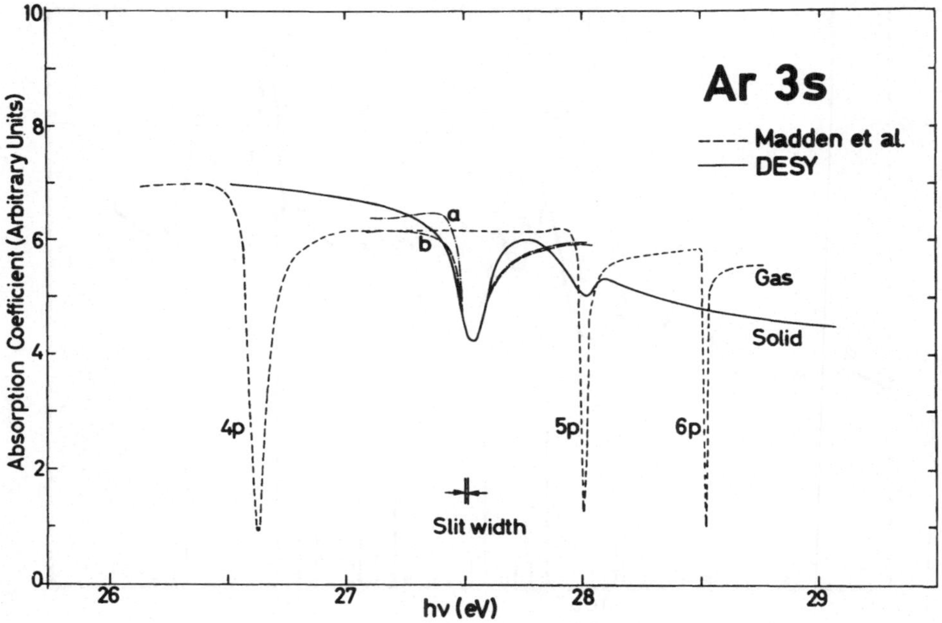

Fig. 8 3s Absorption of gaseous and solid Ar (ref. 47)

shifted to higher energies but the characteristic line shapes remain unchanged. The epxerimental results for the 3s absorption of gaseous and solid Ar are shown in Fig. 8.

Onodera[48] recently discussed the line shape problem for the interference of an exciton with continuum absorption. For the fundamental absorption spectrum of NaBr he showed that the line shape of exciton lines can only be explained by taking configuration interaction into account.

C. Multiple Excitations

Two electron excitations have been discussed for inner shell absorption of solids in different cases. In the Xe 4d spectrum (Fig. 4) near 80 eV in the gas very weak absorption peaks are observed which have been ascribed to the simultaneous excitation of a 4d and 5p electron[49]. In the same range weak structures are also observed in solid Xe[11]. The assumption that these peaks are also due to double excitations is supported by results from measurements on solid rare gas alloys[50]. Simultaneous excitations of one core electron and one valence band electron have been discussed in connection with several alkali halide absorption measurements[22,36,51]. Theoretical predictions for the oscillator strength of double excitation have, so far, given different results and were restricted to the fundamental absorption region[52,53].

Amusia et al.[54] have shown that multiply excited configurations are important in calculating the photoionization cross-sections of the two outer shells of Ar, Kr and Xe. Whereas the one electron model gives a good qualitative description of the absorption spectra, the calculated cross-sections, taking into account multi-electron correlations, are in good agreement with the experiments.

Relatively broad, periodic structures have been observed above the 2p-absorption threshold in the light metals Na, Mg and Al[55] and also in inner shell absorption spectra of some alkali halides[56-58]. These structures have been interpreted as being due to the simultaneous excitation of one electron and one or several plasmons. No theoretical calculations on the probability of such processes are available.

ACKNOWLEDGMENTS

The authors would like to thank F.C. Brown, J. Sugar and U. Rössler for permitting the reproduction of Figs. 2, 3, 6 and 7 from their papers. Thanks are also due to C. Kunz for a critical reading of the manuscript and to N. Kosuch and M. Skibowski for technical assistance and fruitful discussions during the preparation of the manuscript.

REFERENCES

1. For example A. Sommerfield, Elektrodynamik (Akademische Verlagsgesellschaft Leipzig, 1961)
and J.D. Jackson, Classical Electrodynamics, (John Wiley & Sons, New York, 1962)
2. J. Schwinger, Phys.Rev. 75, 1912 (1949)
3. A.A. Sokolov and I.M. Ternov, Synchrotron Radiation (Akademie-Verlag, Berlin, 1968)
4. P.L. Hartman and D.H. Tomboulian, Phys.Rev. 91, 1577 (1953)
5. R. Haensel and C. Kunz, Z. Angew. Phys. 23, 276 (1967)
6. R.P. Godwin, in Springer Tracts in Modern Physics, ed. G. Höhler (Springer Verlag, Berlin, 1969) Vol. 51 p. 1 and references therein
7. C. Gähwiller, F.C. Brown and H. Fujita, Rev.Sci.Instr. 41, 1275 (1970)
8. J.A.R. Samson, Techniques of Vacuum Ultraviolet Spectroscopy (John Wiley & Sons, New York, 1967)
9. J.A. Bearden and A.F. Burr, Rev.Mod.Phys. 39, 125 (1967)

10. R. Haensel, G. Keitel, P. Schreiber and C. Kunz, Phys.Rev. Letters 22, 398 (1969)
11. R. Haensel, G. Keitel, P. Schreiber and C. Kunz, Phys.Rev. 188, 1375 (1969)
12. R. Haensel, G. Keitel, N. Kosuch, U. Nielsen and P. Schreiber, J. de Physique (to be published)
13. U. Rössler, phys.stat.sol. 42, 345 (1970)
14. U. Rössler, phys.stat.sol. (b) 45, 483 (1971)
15. F.C. Brown, C. Gähwiller, H. Fujita, A.B. Kunz, W. Scheifley and N. Carrera, Phys.Rev. B 2, 2126 (1970)
16. see U. Rössler's paper in this volume
17. D.L. Ederer, Phys.Rev. Letters 13, 760 (1964)
18. J.W. Cooper, Phys.Rev. Letters 13, 762 (1964)
19. U. Fano and J.W. Cooper, Rev.Mod.Phys. 40, 441 (1968)
20. S.T. Manson and J.W. Cooper, Phys.Rev. 165, 126 (1968)
21. E.J. McGuire, Phys.Rev. 175, 20 (1968)
22. M. Cardona, R. Haensel, D.W. Lynch and B. Sonntag, Phys.Rev. B 2, 1117 (1970)
23. M. Cardona and R. Haensel, Phys.Rev. B 1, 2605 (1970)
24. T.M. Zimkina, V.A. Fomichev, S.A. Gribowskij and I.I. Zhukova, Sov.Phys. - Solid State 9, 1128 (1967)
25. R. Haensel, P. Rabe and B. Sonntag, Solid State Comm. 8, 1845 (1970)
26. B. Sonntag, R. Haensel and C. Kunz, Solid State Comm. 7, 597 (1969)
27. R. Haensel, K. Radler, B. Sonntag and C. Kunz, Solid State Comm. 7, 1495 (1969)
28. R.S. Knox, Theory of Excitons (Academic Press, New York, 1963)
29. N.O. Lipari and A.B. Kunz, Phys.Rev. B 3, 491 (1971)
30. D. Blechschmidt, R. Haensel, E.E. Koch, U. Nielsen and M. Skibowski, phys.stat.sol. (b) 44, 787 (1971)
31. C.J. Peimann and M. Skibowski, phys.stat.sol. (b) 46, 655 (1971)
32. D. Blechschmidt, V. Saile, M. Skibowski and W. Steinmann, Phys.Letters 35A, 221 (1971)
33. H. Saito, M. Watanabe, A. Ejiri, S. Sato, H. Yamashita, T. Shibaguchi, H. Nishida and S. Yamaguchi, Solid State Comm. 8, 1861 (1970)
34. H. Saito, S. Saito and R. Onaka, J.Phys.Soc. Japan 28, 699 (1970)
35. R. Haensel, C. Kunz, T. Sasaki and B. Sonntag, Phys.Rev. Letters 20, 1436 (1968)
36. S. Nakai and T. Sagawa, J.Phys.Soc. Japan 26, 1427 (1969)
37. D. Blechschmidt, M. Skibowski and W. Steinmann, phys.stat. sol. 42, 61 (1970)
38. H. Sugawara, T. Sasaki, Y. Iguchi, S. Sato, T. Nasu, A. Ejiri, S. Onari, K. Kojima and T. Oya, Opt.Comm. 2, 333 (1970)
39. Y. Iguchi, T. Sasaki, H. Sugawara, S. Sato, T. Nasu, A. Ejiri, S. Onari, K. Kojima and T. Oya, Phys.Rev. Letters 26, 82 (1971)

40. R. Haensel, G. Keitel, G. Peters, P. Schreiber, B. Sonntag and C. Kunz, Phys.Rev. Letters 23, 530 (1969)
41. C. Moore, Natl.Bur.Std. (U.S.) Circ. No. 467 (U.S. GPO Washington, D.C.) Vol. I (1949); Vol. II (1958); Vol. III (1958)
42. V.A. Fomichev, T.M. Zimkina, S.A. Gribowskij and I.I. Zhukova, Sov. Phys. - Solid State 9, 1163 (1967)
43. J.L. Dehmer, A.F. Starace, U. Fano, J. Sugar and J.W. Cooper, Phys.Rev. Letters 26, 1521 (1971)
44. J. Sugar, Phys.Rev. A (to be published)
45. K. Codling, R.P. Madden and D.L. Ederer, Phys.Rev. 155, 26 (1967)
46. R.P. Madden, D.L. Ederer and K. Codling, Phys.Rev. 177, 136 (1969)
47. R. Haensel, G. Keitel, C. Kunz and P. Schreiber, Phys.Rev. Letters 25, 208 (1970)
48. Y. Onodera, paper given at the III. International Conference on Vacuum Ultraviolet Radiation Physics, Tokyo 1971 and Phys. Rev. B (to be published)
49. K. Codling and R.P. Madden, Appl.Opt. 4, 1431 (1965)
50. R. Haensel, N. Kosuch, U. Nielsen, B. Sonntag and U. Rössler (to be published)
51. Y. Iguchi, T. Sagawa, S. Sato, M. Watanabe, H. Yamashita, A. Ejiri, M. Sasanuma, S. Nakai, M. Nakamura, S. Yamaguchi, Y. Nakai and T. Oshio, Solid State Comm. 6, 575 (1968)
52. T. Miyakawa, J.Phys.Soc. Japan 17, 1898 (1962)
53. J.C. Hermanson, Phys.Rev. 177, 1234 (1969)
54. M.Y. Amusia, N.A. Sherepkov and L.V. Chernysheva, Zh. Eksp. Teor.Fiz. 60, 160 (1971) (Engl. transl.: Sov.Phys. - JETP 33, 90 (1971))
55. R. Haensel, G. Keitel, B. Sonntag, C. Kunz and P. Schreiber, phys.stat.sol. (a) 2, 85 (1970)
56. C. Gähwiller and F.C. Brown, Phys.Rev. B 2, 1918 (1970)
57. F.C. Brown, C. Gähwiller, A.B. Kunz and N.O. Lipari, Phys. Rev. Letters 25, 927 (1970)
58. M. Elango and A. Saar, paper given at the III. International Conference on Vacuum Ultraviolet Radiation Physics, Tokyo 1971

POSITRON ANNIHILATION EXPERIMENTS AND THE BAND STRUCTURE OF V_3Si*

S. Berko

Brandeis University, Waltham, Massachusetts 02154, USA

M. Weger

The Hebrew University, Jerusalem, Israel

1. THE PRINCIPLES OF THE POSITRON ANNIHILATION METHOD

In the positron annihilation method[1] the angular correlation between the two γ-rays emitted as a result of a two-photon positron-electron annihilation is measured. The angular correlation is directly connected to the momentum distribution of the annihilating electron-positron pair and depends mainly on the momenta of the electrons in the solid to be studied, since the positron thermalizes prior to annihilation. The probability of the annihilation γ-rays carrying away momentum $\underline{p}$ is given, in the independent particle model, by

$$\rho(\underline{p}) = \sum_{\underline{k},\ell} \left| \int d^3\underline{r}\, \exp(-i\underline{p}.\underline{r})\psi_+(\underline{r})\psi_{\underline{k},\ell}(\underline{r}) \right|^2 \qquad (1)$$

where $\psi_+(\underline{r})$ is the ground state positron wavefunction and $\psi_{\underline{k},\ell}$ is the electron wavefunction with wave number $\underline{k}$ and band index ℓ; the summation is over all occupied states $\underline{k},\ell$. In the actual experiment, the angular distribution $N(\theta)$ is obtained using long slits (fig. 1), so that only one component of the momentum is being measured

$$N(p_z) = \iint \rho(p_x p_y p_z)\, dp_x dp_y \quad ; \quad p_z = mc\,\theta \qquad (2)$$

Let us note that the density in momentum space is not the Fourier transform of the density in ordinary space (for example, for free electrons the density in ordinary space is constant, its Fourier transform is a δ-function, while the momentum distribution

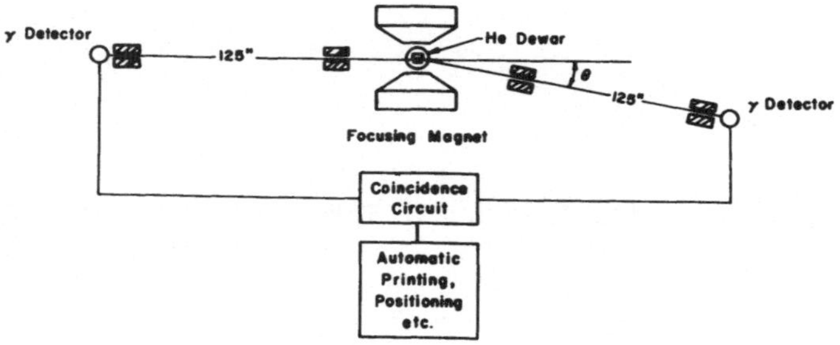

Fig. 1. Sketch of the experimental setup for measuring the 2γ correlation N(θ) from positrons annihilating in solids. The cylindrical γ ray NaI detectors are 10" long, with axes perpendicular to the plane of the drawing.

is the Fermi sphere). Therefore this method provides information independent of methods like X-ray diffraction. In principle the method can determine the Fermi surface of any metal, since $\rho(\underline{p})$ will exhibit breaks at the Fermi momentum, independent of the exact nature of the wavefunctions. In addition, the exact shape of $\rho(\underline{p})$ depends on the details of the electronic wavefunctions, since $\underline{p}$ is not equivalent to the crystal momentum k, thus leading to additional information on the band structure.[2] This method possesses advantages over galvanomagnetic methods (such as dHvA) more commonly employed to determine shapes of Fermi surfaces, in that it does not require long electron mean free paths (and thus low temperatures and very pure materials), and can even be applied to alloys[3]; it is not disturbed by open orbits; and it is less sensitive to hybridization, since if two energy surfaces intersect, the electron orbits are radically altered, while the momentum distribution is affected only locally. But this method also has severe drawbacks: it is expensive; it is very time consuming, since to obtain good counting statistics with positron sources of reasonable intensity, counting times of order of months may be needed. It is also a low-resolution method (for example, the Fermi momentum in Al can be determined by the positron method to an accuracy of order 10^{-2}, while dHvA yields accuracies of order 10^{-4}). The method requires good single crystals, since the positrons can be disturbed (even trapped) by dislocations and vacancies.[4] Because the positron disturbs the electronic wavefunction, the quantitative determination of the momentum distribution (in contrast to the determination of the points of discontinuity, which give the FS) is subject to complex many-body corrections.[5] In this respect, the use of the Compton effect[6] is theoretically better suited, but it suffers at the present of an even lower resolution capability. Also, the determination of the momentum distribution $\rho(p_x p_y p_z)$ from $\int\rho(p_x p_y p_z)dp_x dp_y$ is difficult and not always unique.[7] The positron annihilation method has been applied successfully to a few systems, such as Be,[8] Graphite,[9] the alkali metals,[10] Si and Ge,[11] Cu[12] and its alloys, the rare earths,[13] and even to some ferromagnetic problems.[14] However, there still exists a gap between what the positron method can do in principle, and the extent to which it has been applied in practice.

2. THE β-W SYSTEM

A number of alloys with the β-W structure, such as Nb_3Sn, Nb_3Al, V_3Si, V_3Ga, etc., are interesting because of their high superconducting transition temperature T_c (17°-20°K, roughly). The V_3X (X = Si, Ga, Ge, Pt, Au, etc.) system possesses in addition anomalies in the electronic properties, such as the electronic specific heat[15], the susceptibility χ, the Knight-Shift[16] and nuclear relaxation time T_1,[17] the elastic constants c_{ij},[18] and other properties. These anomalies manifest themselves as a strong temperature dependence of χ, $\Delta H/H$, T_1T, c_{ij}, etc., which in "normal" metals are temperature independent. These anomalies are strongly correlated with T_c. In the Nb_3X system, sometimes these anomalies are also present (Nb_3Sn), but in Nb_3Al[19] and $Nb_3Al_{.75}Ge_{.25}$, which have the highest values of T_c, they are not very strong. These anomalies have been attributed by Clogston and Jaccarino[20] to a sharp peak in the density of states function, about 2 mRy wide, at the Fermi level. Of particular interest here are two special types of anomalies, namely the effect of uniaxial stress on T_c, and the tunneling gap into surfaces of different orientations. Uniaxial stress has a very strong and anisotropic effect on T_c.[21] Stress in the [100] direction reduces T_c drastically (about 0.5°K/kbar for V_3Si and Nb_3Sn), while stress in the [111] direction has only a very small effect, actually slightly increasing T_c. Hydrostatic stress also has a small effect only. The dependence of T_c on stress can be very roughly described by a relation like $\Delta T_c = a\,(|s_{xx}-s_{yy}|+|s_{yy}-s_{zz}|+|s_{zz}-s_{xx}|)$ where s_{ij} is the stress tensor. This indicates a strongly singular and anharmonic dependence.[22] The tunneling gap in Nb_3Sn depends strongly on the surface,[23] being large (almost the bulk value) for tunneling into (100) surfaces and very small (about 1/3 of this value) for tunneling into (111) surfaces, with an intermediate value for (110) surfaces. These anomalies do not occur generally in superconductors.

At a rather early stage, a linear chain model was proposed to account for the anomalies of the electronic properties of this system.[24] The vanadium (or niobium) atoms in the β-tungstens are arranged in three families of linear chains, such that the distance between neighbors belonging to different chains is larger than the

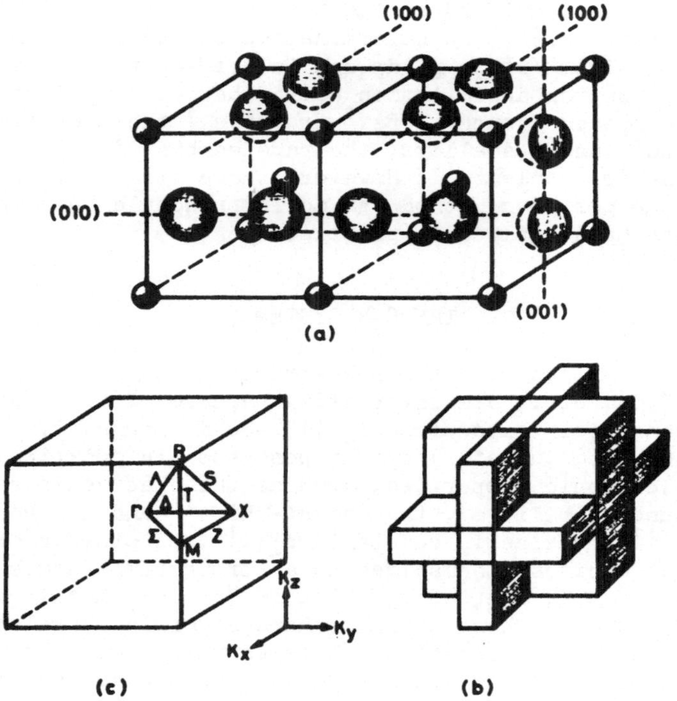

Fig. 2a. The crystal structure of the β-tungstens. The large spheres represent vanadium or niobium atoms. The smaller spheres are the X atoms occupying a bcc lattice.

2b. The Fermi surface of some 3d sub-band in the nearest-neighbor, tight-binding approximation.

2c. The Brillouin zone with its symmetry points.

distance between neighbors belonging to the same chain by a factor of $\sqrt{3}/2$ (fig. 2a). Therefore, in the tight-binding approximation, if we consider only nearest neighbor interactions, the band structure is one-dimensional: There are three families of bands, $E_1(k)$, $E_2(k)$, and $E_3(k)$, where $E_1(k)$ consists of states localized on chains in the [100] direction, with $E_1(k) = E(k_x)$, independent of k_y and k_z; $E_2(k)$ consists of states on chains in the [010] direction, $E_2(k) = E(k_y)$, and $E_3(k) = E(k_z)$. The Fermi surface consists of planes in the [100] directions (fig. 2b). One dimensional metals had previously been shown to possess strong anomalies in their electronic properties[25]; for example, the Kohn anomaly is particularly strong in such a system, resulting in a second-order Jahn-

Teller effect and a lattice distortion as shown by Frohlich,[25] so that the presence of anomalies in the β-tungstens was understandable. Subsequently, Labbe and Friedel[26] suggested that instead of a Kohn anomaly (with $q = 2k_F$), the anomaly in these systems is associated with the $1/\sqrt{E}$ divergence of the density of states characteristic of one-dimensional systems, and this divergence results in an elastic anomaly with $q = 0$, and this accounted for the softening of the elastic constants and the martensitic transformation.[27]

Actually, the transfer integrals between atoms belonging to different chains are not weak compared with nearest-neighbor integrals, since the ratio of the distances ($\sqrt{3/2}$) is not very large. Therefore, the three families of chains cannot be regarded as independent. The coupling between the chains, for a general point of the Brilluoin Zone, was shown[28] to result in displaced planar sections of the constant-energy surfaces ("torn out" planes), while at symmetry planes ($k_x = 0$, or the zone boundary) interaction between different chains may vanish due to the symmetry of the lattice for particular 3d states. For these particular symmetries, very sharp Van-Hove singularities in the density of states produce the peaks needed to account for the experimental anomalies. In this coupled-chain model, the high transition temperature is due in part to coupling between the chains by electron-phonon interaction.[29] When this coupling is broken, T_c decreases somewhat. The anisotropy in the tunneling gap in Nb_3Sn($T_c \sim 18°K$) is accounted for by noting that a (100) surface cuts one family of chains (in the [100] direction), leaving two coupled families intact (and $T_c \sim 15°K$); a (110) surface cuts two families (in the [100] and [010] directions), leaving one (in the [001] direction), which has a lower T_c($\sim 11°K$), while a (111) surface cuts all chains, resulting in a very low T_c($\sim 5°K$). The uniaxial stress experiment is accounted for by assuming that the stress "squeezes" electrons (of a given 3d sub-band, which contributes significantly to superconductivity) out of one, or two, families of chains, leaving eventually one family only.

Thus the concept of electrons of a given 3d sub-band moving along chains of V (or Nb) atoms, with some coupling between the chains, appears to be a useful model for these substances. The problem with the linear chain model is that it is a qualitative model, rather than a quantitative theory, and therefore it is necessary to verify whether a quantitative band calculation indeed demonstrates this property.

An APW calculation for several V_3X compounds was carried out by Mattheiss.[30] The band structure appeared to be very complicated, and in complete disagreement with a nearest-neighbor, two-center, tight-binding calculation. Therefore a critical experiment was needed to find out whether the linear-chain model has merits; and the positron annihilation technique appeared to be ideal for such an experiment, as was suggested some time ago by W. Kohn.

3. A POSITRON ANNIHILATION EXPERIMENT ON V_3Si

A planar FS like the one predicted by the linear chain model is particularly suitable for investigation by the positron annihilation technique, since for the component of the momentum perpendicular to the surface, p_z, $N(p_z)$ (Eq. 2) has a finite discontinuity at $p_z = p_F$. (For a spherical FS, $n(p_z)$ has a discontinuity in the derivative at $p_z = p_F$, and for a cylindrical FS, it has an infinite derivative, but is continuous.) This step should be detectable even in a complicated system like V_3Si, which has 38 conduction electrons per unit cell, and only 1-2 can be expected to contribute to a given band. Since the 3d function deviates strongly from a plane wave, many higher Fourier components should be present, thus the angular correlation curve for a (100) surface should have a step-function component, with steps centered around 0, $2\pi/a$, $4\pi/a$, etc., of a magnitude of order 2-5% (fig. 3). For other surfaces,

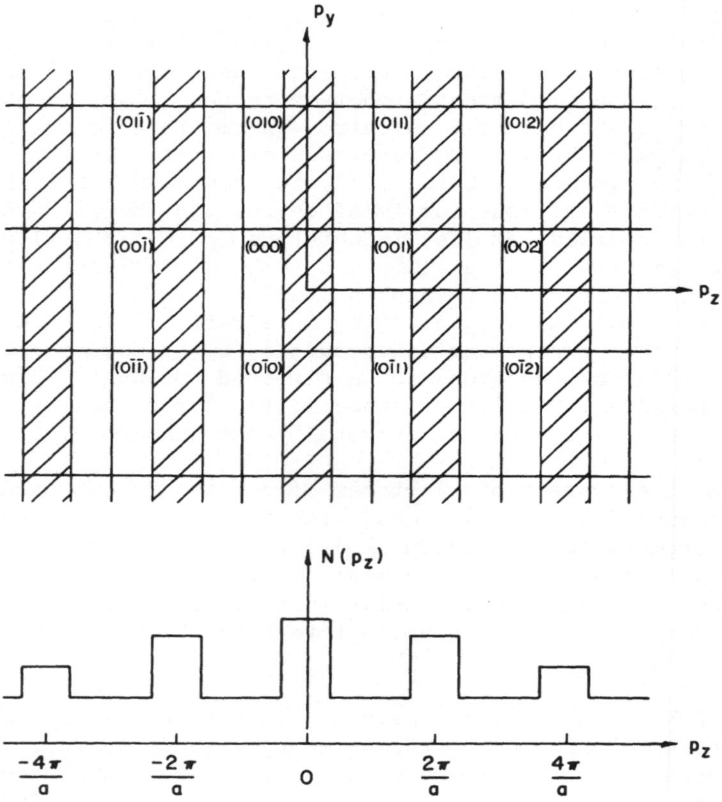

Fig. 3. The pattern expected for the angular-correlation curves for p_z along [100], for the Fermi surface of fig. 2b.

such as (111), the planar structure should be smeared out and the curve should be more continuous. Thus, the small effect on the (100) curve should become visible when the (111) curve is subtracted. We have performed such an experiment[31] on V_3Si single crystals using a long slit geometry and an angular resolution of 0.5 mrad (1 mrad corresponds to a momentum of 0.137 atomic units). The predicted anisotropy was indeed observed[32] (fig. 4). There are maxima in $N_{100}(p_z) - N_{111}(p_z)$ at $p_z = 0$, $2\pi/a$, and $4\pi/a$ (roughly) and minima in between, and the size of the steps is of order 5%. This is a very high anisotropy for a cubic system with so many bands. The excellent agreement between the experimental difference curve and the expected one is most probably fortuitous.

The discontinuities in $N(p_z)$ should be particularly conspicuous when the derivative $dN(p_z)/dp_z$ is plotted, since a discontinuity manifests itself as a δ-function in the derivative. The derivate is shown in fig. 5. Many elements of structure are indeed observed. Since the planes are expected to extend over several Brilluoin zones, they are expected to be smeared out when the crystal is rotated by a few degrees. (If the width of the structure is given by Δp_z, then it should smooth out when the crystal is rotated by an angle $\alpha \sim \Delta p_z/(2\pi m/a)$ where m is an integer of order 2-3.) The structure was indeed observed to be smeared out when viewed at an angle of about 3° from the "true" [100] direction. Cylindrical sections, parallel to the [100] direction, may in principle also provide structure similar to that of planar sections (only quite weaker); however, for cylinders, this structure should be present also when the crystal is viewed in the [110] direction, since then cylindrical sections belonging to different Brilluoin zones are again in line; experimentally, the derivative curve for a crystal cut in the (110)

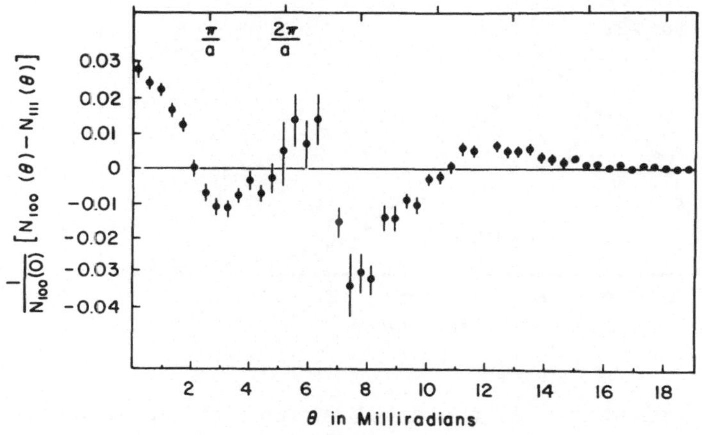

Fig. 4. The difference between the angular correlation curves of V_3Si for the [100] and [111] directions.

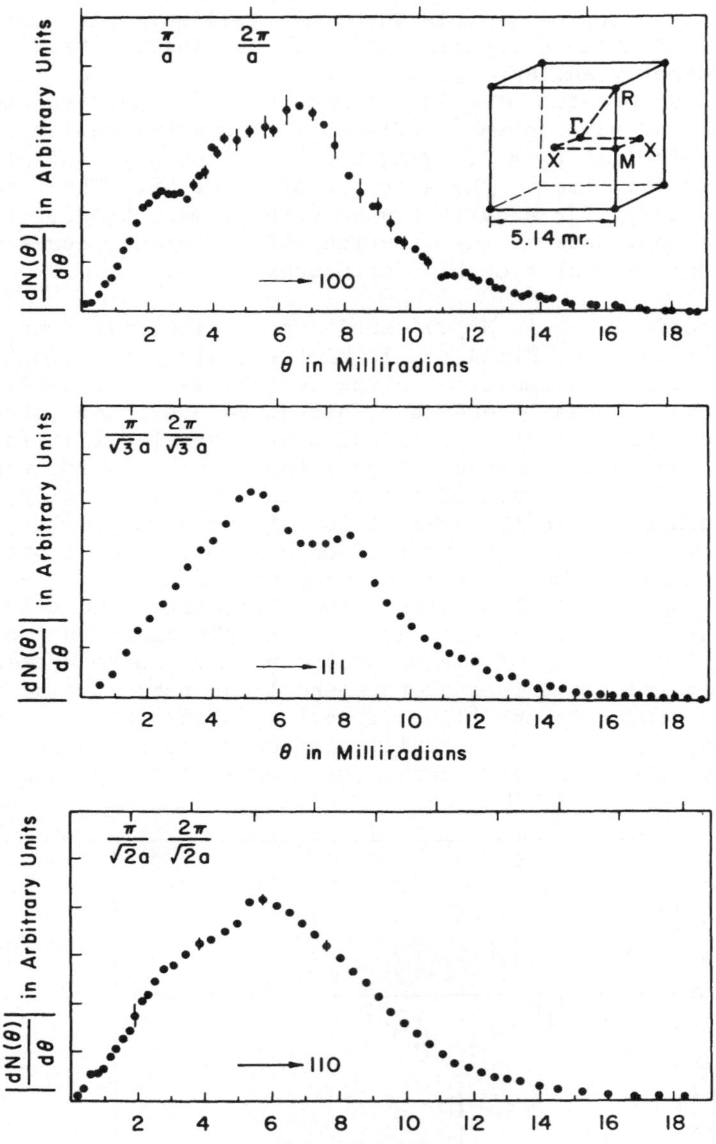

Fig. 5. The derivatives of the angular correlation curves for [100], [110], and [111] directions (V_3Si at 80°K).

plane is quite smooth; therefore it appears that the structure in the [100] derivative is due to planar sections. Following these considerations the derivative curve in the [111] direction should be smooth. Actually the derivative curve in that direction displays two broad maxima (fig. 5). Such a structure cannot be accounted for by the linear chain model. A maximum in the derivative curve corresponds to a sharp fall in $N(p_z)$ in this region, which may be due to an approximately flat region of the FS. The simplest way to account phenomenologically for such flat regions at 4.5 and 7.5 mrad is by an "electron" pocket strong in the Brillouin zones around points (100) and (110) of the reciprocal lattice, or by a "hole" pocket strong in zones (200) and (111) (fig. 6). The "electron" structure has a larger volume in momentum space and therefore is likely to give rise to larger peaks. As seen from the figure, the electron structure possesses necks in the [100] directions. This is not the only structure that can account for the peaks in the [111] derivative curve, but probably the simplest one. We have also studied the dependence of these structures on temperature; within the present experimental statistics no change was observed in the derivative curves when going through the superconducting transition.

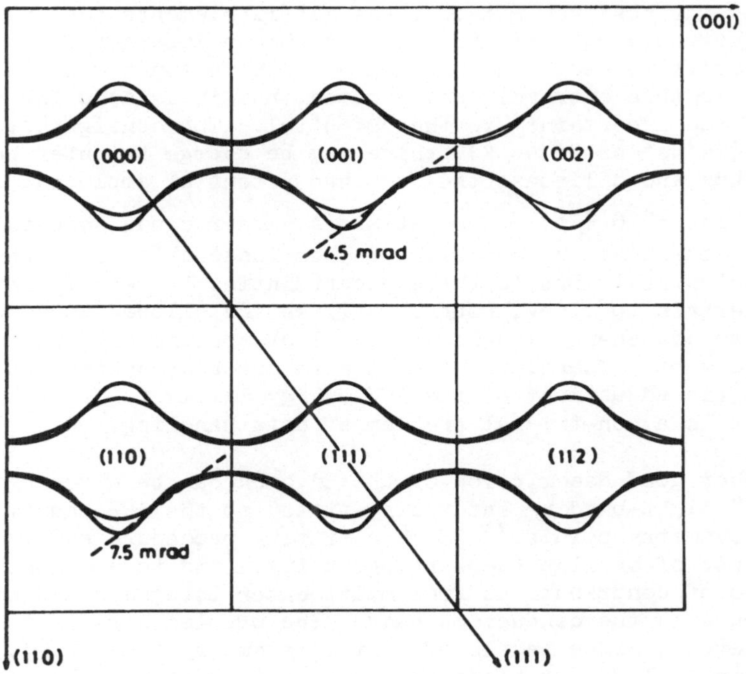

Fig. 6. An attempt to derive a "phenomenological" Fermi surface from the [111] derivative curve (fig. 5).

Further experiments are in progress to study the possible influence of the martensitic transformation on the angular correlations. Preliminary work on single crystals of V_3Ge and Nb_3Sn indicates very similar anisotropies to those of V_3Si.

Thus, we have seen that the positron annihilation experiment shows that in V_3Si some constant energy surfaces are indeed in the form of planar sections perpendicular to the [100] directions; the experiment also provides additional data about the band structure which must be accounted for theoretically.

4. THE APW CALCULATION

At first sight, Mattheiss' APW calculation[30] appears to be hopelessly complicated. Let us see whether we can analyze this calculation so that we can determine from it shapes of constant energy surfaces (to see whether they are planar), wavefunctions, the density of states function (to see whether it possesses a sharp peak at the Fermi level, as proposed by Clogston and Jaccarino,[20] and whether the peak is due to the $k_z = 0$ states of the $m_\ell = \pm 2$ band as proposed by Labbe and Friedel[26]), the momentum distribution (as observed by the positron method), the matrix elements responsible for the electronic part of the electron-phonon interaction causing superconductivity, etc. The problem here is that not only is the APW band structure extremely complicated, but it is also inaccurate, because of the uncertainty in the potential. The configuration is not known ($3d^3 4s^2$ or $3d^4 4s$?), there may be charge transfer between the vanadiums and silicons; the exchange potential employed, namely the "full Slater" $6(\frac{4\pi}{8}\rho)^{1/3}$ potential is rather crude, and the muffin-tin potential is probably a rather inaccurate description of the real potential. Due to these uncertainties the energy levels may be uncertain to ±1 eV, roughly. If we are allowed to adjust the 80 or so odd energy levels within ±1 eV arbitrarily, we can get anything we want. Therefore we must work out some systematic procedure for the adjustment of the APW energy levels to fit experiment. This is a non-trivial problem of data handling.

A rather good description of the 3d band can be given by "effective" tight-binding integrals, fitted to the APW energy levels at symmetry points.[33] Such a fitting procedure automatically takes account of overlap (non-orthogonality), and to a certain extent also of contributions from multicenter integrals and of hybridization with the conduction band. The problem here is to identify the levels, since levels of a given symmetry (say, Γ_{25}), may belong to several 3d sub-bands (here, $m_\ell = 0$, or the σ-sub-band, $m_\ell = \pm 1$, or the π sub-band); to the silicon 3p band, and to the vanadium 4s band. Also, due to hybridization, the APW levels may be mixtures of states of several bands.

The determination of the TB integrals may be carried out with the aid of group theory.[34] It turns out that there are a number of "lonely" levels, namely Γ_1', Γ_{12}', Γ_{15}', M_7, R_1R_2, and R_3, which can belong only to a definite 3d sub-band. (Γ_1' belongs to δ_2, corresponding to d_{xy} functions on chains in the z-direction, and appropriate functions for the other chains. Γ_{15}', M_7, R_1R_2, R_3 belong to π, corresponding to d_{xz} and d_{yz} functions.) Thus, the expressions for the energies of these states in terms of the effective integrals can be written down immediately, and equated with the APW energy; for example, $E(\Gamma_1')_{APW} = K(\delta_2)-2J(\delta_2)+8I(\delta_2)$, where $K(\delta_2)$ is the crystal-field integral of the δ_2 sub-band, $J(\delta_2)$ is the nearest-neighbor transfer integral, and $I(\delta_2)$ is the next nearest neighbor transfer integral. The number of equations for lonely levels (6) is not sufficient to determine all integrals of these two sub-bands (9), but there are several levels that occur in pairs and triples, but belong only to the δ_2 and π sub-bands, namely Γ_{25}', X_2, and M_3. The sum of the energies of all states belonging to each of these symmetries can be written in terms of the integrals of these sub-bands. This yields a sufficient number of equations to determine these integrals (if interactions up to next-nearest-neighbor are considered), and since the equations are linear, the solution is unique. Once these integrals are determined, the procedure is extended to include the δ_1 sub-band ($d_{x^2-y^2}$ for chains in the z-direction). As for the σ sub-band, ($d_{z^2-\frac{1}{3}r^2}$), it has exactly the same symmetry as the 4s band; however, its Γ_1 level can be identified from the wavefunction analysis of Mattheiss, and states Γ_{25} and M_1 of the 4s band are so high that the corresponding states of the σ sub-band can be determined too. Thus, the integrals of the σ sub-band can be determined. A similar procedure applies to the silicon 3s and 3p bands, which, it turns out, can also be described very well by the tight-binding approximation with integrals up to n.n.n Si-Si. Now, in principle, in order to calculate all the energy levels, we must also determine the integrals connecting different sub-bands. These can be determined by making use of the Koster-Slater symmetry considerations, and an interpolation procedure.[34] However, it turns out that to an accuracy of 0.03 Ry, we can ignore them except for the integrals connecting the Si 3p band and V δ_1 sub-band, which are very large and must be included (the lobes of the δ_1 functions point directly towards the Si). The fit between the APW levels and the TB levels obtained by this procedure is shown in fig. 7. The 2-center integrals can also be calculated directly,[35] using the tabulated Herman-Skillman atomic functions.[36] The agreement between these and the effective integrals is about 15% for the larger ones, with a somewhat larger difference for the smaller ones. This close agreement provides extra support to this scheme.

Using these integrals, we can easily determine the shapes of the constant-energy surfaces, and we find that they are quite planar,[34]

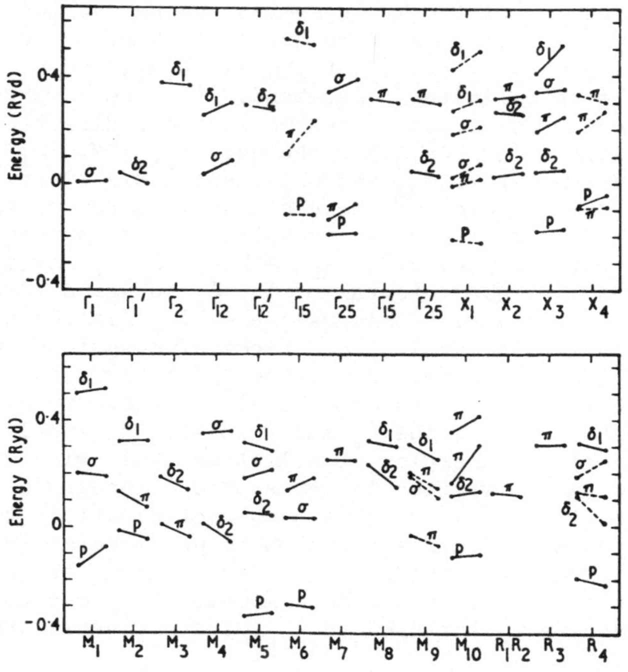

Fig. 7. The agreement between the energy levels at symmetry points calculated by the TB method assuming independent 3d sub-bands (left) to those calculated by the APW method (right).

in particular for the σ sub-band, but also for certain sections of the other sub-bands. They are in the form of "cubes" and "crosses," as suggested by the coupled-chain model.[28] The density of states can also be calculated easily, because the matrices are small. An accuracy of about 2-3% with a resolution of 2 mRy can be obtained in about 10 minutes of a CDC 6400, even without the Gilat-Raubenheimer[37] algorithm; use of this algorithm reduces the time by a factor of 3, roughly. The density of states curve is shown in fig. 8. It possesses many peaks which are automatically identified by this procedure (the sub-band and region in the BZ responsible for them). It is seen that the largest and sharpest peak is due to the δ_2 sub-band, in the vicinity of the Γ_{25}'-X_3-M_5 plane. This peak is about 2-4 mRy wide, and contains roughly one electron per unit cell. In Mattheiss' calculation, it is about 1 eV below E_F. The size and sharpness of this peak are rather insensitive to the assumed potential, as can be seen by comparing the calculation for V_3Ga ($3d^3 4s^2$) with the ones for

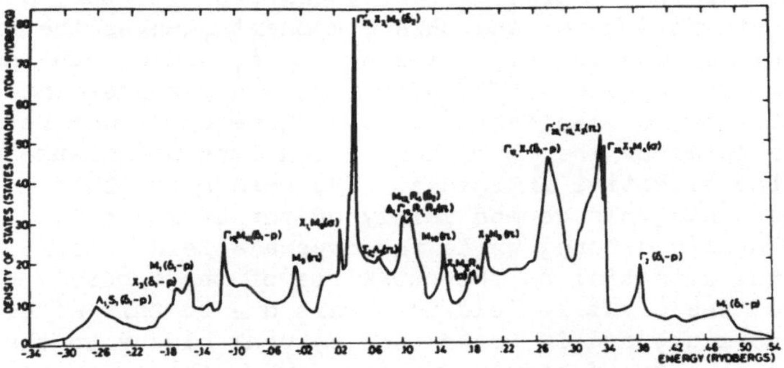

Fig. 8. The density of states curve following Mattheiss'calculation.

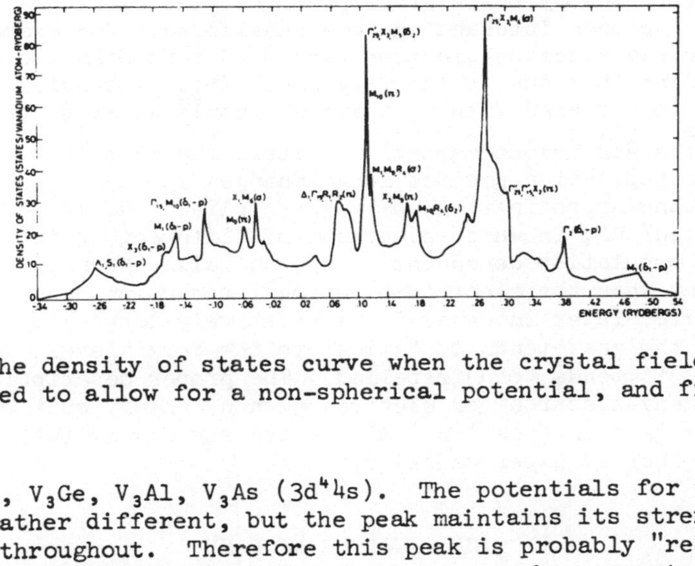

Fig. 9. The density of states curve when the crystal field integrals are adjusted to allow for a non-spherical potential, and fit to experiment.

V_3Ga, V_3Si, V_3Ge, V_3Al, V_3As ($3d^4 4s$). The potentials for these systems are rather different, but the peak maintains its strength and sharpness throughout. Therefore this peak is probably "real," at least in 1-electron theory. Its sharpness can be accounted for, at least partially, by the absence of inter-chain coupling due to symmetry, and the absence of hybridization with the 4s and Si 3s, 3p bands also due to symmetry (none of these bands contains a Γ_{25}' state).

Now we can attempt to apply corrections to the APW calculation. It appears that the largest error is due to the use of the muffin-tin potential. The assumption of spherical symmetry inside the atomic sphere is justified when the local symmetry is high, for example cubic, as in fcc, bcc, sc, and perhaps even diamond lattices. However, here the local symmetry is very low ($\bar{4}2m = D_{2d}$). Each vanadium atom has two nearest neighbors very close to it on a line. Therefore, there is a large $P_2(\cos\theta)$ component in the potential inside the atomic sphere, and this component changes the crystal-field integrals considerably, raising the δ_1 and δ_2 sub-bands and lowering the σ sub-band. (The effect on the 2-center integrals is probably smaller, as manifested by the close agreement between the values calculated by the TB method, which does not assume a MT potential, and the effective integrals. The reason for this may be that the 2-center integrals depend mainly on the atomic potential deep inside the atomic sphere, while the crystal-field integrals depend mainly on the potential on the outskirts of the atomic spheres.) The shift of the crystal-field integrals due to this effect may be of order 0.5-1 eV, and it brings the δ_2 peak closer to the Fermi level. The density of states function, when this adjustment is made, is shown in fig. 9.

The existence of this sharp peak close to the Fermi level in the 1-electron theory poses grave problems when electron-electron and electron-phonon interactions are considered. For example, when we consider the electron-electron repulsion term U in the Anderson model, we find that due to the very small "V_{sd}" (actually here $V_{\delta_2\pi}$ or $V_{\delta_2\sigma}$) compared with U, the peak should split to spin ↑ and spin ↓ states and become magnetic. Within the HF theory, changes in electron population produce large changes in the HF energies and the rigid-band approximation no longer applies. We must do "unrestricted" HF, in which electrons of different 3d sub-bands see different, population dependent, HF potentials. When we do this,[38] we find that when the electron-electron interaction term described by the atomic Slater integral F_2 is relatively large compared with F_o (or U), the peak tends to "stick" to the Fermi level, while if U is large, it tends to be repelled. The proper description of such a peak in many-electron (or electron-phonon) theory must still be regarded as an open question. All we can say now is that this peak is indeed observed experimentally.

5. THE VANADIUM 4s BAND

Once the energy levels of the 3d band have been identified, the energy levels of the V 4s band at symmetry points can be determined by elimination. The energies at a general point in the BZ must be determined by some interpolation procedure. The TB scheme was found not to work at all for the 4s band (there is no convergence even with a very large number of effective integrals). On the other hand, the OPW scheme is found to work if also the Si 3s and 3p bands are included.[34] With 7 OPW's, a fit of 0.2 Ry over a total bandwidth of 1.5 Ry is obtained. Levels near E_F are given greater weight in the least-squares fitting procedure, so that the fit near E_F is quite good. This type of behavior is typical for the OPW

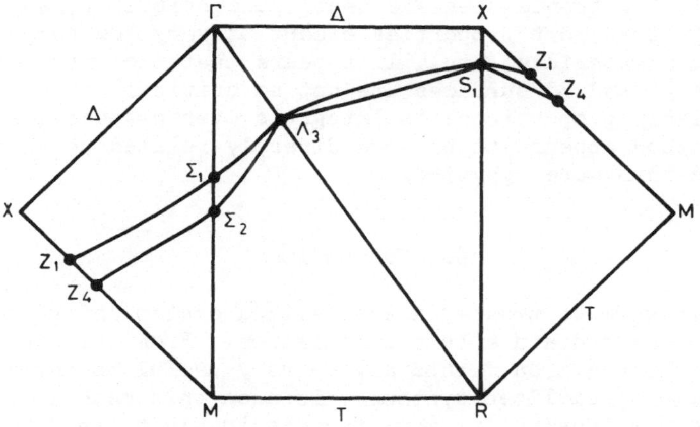

Fig. 10. The 4s Fermi surface derived from the APW calculation.

approximation with a local, energy-independent potential, which cannot be an extremely good approximation for a dense structure, with an unfilled d-shell, over a wide energy range. The Fermi surface determined this way is shown in fig. 10. It is seen that it agrees extremely well with the surface determined phenomenologically from the positron experiment (section 3). It is doubly degenerate, and the OPW's have particularly strong components in the zones centered around points (100) and (110) of the reciprocal space, in excellent agreement with experiment. This is because the lowest state of the 4s band is a Γ_{12} state, which has no component with $p = 0$. The FS contains only about 1/3 of an electron per vanadium atom. Probably the rest of the 4s contribution is included in the "effective" 3d and Si 3s and 3p bands by hybridization.

It should be noted that the 4s FS should be cut by the 3d FS at many points, and therefore even minute hybridization will cause dHvA orbits to be entirely different from what fig. 10 suggests. However, these hybridizations need not affect the positron annihilation curves very much. The reason for this is that the magnitude of some structure seen in the positron annihilation experiment depends (among other factors) on the volume in momentum space producing this structure. Two surfaces (such as a 3d and a 4s FS) cross at a line (which has zero volume in p-space), and if the hybridization is small, wavefunctions and energy levels are affected only in the close vicinity of this line, which has a small volume in p-space. On the other hand, galvanomagnetic effects depend on electron orbits which are altered altogether even if two orbits cross at a point (which has virtually zero volume in p-space). The question arises, what is more "physical," the galvanomagnetic methods or the positron annihilation method. For most electronic properties, such as cohesive energy, electronic specific heat, susceptibility, superconductivity, NMR, transport properties except at very low temperatures and very pure materials, etc., it appears that the intersections between the 4s and 3d surfaces are not so critical (since these too are effectively proportional to integrals over p-space), so the positron method appears to be more directly related to these properties, and thus more "physical."

6. CONCLUSION

In this work we treated a specialized system using the positron-annihilation method and a band calculation. From this work, it appears that the positron method may be very useful to investigate at least some specialized systems. Because this method is so time-consuming and expensive, it may prove desirable to perform a computer-simulated "positron experiment" prior to an actual experiment.

As for the band calculation, the actual computation does not appear to be particularly difficult with present day computers, since by now several methods, such as APW, KKR, and LCAO, have been developed to be standardized and reliable. The problem is now mainly how to handle the "input" and "output" to the computer. The input consists of the approximations made (muffin-tin potential? restricted or unrestricted HF?) which may have to be adjusted to the particular system under investigation. The "output" consists of the quantities calculated explicitly (phase shifts? energy levels? effective parameters describing the band? shapes of constant energy surfaces? wavefunctions? momentum distributions? deformation potentials? etc.). For complicated systems, that possess some special physical features (such as high-temperature superconductivity, metal to insulator transitions, magnetic anomalies, etc.), the quantities to be calculated may have to be adapted especially to the system, and the "standard" outputs such as energy levels along symmetry lines, may prove to be inappropriate and too difficult to handle directly.

*The experimental work was supported at Brandeis University by the National Science Foundation and the U. S. Army Research Office, Durham, N. C. The theoretical computations were performed at the Hebrew University.

REFERENCES

[1] *Positron Annihilation*, edited by Stewart, A. T., and L. O. Roellig (Academic Press Inc., New York, 1967).

[2] Berko, S., and J. S. Plaskett, Phys. Rev. 112, 1877 (1958).

[3] Stewart, A. T., Phys. Rev. 133, A 1651 (1964); K. Fujiwara, O. Sueoka, T. Imura, J. Phys. Soc. Japan 24, 467 (1968); B. W. Murray and J. D. McGervey, Phys. Rev. Letters 24, 9 (1970).

[4] Dekhtyar, I. Ya, D. A. Levina and V. S. Mikhalenkov, Dokl. Akad. Nauk SSSR 156, 765 (1967); S. Berko and J. C. Erskine, Phys. Rev. Letters 19, 307 (1967); I. K. MacKenzie, et al, Phys. Rev. Letters 19, 946 (1967).

[5] See for example J. P. Carbotte and A. Salvadori, Phys. Rev. 162, 290 (1967) and references quoted therein.

[6] Phillips, W. C., and R. J. Weiss, Phys. Rev. 171, 790 (1968).

[7] Mijnarends, P. E., Phys. Rev. 160, 512 (1967).

[8] Stewart, A. T., et al, Phys. Rev. 128, 118 (1962); S. Berko, Phys. Rev. 128, 2166 (1962).

[9] Berko, S., R. E. Kelly and J. S. Plaskett, Phys. Rev. 106, 824 (1957).

[10] Donaghy, J. J., and A. T. Stewart, Phys. Rev. 164, 391, 396 (1967).

[11] Erskine, J. C., and J. D. McGervey, Phys. Rev. 151, 615 (1966); D. Stroud and H. Ehrenreich, Phys. Rev. 171, 399 (1968).

[12] See for example S. Cushner, J. C. Erskine, and S. Berko, Phys. Rev. 1B, 2852 (1970) and references therein.

[13] Williams, R.W., and A. R. Mackintosh, Phys. Rev. 168, 679 (1968).

[14] Mijnarends, P. E., and M. H. H. Höfelt, J. Physique 32, C1-284 (1971); S. Berko and A. P. Mills, J. Physique 32, C1-287 (1971).

[15] Morin, J. F., and J. P. Maita, Phys. Rev. 129, 1115 (1963).

[16] Blumberg, W. E., J. Eisinger, V. Jaccarino, B. T. Matthias, Phys. Rev. Letters 5, 149 (1960); A. M. Clogston, A. C. Gossard, V. Jaccarino, Y. Yafet, Phys. Rev. Letters 9, 262 (1962); Rev. Mod. Phys. 36, 170 (1964).

[17]Silbernagel, B. G., M. Weger, W. G. Clark, J. H. Wernick, Phys. Rev. 153, 535 (1967).

[18]Testardi, L. R., R. R. Soden, E. S. Greiner, J. H. Wernick, V. G. Chirba, Phys. Rev. 154, 399 (1967); K. R. Keller and J. J. Hanak, Phys. Rev. 154, 628 (1967); M. Rosen, H. Klimker, M. Weger, Phys. Rev. 184, 466 (1969).

[19]Willens, R. H., T. H. Geballe, A. C. Gossard, J. P. Maita, A. Menth, G. W. Hull, R. R. Soden, Solid State Comm. 7, 837 (1969).

[20]Clogston, A. M., and V. Jaccarino, Phys. Rev. 121, 1357 (1961).

[21]Weger, M., B. G. Silbernagel, E. S. Greiner, Phys. Rev. Letters 13, 521 (1964); J. P. McEnvoy, Superconductivity Conference, Stanford (1969).

[22]Testardi, L. R., J. E. Kunzler, H. J. Levinstein, J. P. Maita, J. H. Wernick, Phys. Rev. 3B, 107 (1971).

[23]Hoffstein, V., and R. W. Cohen, Phys. Letters 29A, 603 (1969).

[24]Weger, M., Rev. Mod. Phys. 36, 175 (1964).

[25]Frohlich, H., Proc. Roy. Soc. A223, 296 (1954); A. W. Overhauser, Phys. Rev. 128, 1437 (1962).

[26]Labbe, J., and J. Friedel, J. Physique 27, 153, 303 (1966); Phys. Rev. 158, 647 (1967).

[27]Batterman, B. W., and C. S. Barrett, Phys. Rev. 145, 296 (1966).

[28]Weger, M., J. Phys. Chem. Solids, 31, 1621 (1970).

[29]Weger, M., Solid State Comm. 9, 107 (1971).

[30]Mattheiss, L. F., Phys. Rev. 138, A112 (1965).

[31]Berko, S., and M. Weger, Phys. Rev. Letters 24, 55 (1970).

[32]An earlier positron experiment on V_3Si failed to exhibit any anisotropy within statistics - T. W. Mihalisin and R. D. Parks, in *Proceedings of the Ninth International Conference on Low Temperature Physics, Columbus, Ohio, 1964,* edited by J. G. Daunt, D. O. Edwards, F. J. Milford, and M. Yaqub (Plenum Press, Inc., New York, 1965), Pt. A, p. 487.

[33]Hodges, L., H. Ehrenreich, N. D. Lang, Phys. Rev. 152, 505 (1966).

[34]Goldberg, I. B., and M. Weger, "The Electronic Structure of V_3Ga and V_3Si, I" to be published in Solid State Physics and J. Physique.

[35]Ashkenazi, J., and M. Weger, "Simplified Procedure for Calculation and Parametrisation of Tight Binding Integrals," to be published.

[36]Herman, F., and S. Skillman, *Atomic Structure Calculations* (Prentice Hall, N. J., 1963).

[37]Gilat, G., and L. J. Raubenheimer, Phys. Rev. 144, 390 (1966).

[38]Goldberg, I. B., and M. Weger, J. Phys. C. 4, L188 (1971).

II

Theoretical Studies of the Electronic Structure of Crystals

INTRODUCTORY REMARKS

N. W. Dalton[†]

IBM Research Laboratory, San Jose, California 95114

The problem of calculating electronic band structures and properties has remained at the forefront of solid state research from the time when Wigner and Seitz first attempted to calculate the band structure and cohesive energy of metallic sodium. An indication of the scope and size of the band-structure problem can be gained from a glance at the published proceedings of a recent conference[1] devoted solely to this topic. However, in spite of the enormous effort and attention accorded to this problem, the truth still remains that no single method for calculating band structures has been sufficiently developed to apply equally well to all materials.

This is not to say that little progress has been made during the last few years. On the contrary, the relationship between the different methods has been clarified and the limitations of each method investigated in detail.[2] Whereas ten years ago only a few isolated materials had been studied in depth, nowadays the electronic properties of whole groups of compounds have been determined. Our understanding of the nature of the chemical bond and of the properties of amorphous materials has also been considerably advanced in recent times. The papers appearing in the following pages reflect much of this mainstream activity.

The trend of current band-structure work has been to develop empirical methods, e.g., the pseudopotential method, or to refine and extend the classic first-principles methods, e.g., APW, OPW

[†] Permanent address: Theoretical Physics Division, A.E.R.E., Harwell, Berkshire, England.

and KKR. The first two papers in this section are representative of the empirical approach to the calculation of the electronic properties of solids. In his article, Treusch presents a general review of the pseudopotential method and summarizes several applications including the calculation of crystal charge densities and the optical properties of disordered materials. On the other hand, Schneider and Stoll consider a particular application of the pseudopotential method to metallic lithium, but discuss in detail the calculation of phonon dispersion curves, ion-ion potentials, and dielectric functions. Both papers, considered together, underline the importance of empirical schemes as a bridge between first-principles methods and experiment.

Interest in the classic first-principles band-structure methods remains strong as evidenced by the contributions from Dalton (OPW), Bross and Stöhr (APW), Christensen (APW), and Rössler (KKR). The paper by Dalton contains a general method for calculating relativistic band structures which, although based upon the OPW method, also incorporates the LCAO method. The importance of using overlapping orbital basis functions in this method is demonstrated by a calculation of the 3d bands in copper the germanium.

An investigation of many-body effects in copper is reported by Bross and Stöhr using a modified version of the APW method. The results of another application of the APW method directed at choosing potentials for silver, molybdenum and vanadium which give closest agreement with experiment are discussed by Christensen. Finally, extensive applications of the KKR method to semiconductors and ionic crystals are surveyed by Rössler.

Unfortunately, nature does not always provide us with perfect crystal lattice structures, as is the case with glasses and amorphous materials. Our theoretical understanding of the properties of such complex materials is primitive at the present time. However, it is generally agreed that the principal feature which distinguishes amorphous materials from crystalline materials is the absence of long-range order in the former. Effects due to differences in short-range order are expected to be small although not insignificant. Thus the contributions of Henderson, relating to the atomic arrangements in amorphous silicon and germanium, and of Rudge and Ortenburger, on the optical properties of the polytypes of germanium, are of particular interest in assessing the relative importance of effects due to short- and long-range order.

This section concludes with an instructive application of computer graphics to solid state physics by Dalton and Schreiber. Using computer graphics techniques they have generated a color motion picture which explores the three-dimensional character of the electronic charge distribution in the gallium arsenide crystal.

REFERENCES

1. Marcus, P. M., Janak, J. F. and Williams, A. R., editors (1971) Computational Methods in Band Theory (Plenum Press, New York).

2. Ziman, J. (1971) Solid State Physics 26, 1.

EMPIRICAL BAND CALCULATIONS

Joachim Treusch

Institut für Physik der Universität Dortmund

INTRODUCTION

Much has been said in the very recent past about philosophy and technique of the pseudopotential method, which we regard to be the fundamental method in the area of empirical band calculations. Authoritative writers like Marvin Cohen, Volker Heine, J. C. Phillips, John Ziman, and my late friend Rolf Sandrock gave brilliant reviews (1-5), illuminating the various aspects of the work done in this field during the last ten years. Thus little can be done in the frame of this article to enhance the reputation of the pseudopotential scheme, and we will restrict ourselves to the modest aim of giving a report of recent work and of what we feel can be the impact of this method for the next future. At the beginning, however, we have to recall some of the basic ideas and formulas in order to make clear, what we are speaking about.

THE PSEUDOPOTENTIAL SCHEME

The one-electron Schrödinger- (or Pauli-) equation, poor as it is for a crystal, is the starting point of almost all band calculations, be it self-consistent OPW-, adhoc- (like KKR or APW), semi-empirical (whatever that means), or fully empirical calculations.

$$\{ -\Delta + V(r)\} \psi = E \psi \qquad (1)$$

Using an ansatz wave-function of the OPW-type

$$|\psi\rangle = |\phi\rangle - \sum_1 |\chi_1\rangle\langle\chi_1|\phi\rangle \ , \qquad (2)$$

where ϕ means a symmetrized linear combination of plane waves and χ an atomic eigenfunction of the atoms building up the crystal, we can rearrange equation (1) to

$$\{ -\Delta + V(\vec{r}) + \sum(E-E_1)|\chi_1\rangle\langle\chi_1|\}|\phi\rangle = E|\phi\rangle \ . \qquad (3)$$

Here we have tacitly assumed, that the crystal potential yields atomic eigenvalues when acting upon atomic eigenfunctions. Formally we can interpret equation (3) as a pseudo-wave equation for the pseudo-wave function ϕ, the Hamiltonian consisting of the kinetic energy term and a pseudopotential

$$V_{Ps} = V(\vec{r}) + \sum(E-E_1)|\chi_1\rangle\langle\chi_1| \ , \qquad (4)$$

which is nonlocal, dynamic, and depends on angular momentum. Obviously little progress has been made up to here. There acts, however, a strong "cancellation theorem", which causes a rather effective weakening of V_{Ps} as compared to the real crystal potential. Moreover, since the set of trial functions (2) is overcomplete, the pseudopotential given in equation (4) is not unique. It can be shown, that a rather general form of a pseudopotential yielding a pseudo-wave equation equivalent to the original Schrödinger equation is given by

$$V_{Ps} = V(\vec{r}) - \sum|\chi_1\rangle\langle F_1| \ , \qquad (5)$$

where $F_1(\vec{r})$ is a regular function to be chosen at will. If we take $F_1 = (E-E_1)\chi_1$ we get formula (4), the so-called Phillips-Kleinman family of pseudopotentials. If we take $F_1 = V\cdot\chi_1$ we get the Austin-Heine-Sham family of pseudopotentials, from which direct evidence of the cancellation effect can be drawn. If the atomic eigenfunctions were a complete set of functions, the pseudopotential would in fact equal zero. Still wider classes of pseudopotentials than those of equation (5) are described in Ziman's paper (4); we will not regard them.

In practice the procedure is not to construct a pseudopotential according to formulas (4) or (5). One uses rather the cancellation theorem as a motivation to invent model potentials, mostly local and static ones, that combine two essential features:
1. they are weak enough to ensure reasonably rapid convergence of a plane wave expansion of the pseudo-wave function,
2. they are close enough to reality, so that there is no total arbitrariness, and trends that can be followed in measurable physical properties of e.g. neighboring substances of the periodic system are reflected by the respective model potentials.
With the use of these model potentials one then tries to solve the pseudo- wave equation. As usual a variational procedure leads to a secular equation from which eigenvalues E(k) and eigenfunctions may be determined:

$$\det \left| \{(\vec{k}+\vec{K}_n)^2 - E\}\delta_{nn'} + V^{Ps}_{nn'} \right| = 0 . \qquad (6)$$

The main problem is how to provide the "form factors"

$$V^{Ps}_{nn'} = \int V^{Ps}(\vec{r})\exp\{i(\vec{K}_n-\vec{K}_{n'})\vec{r}\}d^3\vec{r}. \qquad (7)$$

In crystals containing more than one atom per unit cell the form factor reads

$$V^{Ps}_{nn'} = S_{nn'} \cdot \overline{V}_{nn'} , \qquad (8)$$

where now $\overline{V}_{nn'}$ is the pure form factor of equation (7). $S_{nn'}$, the so-called structure-factor is given as

$$S_{nn'} = \frac{1}{M}\sum_{j=1}^{M} \exp\{i(\vec{K}_n-\vec{K}_{n'})\vec{\alpha}_j\} . \qquad (9)$$

M is the number of atoms per unit cell, the vector $\vec{\alpha}_j$ describes the position of the j-th atom.

MODEL POTENTIALS AND FORM FACTORS

The historical development of the art to construct model potentials and form factors shows two "schools" of artistry. The best example of the "fully empirical" school is the famous article by Cohen and Bergstresser (6), where band structures of fourteen semiconductors with Zincblende structure were presented. They had been attained through a fit to essentially optical data of these semiconductors.

The "semiempirical" school tries to construct model potentials, which are fitted to atomic or ionic spectroscopic data instead of solid state data. These model potentials are thentransferred to band structure calculations. The most representative name of this school is that of Volker Heine. The extremely useful model potential proposed by him and his coworkers (7) shall be the only one described here, since it influenced most of the recent pseudopotential band calculations. The so-called Heine-Abarenkov potential

$$V_{HA}(r) = \begin{cases} \sum A_l P_l & \text{for } r < R_M \\ -Ze^2/r & \text{for } r > R_M \end{cases} \qquad (1o)$$

consists of an l-dependent inner part (P_l is a projection operator acting only on the l-component of the wave function) and an ionic potential outside a sphere with radius R_M. The parameters A_l and R_M are fitted to spectroscopic data of free ions. Screening is taken into account by Heine and Animalu (7) on the basis of a free-electron model. They published model potentials and form factors for as much as 25 elements.

Free-electron like screening is of course inadequate for semiconductors and insulators. The dielectric constant $\varepsilon(q)$ which is infinite for $q \to o$ for metals cancels the singularity of the fourier transform of the ion potential.

$$V^{eff}(q) = \frac{V_{HA}(q)}{\varepsilon\,(q)}$$ is finite for $q=o$. In a semiconductor the screening function $\varepsilon(q)$ tends to a constant value ε_o for $q \to o$. Thus we would have a remaining singularity in the screened potential; the crystal would not be neutral as a whole. Phillips, in his by now rather well known "dielectric theory of covalent bonding" (8) postulated a localised negative point charge in between

neighboring atoms, a "bond charge", in order to cure this failure. The total pseudopotential then results as

$$V(q) = S_a(q) \cdot V_a(q) + S_b(q) \cdot V_b(q) . \qquad (11)$$

Here $S_{a,b}$ means the atomic and bond charge structure factor, respectively, $V_a(q)$ is the fourier transform of a Heine-Abarenkov potential screened with Penn's dielectric function (9), $V_b(q)$ is the fourier transform of a point charge. The form factors calculated along these lines are indeed able to yield realistic band structures, or, to be more specific, they agree well (within o.leV) with empirical form factors of Cohen and Bergstresser (6), which were known at the time, when the dielectric theory was born.

BAND CALCULATIONS

As has been mentioned the great break-through of the pseudopotential method in the field of band calculations was the Cohen and Bergstresser paper of 1966. Since then band structures of some forty semiconductors have been calculated, and lots of Fermi surfaces of simple and even of transition metals have been determined. Ample citation can be found in the article by Cohen and Heine (1). Two main conclusions can be drawn from the abundance of results:

1. the empirically determined form factors fit rather well the V(q)-curves calculated by Heine and Animalu, modified in the case of semiconductors according to Phillips theory,

2. the form factors of similar substances are similar; well defined trends can be followed going through the periodic table.

In order to prove the significance of these two statements let us look at some examples. The first example is the Cohen-Bergstresser letter itself. It shows clearly the transferability of form factors and their connexions for different substances. E.g. the form factors of Germanium were directly applicable as the symmetric part of the GaAs form factors. Later Cohen-Bergstresser's empirical form factors were used in quite different circumstances. Liu and Falicov (1o) - our second example - calculated the Fermi surfaces of Arsenic and Antimony with the help of form factors constructed from those which Cohen and Bergstresser had used for GaAs and InSb, respectively. Sandrock, to give a final example, succeeded in a band structure calculation of Selenium (11) taking Se-formfactors from empirical ZnSe results. Small corrections had to be added, since the type of screening and the coordination number is different in Selenium as compared to Zincselenide. Here the model potentials of Heine and Animalu could serve as a guideline, limiting the range of arbitrariness in the determination of V(q) to tenths of an eV, and allowing for an inter- or extrapolation of V(q) to values of reciprocal lattice vectors characteristic for trigonal Selenium but not calculated in the case of cubic ZnSe.

The example of Selenium seems to us to be characteristic for the present situation. In contrast to the first pseudopotential band calculations where one could not be sure whether a particular set of formfactors was unique, we now have some sound arguments to decide. Transferability of form factors from one compound to another one, to an element, or to the same compound in another crystalline structure (e.g. cubic and hexagonal ZnS (12)) has been proven to hold. Since adjusted formfactors are known for quite a lot of elements, one can in principle calculate band structures of a huge number of compounds, if one uses the additional possibility to extrapolate or interpolate V(q)-curves following the functional form of Heine-Animalu model potentials. Of course there are computational restrictions: the higher the number of atoms per unit cell is, the higher is the density of reciprocal lattice vectors. Since the convergence is roughly determined by the absolute value of the largest K-vector taken into account, the dimension of the secular determinant grows rapidly for non-Bravais lattices thus leading beyond the limits of the present computer generation's capabilities. The most complex crystals treated up to now in the framework of the pseudopotential theory are, at least as to our knowledge, the different polytypes of SiC (13). For the type 6H-SiC with 12 atoms per unit cell 37o plane waves had to be taken into account to obtain convergence within o.3 eV for the highly symmetric points of the Brillouin zone.

We are not sure whether it is reasonable at this point to make propaganda for still faster computers in order to enable us to calculate still more complex crystals with still more plane waves and improved convergence but without much hope to get new physical insight, d-BANDS, RELATIVISTIC CORRECTIONS, NON-LOCAL POTENTIALS

More interesting extensions of the simple pseudopotential scheme are concerned with relativistic effects, the occurence of high-lying d-bands or resonances, and the improvements possible through the use of non-local pseudopotentials.

If formfactors are fitted to interband transition energies taken from experiment they automatically contain the relativistic corrections due to mass-velocity- and Darwin-term, since these terms transform according to single group representations and cause no additional splitting as compared to non-relativistic calculations. Taking into account the spin-orbit coupling operator

$$- \frac{i\alpha^2}{4}(\nabla V x \nabla)\vec{\sigma} = H_{soc} \tag{12}$$

is not as straightforward as e.g. in a tight-binding or Green's function formalism (see following papers of this volume). Matrix elements of H_{soc} cannot be calculated simply with pseudo-wave functions, since the largest contributions stem from the close neighborhood of the nucleus, where H_{soc} is singular. Assuming H_{soc} to be spherically symmetrical in this region Weisz (14) calculated the matrix element in plane wave representation:

$$<\vec{k}'s'|H_{soc}|\vec{k}s> = S(\vec{k}-\vec{k}')\{(\alpha^2/4)V(|\vec{k}'-\vec{k}|) - \lambda_1 - \lambda_2(\vec{k}'\cdot\vec{k})\}(\vec{k}'x\vec{k})\cdot\vec{\sigma}_{ss'} \tag{13}$$

where S is the structure factor, U the Fourier transform of the crystal potential, and λ_1 and λ_2 positive constants that account for the contribution of the core p- and d-states, respectively. Usually these core-core contributions to the matrix element exceed the plane-wave contribution by far, so that the term involving the crystal potential can be neglected. Actual calculations on the basis of Weisz' model were performed for e.g. SnTe, GeTe, and PbTe (15); grey tin, InSb, and CdTe (16).

The pseudopotential scheme in its plane wave form fails if there are d-bands in the neighborhood of the Fermi energy in a metal or of the gap in a semiconductor, since these d-electrons are strongly localised and cannot be adequately described by few plane waves. If one adds to the overcomplete set of trial functions given in equation (2) the atomic d-functions, which are not eigenfunctions of the crystal Hamiltonian as had been assumed for the lower lying core states, an s-d-hybridization occurs. On the basis of this combined scheme the band structures of noble and transition metals can be calculated (see e.g. 17).

Another way to include d-bands is to use an l-dependent pseudopotential. Fong and Cohen (18) calculated the band structure of copper using an empirical s-p-pseudopotential and a strong square-well potential acting on d-electrons alone. A very similar procedure was followed by Hemstreet, Fong, and Cohen (19) in the case of diamond. Here the authors introduced an l=1 nonlocal potential term in order to account for the missing cancellation of p-electrons (no l=1 core state exists).

It is anticipated by Cohen and his coworkers - and by the present author - that this nonlocal version of the empirical pseudopotential scheme will help to clarify the bandstructures of noble and transition metals and their compounds, which gain increasing importance.

OPTICAL CONSTANTS, EFFECTIVE MASSES, CHARGE DENSITIES

Hitherto we have discussed the various ways to calculate band structures, i.e. eigenvalues by means of the pseudopotential method. By now one has learnt that the eigenfunctions of the pseudo Schrödinger equation, which describe only the smooth part of the real electronic wave function in the crystal, can be widely and successfully used.

Optical transition matrix elements were computed and used to calculate the imaginary part of the dielectric function $\varepsilon_2(\omega)$ and from that reflectivity and absorption coefficients of a given substance. Brust's early work on Germanium (2o) showed that dipole matrix elements of pseudo-wave functions fairly well described the absolute height of structures in the optical constants, the energetic position of the peaks being described mainly by the joint density of states. Sandrock's calculation (11) of the optical constants of Selenium demonstrated the usefulness of pseudo-wave functions also for strongly anisotropic substances. The classical critical-point assignment has been replaced by a detailed

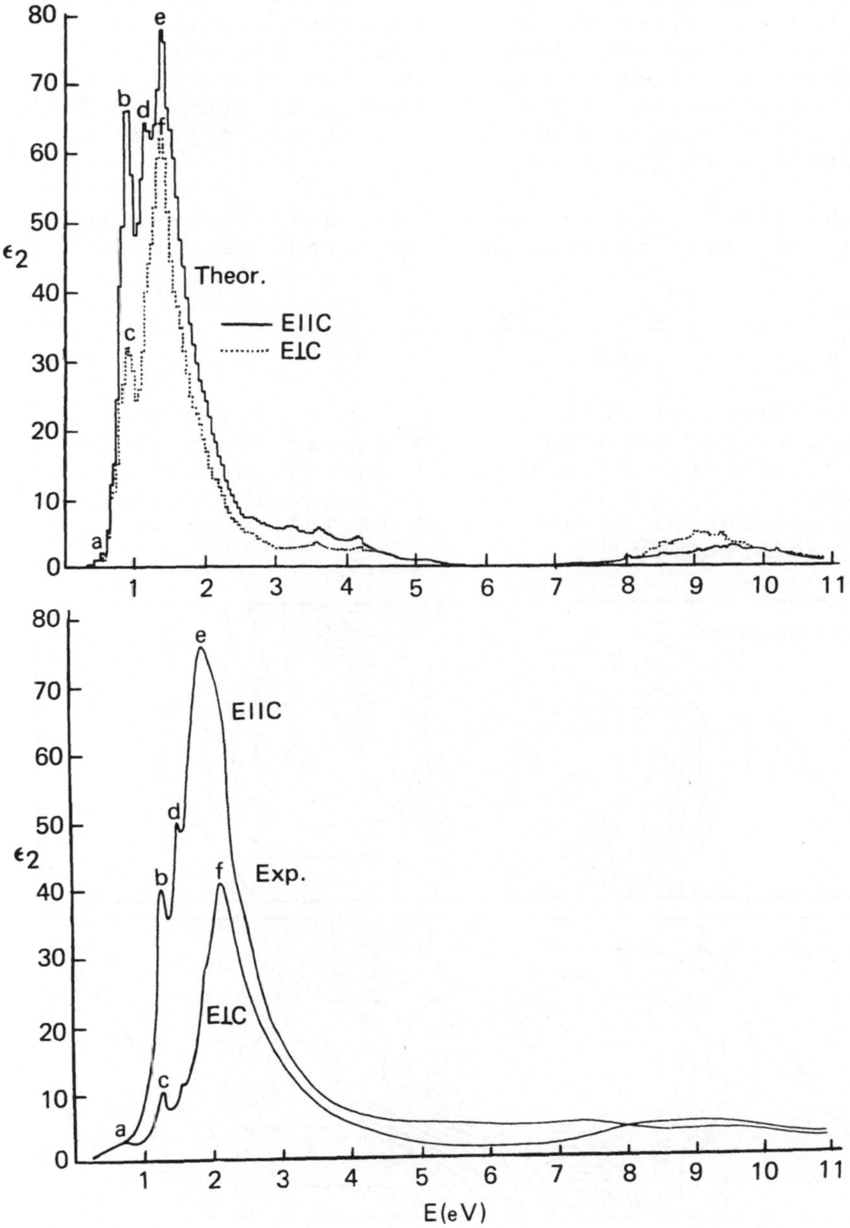

Figure 1: Imaginary part of the dielectric constant of trigonal Tellurium taken from reference 21.

calculation of what parts of the Brillouin zone contribute to certain structures in the optical spectrum, and it is the existence of selection rules for the matrix elements that causes the polarization dependence of the dielectric constant. As an example we show Maschke's recent calculation of the dielectric constant of Tellurium (21), which was performed in close analogy to Sandrock's work (see fig.1!).

The dipole matrix element can also be used to calculate effective masses. The k·p-formalism yields

$$\frac{m}{m^{+}_{\alpha,i}} = 1 + \frac{2}{m} \sum_{\alpha \neq \beta} \frac{|\langle\alpha|p_i|\beta\rangle|^2}{E_\alpha - E_\beta} \qquad (14)$$

for the effective mass of band α in i-direction. Effective conduction band masses for eleven cubic semiconductors have been calculated by Bowers and Mahan (22) with the aid of pseudo-wave functions. The deviations from experimental values are about 1o%, what we feel is quite satisfactory.

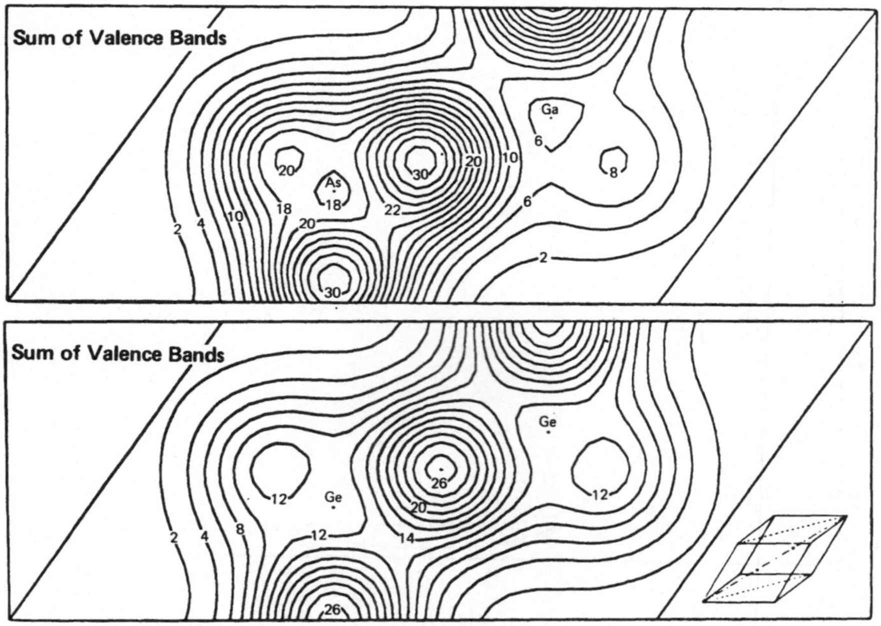

Figure 2: Valence-electron-density contour map for GaAs and Ge in the (1,-1,o)plane taken from ref.23

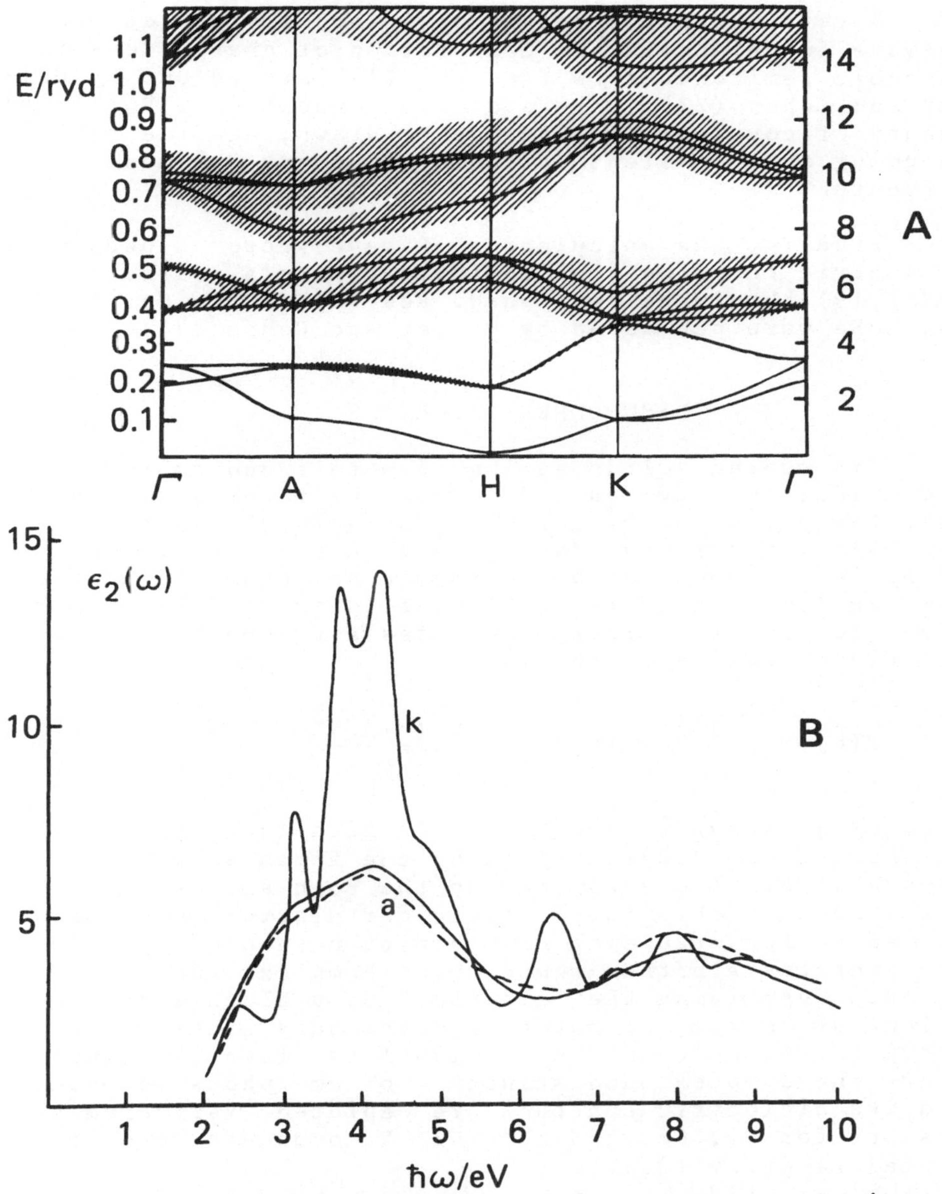

Figure 3: Complex band structure of amorphous Selenium calculated with α=o.1. The imaginary part of the energy is shown as a blurring of the real part. (b) Imaginary part of the dielectric constant. Curve k: isotropic spectrum of crystalline Se; curves a, spectrum of amorphous Se, calculated (solid line) and experimental (dotted line). Taken from reference 25

A very intuitive result of an application of pseudo-wave functions is the computation of charge densities in cubic semiconductors (see fig.2!) carried out by Walter and Cohen (23). Here a direct connexion to Phillips' theory of covalent bonding can be drawn, and the trends in going from covalent to ionic compounds are made very perceptual.

Finally, the calculation of wave-vector dependent dielectric functions, which improve Penn's isotropic model (9) should be mentioned. Results on Si, Ge, GaAs, and ZnSe were published by Walter and Cohen (24).

DISORDERED MATERIALS

Disordered solids as liquid metals and amorphous semiconductors have gained increasing interest in the last few years. We will confine ourselves here to mentioning one recent development in the description of optical properties of amorphous semiconductors: the Complex Band Structure (CBS)-model suggested by Kramer (25). To describe disorder he uses a two-particle correlation function of the form

$$P(\vec{q}) = \pi^{-3/2} \sum_n (\alpha|\vec{K}_n|)^{-3} \exp\frac{(\vec{q}+\vec{K}_n)^2}{\alpha^2 K_n^2} , \quad (15)$$

i.e. if α is not to large (α<o.1) then there are relics of short-range order. Utilizing the flatness of the pseudopotential one can sum up the Born series constructed with this two-particle correlation function and solve the resulting non-hermitian secular equation. The energies are then found to be complex, the imaginary part describing the "lifetime" of a Bloch state, which, of course, cannot be a stationary state if $\alpha \neq$o. A typical example of the CBS-model is given in figure 3, where the complex band structure of amorphous Selenium and its dielectric constant are depicted. Very promising results for Si, Ge, and some III-V compounds were reported recently (26).

PERSPECTIVES

What is the impact now of what has been said on the last few pages? We feel that the time of pure band calculations is seeing its horizon. That does not at all mean, however, that the time of pseudopotentials has gone.

At first there are the direct extensions of the simple pseudopotential scheme that deserve further attention; namely the inclusion of d-bands, relativistic, and nonlocal effects and the calculation of optical constants, effective masses, and charge densities. Then there is the dielectric theory of bonding which is based on empirical band calculations, but has become fruitful also in the field of impurities (27) and phonon dispersion calculations (28). The theory of cohesion has got new impulses from the pseudopotential scheme (2,29), binding energies of excitons can be calculated with the help of pseudopotentials (3o). To conclude this enumeration, the vast field of amorphous semiconductors is a great challenge for those who have by now learnt to understand a good deal of the properties of the crystalline state.

REFERENCES

(1) M.L. Cohen and V. Heine, Solid State Physics, Vol.24, 38 (197o)
(2) V. Heine, Solid State Physics, 24, 1 (197o)
V. Heine and D. Weaire, Solid State Physics 24,25o
V. Heine, in "The Physics of Metals" Vol.I, Electrons (J.M. Ziman, ed.), Cambridge Univ.Press 1968
(3) J.C. Phillips, Advances in Physics 17,79 (1968)
(4) J.M. Ziman, Solid State Physics 26, 1 (1971)
(5) R. Sandrock, in Festkörperprobleme X (O. Madelung, ed.) p.283, Vieweg, Braunschweig 197o
(6) M.L. Cohen and T. K. Bergstresser, Phys.Rev. 141, 789 (1966)
(7) V. Heine and I.V. Abarenkov, Phil.Mag. 9,451(1964)
I.V. Abarenkov and V. Heine, Phil.Mag.12,529(1965)
A.O.E. Animalu, Phil.Mag.11,379(1965)
A.O.E. Animalu and V. Heine, Phil.Mag.12,1249(1965)
(8) J.C. Phillips, Phys.Rev. 168, 9o5,912,917 (1968)
(9) D.R. Penn, Phys.Rev. 128, 2o93 (1962)
(1o) L.M. Falicov and P.J. Lin, Phys.Rev.141,562 (1966)
P.J. Lin and L.M. Falicov, Phys.Rev.142,441 (1966)
(11) R. Sandrock, Phys.Rev. 169, 642 (1968)
(12) T.K. Bergstresser and M.L. Cohen, Phys.Rev. 164, 1o69 (1967)
(13) H.-G. Junginger and W. van Haeringen, phys.stat. sol. 37, 7o9 (197o)
(14) G. Weisz, Phys.Rev. 149, 5o4 (1966)
(15) Y.W. Tung and M.L. Cohen, Phys.Rev. 18o,823(1969)
(16) S. Bloom and T.K. Bergstresser, Sol.State Commun. 6, 465 (1968)

(17) F.M. Mueller, Phys.Rev. 153, 659 (1967)
F.M. Mueller and J.C. Phillips, Phys.Rev. 157,6oo (1967)
(18) C.Y. Fong and M.L. Cohen, Phys.Rev.Letters 24,3o6 (197o)
(19) L.A. Hemstreet Jr., C.Y. Fong, and M.L. Cohen, Phys.Rev. B2, 2o54 (197o)
(2o) D. Brust, Phys.Rev. 134, 1337 (1964)
(21) K. Maschke, phys.stat.sol (in press)
(22) R.L. Bowers and G.D. Mahan, Phys.Rev. 185,1o73(1969)
(23) J.P. Walter and M.L. Cohen, Phys.Rev.Letters 26, 17 (1971)
(24) J.P. Walter and M.L. Cohen, Phys.Rev. B2,1821(197o)
(25) B. Kramer, K. Maschke, P. Thomas, and J. Treusch, Phys.Rev.Letters 25, 1o2o (197o)
B. Kramer, phys.stat.sol., 41, 649, 725 (197o)
(26) B. Kramer, in Proc.Internatl.Conf.on Amorphous and Liquid Semicond.,Ann Harbor, Mich., 1971, in press
(27) J.C. Phillips, Phys.Rev. B1 154o, 1545 (197o)
(28) R.M. Martin, Phys.Rev. B1, 4oo5 (197o)
(29) J.C. Phillips and J.A. van Vechten, Phys.Rev. B2, 2147 (197o)
(3o) J. Hermanson and J.C. Phillips, Phys.Rev. 15o,652
J. Hermanson, Phys.Rev. 15o, 66o (1966)
U. Rössler, phys.stat.sol. 42, 345 (197o)

ON THE THEORY OF METALLIC LITHIUM

T. Schneider and E. Stoll

IBM Zurich Research Laboratory

CH-8803 Rüschlikon, Switzerland

1. INTRODUCTION

The pseudopotential formalism has been extensively applied to the interpretation of different physical properties of simple metals.[1-3] However there are some well-known drawbacks to this approach. (i) Its use requires an accurate knowledge of both the core-state eigenfunctions and the core-state eigenvalues. Unfortunately these are not known very accurately. In fact the core-state eigenvalue shifts in the metal as compared to the free ion.[4] For this reason model potentials have been introduced.[4-7] However, Heine and Abarenkov,[5] Ashcroft[6] and Shaw[7] make no attempt to base the shape of their model potentials on any physical principle. This and the missing information about the core shift show the importance of first-principle pseudopotentials. (ii) In reality, the electron gas screens the ionic pseudopotentials. This effect requires a knowledge of the dielectric function of the electron gas. However, there is still some uncertainty about the best approximation to include the effects of exchange and correlation.[8-12]

In this paper we apply the pseudopotential formalism in a consistent manner to metallic lithium. To determine the parameters of the ion entering the Phillips and Kleinman pseudopotential,[13] two methods are used: a first-principles method, and a phenomenological method which adjusts the core-state eigenvalue and eigenfunction parameters to fit the phonon frequencies. We note that this fit provides a powerful way to determine the core shift. Other properties depend only on some average of the pseudopotential or on special Fourier components of the potential. Within this frame we shall discuss the phonon dispersion curves and the effective ion-ion potential with special emphasis on the consequences of the approximations to include the correlation effects in the Hartree-dielectric function of the electron gas. Furthermore we investigate in some detail the lattice stability, the martensitic phase transformation and the band structure of b.c.c. lithium.

2. OUTLINE OF THE THEORY

In this section we shall give a brief outline of the important concepts and formulas of the pseudopotential formalism. Let us first consider the calculation of the total energy of a simple metal in the Born-Oppenheimer adiabatic approximation:

$$H = T_I + \Phi(\{\underset{\sim}{R}^\ell\}), \quad \Phi(\{\underset{\sim}{R}^\ell\}) = V_{II}(\{\underset{\sim}{R}^\ell\}) + \varepsilon_o(\{\underset{\sim}{R}^\ell\}). \tag{1}$$

T_I is the kinetic, Φ the potential energy, V_{II} the electrostatic Coulomb energy of the ions and ε_o the ground-state energy of the valence electrons. Following Kohn and Sham[14] the potential energy Φ per ion, including the effects of electron exchange and correlation, can be written as

$$\Phi(\{\underset{\sim}{R}^\ell\}) = \frac{1}{2N} \sum_{\ell,\ell'}{}' \frac{(Ze)^2}{|\underset{\sim}{R}^\ell - \underset{\sim}{R}^{\ell'}|} + \frac{2}{N} \sum_{k \leq k_F} E_{\underset{\sim}{k}} - \frac{e^2}{2N} \int \frac{\rho(\underset{\sim}{r})\,\rho(\underset{\sim}{r}')}{|\underset{\sim}{r} - \underset{\sim}{r}'|} d^3r d^3r'$$
$$+ \frac{1}{N} \int \rho(\underset{\sim}{r}) \left(E_{ec}[\rho(\underset{\sim}{r})] - \mu_{ec}[\rho(\underset{\sim}{r})]\right) d^3r \tag{2}$$

where

$$\rho(\underset{\sim}{r}) = 2 \sum_{k \leq k_F} \psi_k^*(\underset{\sim}{r})\, \psi_k(\underset{\sim}{r}) . \tag{3}$$

The energy levels and eigenfunctions of the valence electrons enter through

$$H\psi_k = \left\{ T + V + V_c(r) + \mu_{ec}\, [\rho(\underset{\sim}{r})] \right\} \psi_k = E_{\underset{\sim}{k}}\, \psi_k \tag{4}$$

where

$$V_c(\underset{\sim}{r}) = \int \frac{e^2}{|\underset{\sim}{r} - \underset{\sim}{r}'|} \rho(\underset{\sim}{r}')\, d^3r' . \tag{5}$$

E_{ec} is the exchange and correlation energy per electron, μ_{ec} the corresponding chemical potential, Z the valence and V the electron-ion interaction which contains exchange and correlation effects between the core and valence electrons.

In the pseudopotential method one introduces the core projector[1,3]

$$P = \sum_c |\psi_c\rangle\langle\psi_c| = \sum_c P_c \tag{6}$$

and writes for the eigenfunction of a valence electron

$$|\psi_k\rangle = (1 - P)|\varphi_k\rangle . \tag{7}$$

Substitution of (7) into (4) leads to a wave equation for $|\varphi_k\rangle$,[12]

$$(H + V_R)|\varphi_k\rangle = E_k|\varphi_k\rangle \tag{8}$$

where

$$V_R = \sum_c (E_{\underset{\sim}{k}} - E_c)\, P_c . \tag{9}$$

In (8) the core energies E_c lie lower than $E_{\underset{\sim}{k}}$, so that V_R behaves like a repulsive potential which has the effect of cancelling some of the negative potential V in H (4). This implies that $|\varphi_k\rangle$ should approximate a combination of plane waves, namely,

$$|\varphi_{\underset{\sim}{k}}\rangle = \sum_{\underset{\sim}{h}=0} a_{\underset{\sim\sim}{hk}}\, |\underset{\sim}{k}+\underset{\sim}{h}\rangle . \tag{10}$$

Substituting this expression into (8) and multiplying by $\langle\underset{\sim}{k}'|$ we find

$$\sum_{\underset{\sim}{h}=0} \langle\underset{\sim}{k}'|(T+V+V_R+V_c+\mu_{ec})\, a_{\underset{\sim\sim}{hk}}\, |\underset{\sim}{k}+\underset{\sim}{h}\rangle = \sum_{\underset{\sim}{h}=0} E_{\underset{\sim}{k}}\, \delta_{\underset{\sim}{k}',\underset{\sim}{k}+\underset{\sim}{h}}\, a_{\underset{\sim\sim}{hk}} . \tag{11}$$

$V + V_R$ is the so-called bare pseudopotential. Provided that the pseudopotential is weak and degeneracies can be neglected, this equation can be solved iteratively. Starting from free electrons

$$E_{\underset{\sim}{k}}^{\,o} = \langle\underset{\sim}{k}|\; T \;|\underset{\sim}{k}\rangle = \frac{\hbar^2}{2m}\, k^2 \tag{12}$$

one obtains in first order

$$E_k^{\,1} = \frac{\hbar^2}{2m}\, k^2 + \langle\underset{\sim}{k}|(V + \sum_c (E_{\underset{\sim}{k}}^{\,o} - E_c)|\psi_c\rangle\langle\psi_c|)|k\rangle + \frac{6}{5}\,\frac{Ze^2}{r_s} + \mu_{ec}(\rho_o) . \tag{13}$$

r_s is the atomic radius and ρ_o the uniform electron density. It should be noted that $E_{\underset{\sim}{k}}$ enters V_R only in zeroth order, because $|\psi_c\rangle\langle\psi_c|$ is of first order. In second order one finds

$$E_{\underset{\sim}{k}}^{2} = E_{\underset{\sim}{k}}^{1}\left(1 + \sum_{c} \langle \underset{\sim}{k}|\psi_c\rangle\langle\psi_c|\underset{\sim}{k}\rangle\right) + \sum_{\underset{\sim}{h}\neq \underset{\sim}{0}} \frac{|\langle \underset{\sim}{k}+\underset{\sim}{h}|\; U_I \;|\underset{\sim}{k}\rangle|^2}{E_{\underset{\sim}{k}}^{o} - E_{\underset{\sim}{k}+\underset{\sim}{h}}^{o}}$$

$$= E_{k}^{1}\left(1 + \sum_{c} \langle \underset{\sim}{k}|\psi_c\rangle\langle\psi_c|\underset{\sim}{k}\rangle\right) + \sum_{\underset{\sim}{h}\neq \underset{\sim}{0}} \frac{|F(\underset{\sim}{h})|^2 \; |U_I(\underset{\sim}{k}+\underset{\sim}{h},\underset{\sim}{k})|^2}{E_{\underset{\sim}{k}}^{o} - E_{\underset{\sim}{k}+\underset{\sim}{h}}^{o}} . \tag{14}$$

U_I is the screened or self-consistent pseudopotential. The last form is obtained by assuming the potentials to be those of overlapping ions situated at the sites $\underset{\sim}{R}^{\ell}$:

$$V(\underset{\sim}{r}) = \sum_{\ell} V(\underset{\sim}{r} - \underset{\sim}{R}^{\ell}) , \tag{15}$$

the matrix elements may be factorized into a structure factor and a matrix element associated with a single ion,[1] namely,

$$\langle \underset{\sim}{k} + \underset{\sim}{Q}|\; V|\underset{\sim}{k}\rangle = F(\underset{\sim}{Q})\; V(\underset{\sim}{k} + \underset{\sim}{Q},\underset{\sim}{k}) \tag{16}$$

where the structure factor is given by

$$F(\underset{\sim}{Q}) = \frac{1}{N}\sum_{\ell} e^{-i\underset{\sim}{Q}\cdot\underset{\sim}{R}^{\ell}} . \tag{17}$$

The self-consistent pseudopotential U_I is a sum of the bare pseudopotential $V + V_R$ and the Coulomb, exchange and correlation potentials of the screening charge density. We note that the matrix elements of the bare and screened pseudopotentials are related through an integral equation.[15] For local exchange and correlation potentials the solution reads

$$U_I(\underset{\sim}{k},\underset{\sim}{k}+\underset{\sim}{Q}) = \frac{V(Q) + \dfrac{16\pi\rho_o e^2}{Q^2}\,[1-G(Q)] \displaystyle\sum_{k'\leq k_F} \frac{V_R(\underset{\sim}{k}',\underset{\sim}{k}'+\underset{\sim}{Q})}{E_{\underset{\sim}{k}'}^{o} - E_{\underset{\sim}{k}'+\underset{\sim}{Q}}^{o}}}{\varepsilon(Q)} + V_R(\underset{\sim}{k},\underset{\sim}{k}+\underset{\sim}{Q}) \tag{18}$$

where $\varepsilon(Q)$ is given in terms of the Hartree dielectric function and the function $G(Q)$ by

$$\varepsilon(Q) = 1 + (1 - G(Q))(\varepsilon_H(Q) - 1) . \qquad (19)$$

The function $G(Q)$ appearing in (18) and (19) describes the corrections due to exchange and correlation and is related to the second functional derivative of the exchange and correlation energy $E_{ec}[\rho(r)]$ [Eq. (2)]. In the last few years several approximate expressions for $G(Q)$ have been proposed.[8-12] Some examples, including their asymptotic behavior, are listed in Table I.

Table I: Approximate expressions for $G(Q)$; p may be calculated from[20]

$$p = \frac{2}{1 + 0.153 \frac{1}{\pi a_o k_F}} .$$

	$G(Q)$	$\lim_{Q\to 0} G(Q)(\frac{k_F}{Q})^2$	$\lim_{Q\to\infty} G(Q)$
Hartree approximation	0	0	0
Sham's adaption of Hubbard's procedure; Ref. 8	$G_H(Q)=\frac{1}{2}\frac{Q^2}{Q^2+pk_F^2}$	$\frac{1}{2p}$	1/2
Kleinman; Ref. 9	$G_K(Q)=\frac{1}{4}(\frac{Q^2}{Q^2+pk_F^2} + \frac{Q^2}{pk_F^2})$	$\frac{1}{2p}$	$\frac{1}{4}\frac{Q^2}{pk_F^2}$
Singwi et al.; Ref. 10	$G_{STLS}(Q)$ see Ref. 10	3/8	1/2
Singwi et al.; Ref. 11	$G_{SSTL}(Q)=A(1-e^{-B(Q/k_F)^2})$	$A\cdot B$	A

We are now ready to calculate the potential energy of the ions. Using Eqs. (2), (3) and (4) one obtains in second order,

$$\Phi(\{\underset{\sim}{R}^\ell\}) = \Phi(\rho_o) + \Phi_e(\{\underset{\sim}{R}^\ell\}) + \Phi_c(\{\underset{\sim}{R}^\ell\}) \qquad (20)$$

where

$$\Phi(\rho_o)=Z\left[\frac{3}{5}\frac{\hbar^2}{2m}k_F^{\,2}-\frac{9}{10}\frac{Ze^2}{r_s}+E_{ec}(\rho_o)+\frac{2}{NZ}\sum_{k\leq k_F}V_R(\underset{\sim}{k},\underset{\sim}{k})\right] \tag{21}$$

$$\Phi_e(\{\underset{\sim}{R}^\ell\}) = \frac{1}{2}\sum_{\underset{\sim}{Q}\neq 0}|F(\underset{\sim}{Q})|^2\, U_e(Q) \tag{22}$$

$$U_e(Q) = -4\sum_{k\leq k_F}\frac{U_I(\underset{\sim}{k}+\underset{\sim}{Q},\underset{\sim}{k})(V(Q)+V_R(\underset{\sim}{k},\underset{\sim}{k}+\underset{\sim}{Q}))}{E^o_{\underset{\sim}{k}+\underset{\sim}{Q}}-E^o_{\underset{\sim}{k}}} \tag{23}$$

$$\Phi_c(\{\underset{\sim}{R}^\ell\}) = \frac{(Ze)^2}{2}\sum_\ell\frac{G(\varepsilon R^\ell)}{R^\ell}-\frac{2\varepsilon}{\pi}+4\pi\rho_o\sum_{\underset{\sim}{Q}\neq\underset{\sim}{0}}|F(\underset{\sim}{Q})|^2\,\frac{e^{-Q^2/4\varepsilon^2}}{Q^2}$$
$$-\frac{\pi\rho_o}{\varepsilon^2}+\frac{9}{5r_s}-Ze^2\rho_o\int\frac{Z(r)-Z}{r}\,d^3r \tag{24}$$

where

$$G(y) = \frac{2}{\sqrt{\pi}}\int_y^\infty e^{-x^2}\,dx\ . \tag{25}$$

$\Phi(\rho_o)$ is the energy per ion of electrons homogeneously distributed in a Wigner-Seitz sphere, and depends only on the mean electron density ρ_o. The second-order contribution to this energy Φ_e depends on ρ_o and the arrangement of the ions. The method to calculate the Coulomb contribution has been developed by Ewald.[16] An optimized convergence is obtained with ε near

$$\varepsilon = 1/r_s\ . \tag{26}$$

$Z(r)/r$ is the bare electron-ion interaction of a single ion.

A transformation of $\Phi_e + \Phi_c$ to express it in real space instructive. Using Eqs. (22) and (24) we find

$$\Phi_e(\{\underset{\sim}{R}^\ell\})+\Phi_c(\{\underset{\sim}{R}^\ell\}) = \frac{1}{2N}\sum_{\ell,\ell'}{}'\ \Phi(|\underset{\sim}{R}^\ell-\underset{\sim}{R}^{\ell'}|)-Ze^2\rho_o\int\frac{Z(r)-Z}{r}\,d^3r$$
$$+\sum_{Q\neq 0}U_e(Q) \tag{27}$$

where

$$\Phi(R) = \frac{1}{(2\pi)^3 \rho_o} \int \left(\frac{4\pi\rho_o (Ze)^2}{Q^2} + U_e(Q)\, e^{-i\underset{\sim}{Q}\cdot\underset{\sim}{R}} \right) d^3Q \qquad (28)$$

represents the effective ion-ion pair potential.

On the basis of these formulas it is now possible to calculate the phonon frequencies, the cohesive energy and the free energy. Such properties will be discussed in the next section.

Finally, we note that the iterative solution of (11) which led to an approximate expression for the one-electron energies (14), may diverge. The form of the energy denominator indicates [Eq. (14)] that this will occur in the neighborhood of Bragg planes. To avoid such divergencies, which become important in the calculation of the band structure, one has to solve (14) numerically. A formal description of such a procedure has been given in Ref. 18. Here one assumes linear screening.

3. NUMERICAL RESULTS AND DISCUSSION

From Eqs. (9), (19) and (23) one expects the phonon frequencies to be sensitive to changes in the core properties and the function G(Q) which approximates the corrections to the Hartree dielectric function due to exchange and correlation. To investigate this conjecture we calculate the phonon dispersion curves in b.c.c. lithium, using for the core energy and the eigenfunction the atomic values (Table II) and for G(Q) the expressions proposed by Kleinman[9] (G_K) and Singwi et al.[11] (G_{SSTL}), respectively. [The different expressions for G(Q) are listed in Table I.] In Figs. 1 and 2 we compare these results with the neutron measurements of Smith et al.[19] It is seen that both corrections to the Hartree dielectric function [$G_K(Q)$, $G_{SSTL}(Q)$] together with the core data of the free atom (Table II) reproduce the measured phonon frequencies surprisingly well. This indicates that the core properties of the free ion shift in the metal close to the free-atom values. Such an effect can be understood in terms of the interaction between the 1s and the 2s electrons.

To investigate the core shift and the corrections to the Hartree dielectric function in more detail, we fitted the core parameters (r_{1s}, E_{1s}) to the measured phonon frequencies, using $G_K(Q)$ and $G_{SSTL}(Q)$, respectively. From Figs. 1 and 2 it is seen that the Kleinman form of G(Q) leads to a somewhat better agreement with experiment. The fitted core parameters (Table II) show that $G_K(Q)$ lead to a stronger binding of the 1s-electrons compared to the free

Table II. Physical constants of Li^7.

Atomic mass	$= 1.164 \cdot 10^{-23}$ g		
Density ρ_o	$= 4.728 \cdot 10^{22}$ cm^{-3}		
Fermi wavenumber k_F	$= 1.1187$ Å^{-1}		
Wigner-Seitz radius r_s	$= 3.24\ a_o$		
Core properties	r_{1s} [Å]	E_{1s} [Ry]	
Free atom	0.1820	- 4.40	Ref. 22
Fit with $G_K(Q)$	0.1772	- 4.63	
Fit with $G_{SSTL}(Q)$	0.1870	- 4.04	
Free ion		- 5.56	

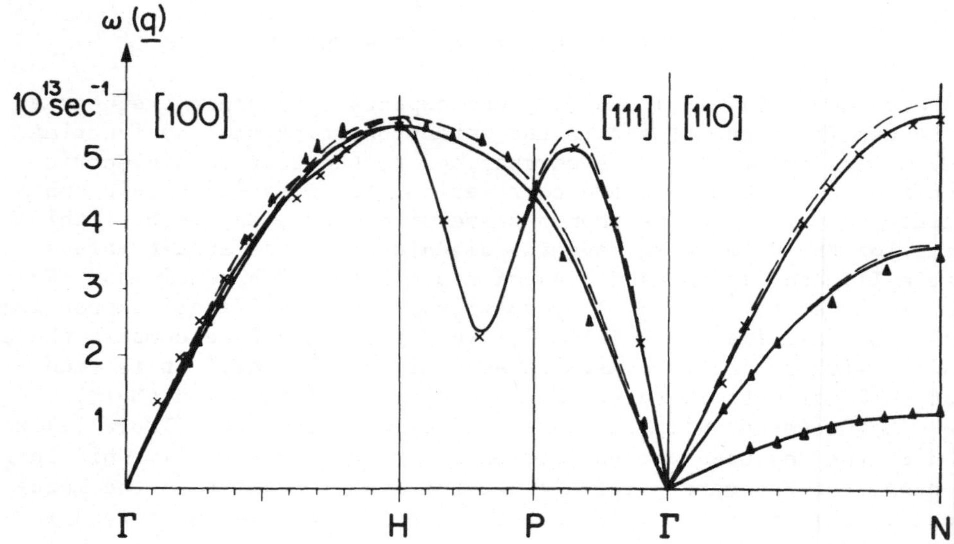

Fig. 1. Phonon dispersion curves in Li.
▲ , × neutron measurements (▲ transverse branch, × longitudinal branch). —— with the aid of $G_K(Q)$ and r_{1s}, E_{1s} from the free atom. —— with the aid of $G_K(Q)$ and r_{1s}, E_{1s} fitted to the phonon frequencies.

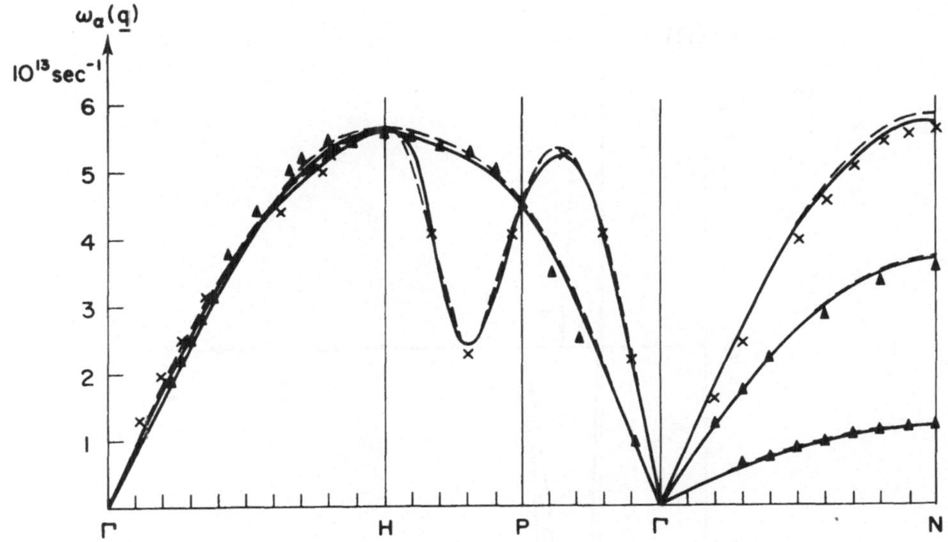

Fig. 2. ——— with the aid of $G_{SSTL}(Q)$ and r_{1s}, E_{1s} from the free atom; —— with the aid of $G_{SSTL}(Q)$ and r_{1s}, E_{1s} fitted to the phonon frequencies.

atom, whereas $G_{SSTL}(Q)$ leads to a weaker binding of the same 1s-electrons. On physical grounds one expects that the binding of the 1s-electron in the metal must lie in between that of the free-ion and the free-atom binding. This would lead one to conclude that the correction to the Hartree dielectric function as proposed by Kleinman is, at least for this case, the more appropriate one.

We turn now to the effective pair potential of the ions, which determines the phonon dispersion curves and the structure-dependent energy of the rigid lattice (27). On the basis of the core data from the fit to the measured phonon frequencies, using $G_K(Q)$, we calculated the effective pair potential for different corrections to the Hartree dielectric function. The results are displayed in Fig. 3. It is seen that close to the position of the first two neighbor shells, $G_K(Q)$ and $G_{SSTL}(Q)$ lead to quite similar potentials. Recalling that in a first approximation the phonon frequencies depend on the shape of this potential at the position of the first two nearest-neighbor shells, this result appears to be consistent with the small discrepancies in the calculated phonon frequencies (Figs. 1 and 2). However, from Fig. 3 one has to expect that the older form for $G(Q)$ by Singwi et al.[10] [$G_{STLS}(Q)$], as well as the Hartree approximation, would lead to rather unrealistic phonon frequencies.

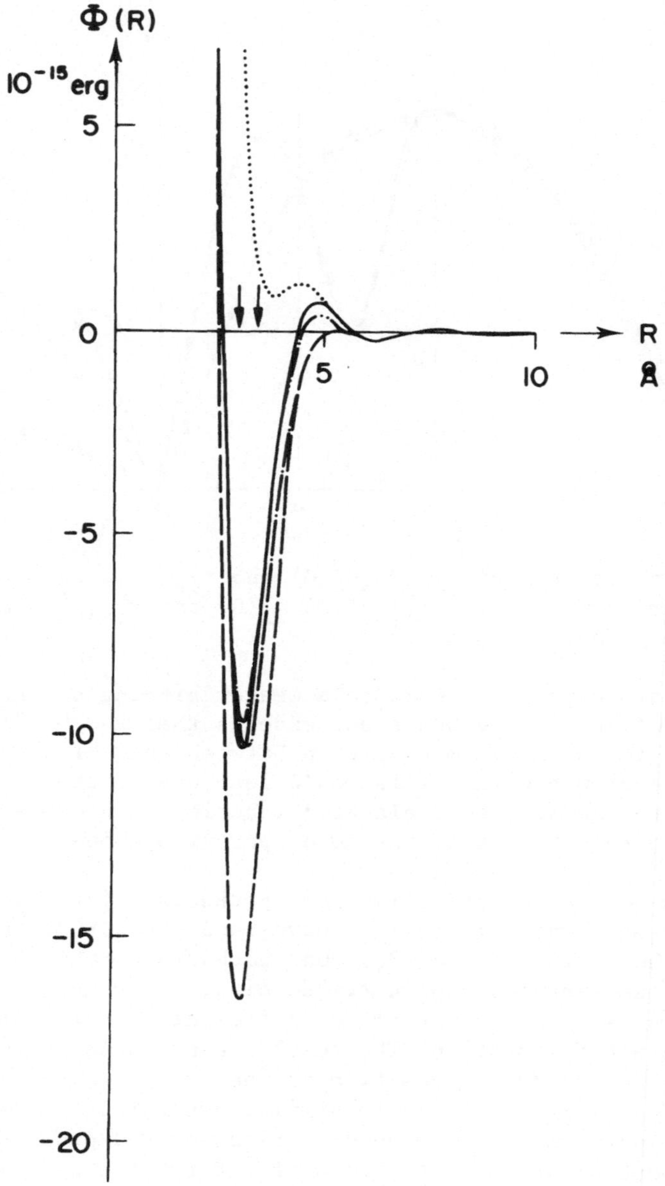

Fig. 3. Effective pair potential; r_{1s}, E_{1s} from the fit with $G_K(Q)$. The arrows indicate the positions of the first two nearest-neighbor shells. —·— $G_H(Q)$ —··— $G_K(Q)$ ····· Hartree dielectric function — — $G_{STLS}(Q)$ —— $G_{SSTL}(Q)$.

Let us now consider the total energy of the system at constant volume. At temperature T the free energy is then given by

$$F(T) = \Phi(\rho_o) + \Phi_e(\{\underset{\sim}{R}^o\}) + \Phi_c(\{\underset{\sim}{R}^o\}) + E_N + E(T) - T\,S(T) . \tag{29}$$

$\underset{\sim}{R}^o$ denotes the equilibrium positions of the ions, E_N the zero point energy, E(T) the temperature-dependent energy of the phonons, and S(T) the entropy. Here we neglect the anharmonic effects and the electronic contribution to the free energy, which should be quite small. In Fig. 4 we have plotted the calculated relative free energy as a function of temperature for the b.c.c., f.c.c. and h.c.p. crystal structures. In these calculations we used the core data fitted to the measured phonon frequencies on the basis of Kleinman's $G_K(Q)$. Our results lead to a phase transition at 134°K. Above 134°K, b.c.c. is stable and below h.c.p. with ideal axial ratio. Actually such a transition has been observed at 78°K.[21] Keeping in mind that the energy differences are extremely small it is surprising that one is even able to predict that a phase transformation should occur. For a more detailed discussion of this result we refer to Ref. 23.

Since $\Phi_e(\{\underset{\sim}{R}^o\})$ is a rather slowly convergent sum in both Fourier and real space, one expects a strong dependence on the relative free

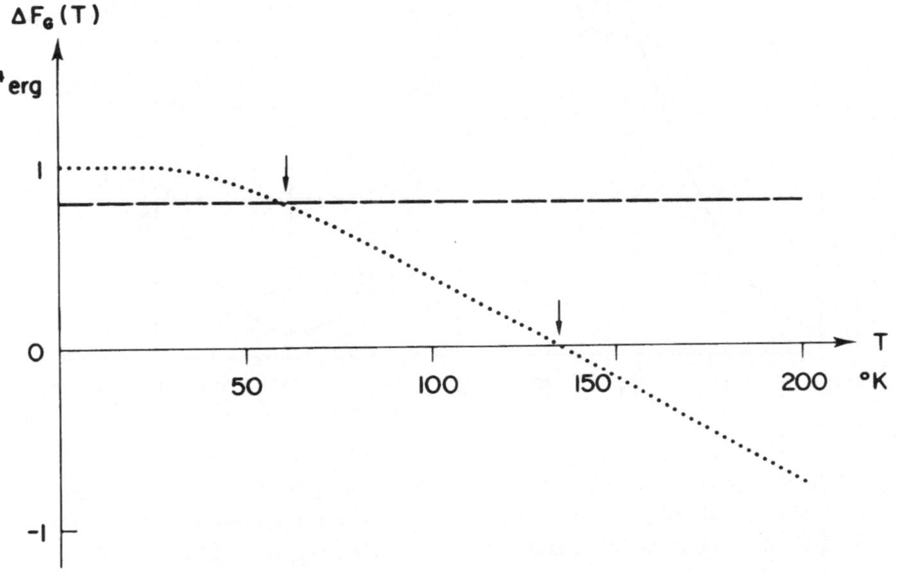

Fig. 4. Relative free energies as a function of temperature. r_{1s}, E_{1s} taken from the fit to the phonon frequencies with $G_K(Q)$. ······ $F_{bcc} - F_{hcp}$, ——— $F_{fcc} - F_{hcp}$.

energy of different crystal structures on small changes in G(Q). So far we have performed these calculations with $G_K(Q)$ only. Similar calculations with $G_{SSTL}(Q)$ are planned. In this way it is hoped to clarify the remaining uncertainties in the core shift and the proposed expressions for G(Q) somewhat further.

Finally, we consider the electronic band structure of b.c.c. lithium. In this calculation we used the procedure proposed in Ref. 18 and assumed linear screening. From Fig. 5 it is seen that the $E(\underset{\sim}{k})$ branches are continuous below the Fermi energy. Consequently,

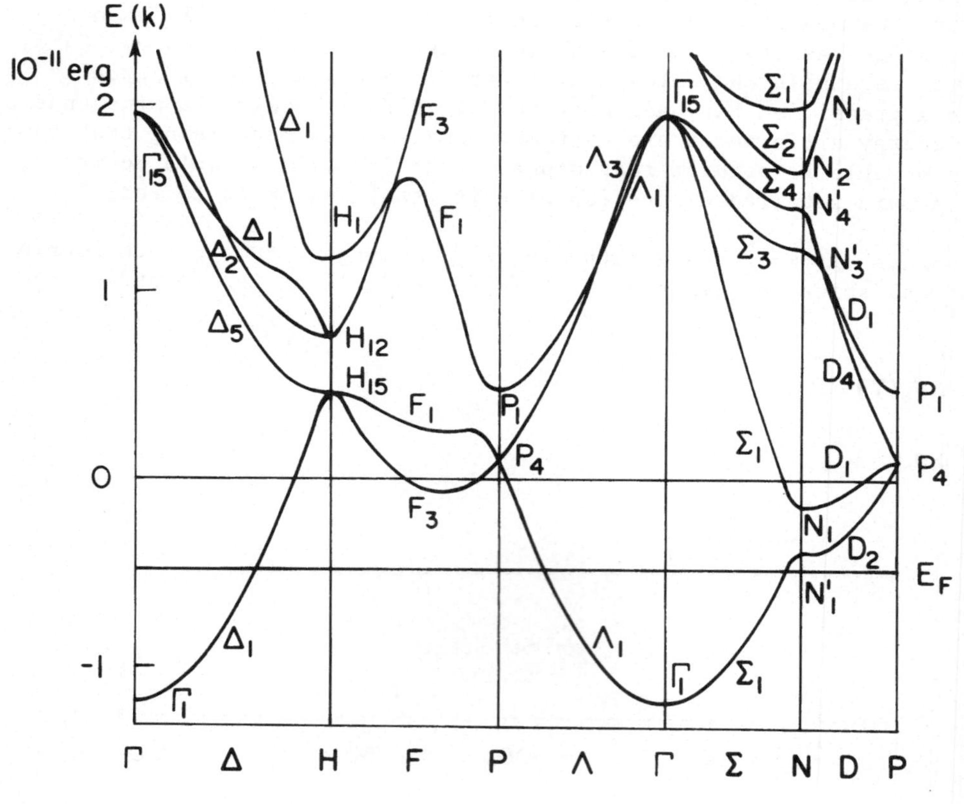

Fig. 5. Electronic band structure of b.c.c. lithium along some symmetry directions. r_{1s}, E_{1s} has been taken from the fit to the phonon frequencies with $G_K(Q)$ (Table II).

the Fermi surface will be closed. In addition, the deviations from the free-electron behavior appear to be rather small. In fact, the deviations of the Fermi surface from a sphere are found to be less than 2%.

4. CONCLUSIONS

We have shown in section 3 that the Phillips-Kleinman pseudopotential used in conjunction with the free-atom values for the core parameters and the dielectric functions of Kleinman[9] and Singwi et al.[11] reproduces the measured phonon frequencies surprisingly well. However, the fit of the core parameters to the phonon frequencies leads one to conclude that the dielectric function of Kleinman[9] is, at least for this case, the more appropriate one. In fact, in this case the binding of the 1s-electrons in the metal lies in between that of the free-ion and the free-atom binding. Using the fitted core parameters and the dielectric function of Kleinman[9] we were even able to predict that a first-order phase transition from b.c.c. to h.c.p. lithium should occur.

REFERENCES

[1] Harrison, W.H., *Pseudopotentials in the Theory of Metals* (Benjamin Inc., New York, 1967).

[2] Joshi, S.K. and A.K. Rajacopal, Solid State Physics 22, 159 (1968).

[3] Heine, V., Solid State Physics 24, 1 (1970).

[4] Lin, P.J. and J.C. Phillips, Ad. Phys. 14, 257 (1965).

[5] Heine, V. and I. Abarenkov, Phil Mag. 9, 451 (1964) and 12, 529 (1965).

[6] Ashcroft, N.W., J. Phys. Chem. 1, 232 (1968).

[7] Shaw, R.W., Phys. Rev. 174, 769 (1968).

[8] Sham, L.J., Proc. Roy. Soc. (London) A283, 33 (1965).

[9] Kleinman, L., Phys. Rev. 160, 585 (1967) and 172, 383 (1968).

[10] Singwi, K.S., M.P. Tosi, R.H. Land and A. Sjölander, Phys. Rev. 176, 589 (1968).

[11] Singwi, K.S., A Sjölander, M.P. Tosi and R.H. Land, Phys. Rev. B1, 1044 (1970).

[12] Schneider, T., Physica 52, 481 (1971).

[13] Phillips, J.C. and L. Kleinman, Phys. Rev. 116, 287 (1959).

[14] Kohn, W. and L.J. Sham, Phys. Rev. 111, 442 (1958).

[15] Ziman, J.M. and L.J. Sham, Solid State Physics 15, 221 (1963).

[16] See Kittel,[17] Appendix A.

[17] Kittel, C., *Introduction to Solid-State Physics*, 2nd ed. (Wiley, New York, 1956).

[18] Waeber, W.B. and E. Stoll, Physica, to be published.

[19] Smith, H.G., G. Dolling, R.M. Nicklow, P.R. Vijayaraghaven and M.K. Wilkinson, Neutron inelastic scattering, Vienna: IAEA 1968, Vol. I, p. 149.

[20] Schneider, T., Helv. Phys. Acta 42, 957 (1969).

[21] Barrett, C.S., Acta Cryst. 9, 671 (1956).

[22] Herman, F., and S. Skillman, *Atomic Structure Calculations*, (Prentice Hall Inc., Englewood Cliffs, N.Y. 1963).

[23] Schneider, T. and E. Stoll, Solid State Commun. 8, 1729 (1970).

GENERALIZATIONS OF THE RELATIVISTIC OPW METHOD INCLUDING OVERLAPPING AND NON-OVERLAPPING ATOMIC ORBITALS

N. W. Dalton[†]

IBM Research Laboratory, San Jose, California 95114 USA

ABSTRACT

The advantages and disadvantages of the standard orthogonalized plane-wave (OPW) method for calculating electronic band structures are briefly examined. Then, following Herring (1940), the possibility of increasing the convergence of the OPW method by including certain 'outer-core' auxiliary tight-binding functions in the OPW basis set is investigated. It is argued that, contrary to much recent work, the overlap of core and outer-core functions on neighboring lattice sites is non-negligible in many materials and must be taken into account (either directly or indirectly) if a modified OPW method is to form the basis for a flexible and rapidly convergent scheme for metals and semi-conductors. Detailed expressions are derived for the relativistic OPW and modified OPW matrix elements which include ℓ-dependent potentials and overlapping orbitals.

1. INTRODUCTION

During the past thirty years or so several methods have been proposed for solving the Schroedinger or Dirac equation for an electron moving through a periodic crystal potential (i.e., electronic band-structure problem). Using conventional notation (see Appendix for explanation of symbols), the problem is to find eigensolutions $\tilde{\Psi}_{\underline{k}}$ (the tilde denotes that the wave function is a four-component spinor) and eigenvalues $E_{\underline{k}}$ of the equation (see Appendix for details)

[†] Permanent address: Theoretical Physics Division, A.E.R.E., Harwell, Berkshire, ENGLAND.

$$[c\tilde{\alpha}\cdot\bar{p} + \tilde{\beta}mc^2 + V_{cr}(\underline{r})\tilde{I}]\tilde{\Psi}_{\underline{k}}(\underline{r}) = E_{\underline{k}}\tilde{\Psi}_{\underline{k}}(\underline{r}) \tag{1}$$

where $V_{cr}(\underline{r})$ is the periodic crystal potential and $\underline{k}$ is the wave vector resulting from Bloch's Theorem. (In the non-relatistic limit this equation reduces to the Schroedinger equation.) For purposes of later discussion it is convenient to regard the crystal wave function as made up of two parts,

$$\tilde{\Psi}_{\underline{k}}(\underline{r}) = \underbrace{\tilde{\Psi}^{<}_{\underline{k}}(\underline{r})}_{\text{near nuclei (oscillatory)}} + \underbrace{\tilde{\Psi}^{>}_{\underline{k}}(\underline{r})}_{\text{interstitial region (smooth)}} \tag{2}$$

where $\tilde{\Psi}^{<}_{\underline{k}}(\underline{r})$ primarily describes the motion of electrons near the nuclei of the crystal and $\tilde{\Psi}^{>}_{\underline{k}}(\underline{r})$ primarily accounts for the behavior of electrons in the interstitial region. In general, $\tilde{\Psi}^{<}_{\underline{k}}(\underline{r})$ will be a rapidly oscillating function since it must contain one or more nodes corresponding to the nodes in the radial atomic functions. Correspondingly, $\tilde{\Psi}^{>}_{\underline{k}}(\underline{r})$ is expected to be reasonably smooth due to the absence of nodes far from the nuclei (i.e., interstitial region). Both regions (near nuclei and interstitial) will be described by an admixture of $\tilde{\Psi}^{<}_{\underline{k}}(\underline{r})$ and $\tilde{\Psi}^{>}_{\underline{k}}(\underline{r})$, but in either region, either $\Psi^{<}_{\underline{k}}(\underline{r})$ or $\Psi_{\underline{k}}(\underline{r})$ is assumed to be dominant. Any method for solving (1) must take into account, either explicitly or implicitly, this dual behavior of the wave function.

The earliest methods proposed for solving the Schroedinger equation in the presence of $V_{cr}(\underline{r})$ were the cellular method (Wigner and Seitz (1933, 1934)), the augmented plane-wave (APW) method (Slater (1937)), and the orthogonalized plane-wave (OPW) method (Herring (1940)). Since all of these methods involve extensive numerical computations, they could not be tested adequately until digital computers had been sufficiently developed, i.e., late nineteen fifties. However, during that time additional schemes were proposed (Korringa (1947), Kohn and Rostoker (1954), (KKR); simplified linear combination of atomic orbitals (LCAO), Slater and Koster (1954); and the pseudopotential method, Phillips and Kleinman (1959)) so that by the early nineteen sixties a great variety of methods were available for calculating electronic band structures. It was clear nevertheless that none of these methods was sufficiently flexible for treating all classes of materials (e.g., simple metals, transition metals, semi-conductors, etc.). This situation prompted many workers to try to extend or modify the available band-structure methods to increase their flexibility and to clarify the relationship between them. In this connection two independent lines of research have developed during the past five or six years directed at understanding transition metal band structures. On the one hand attempts were made to extend the pseudopotential method to transition metals. This led to the concept of the 'model Hamiltonian' (for discussion see Phillips (1968)) which was justified theoretically on the basis of the KKR method by Heine (1967) and Hubbard (1967) (for review of this work, see Dalton (1971)). Implicit in this work is the idea of using mixed basis sets (i.e., OPW's and LCAO's) for transition metal band structures. On the other hand it was also clear to workers in semiconductors that to improve the convergence of the OPW method in certain materials (e.g., C and ZnO) it would be necessary to augment the OPW basis set with suitable tight-binding functions (LCAO's). Thus band theorists are now entering the 'modified' era, i.e., modified OPW schemes, modified APW schemes, modified KKR schemes, etc. This trend is depicted in Fig. 1. The modified OPW method can be thought of as a unification of the OPW and LCAO methods and could equally well be termed the modified LCAO method. Similarly, the modified KKR schemes (Williams et al. (1971), Dalton (1971)) may be regarded as first-principles attempts at constructing suitable mixed basis sets. The main purpose of this paper is to examine the OPW and modified OPW methods as general methods for calculating band structures of metals and semiconductors. Earlier work will be critically discussed, particularly in relation to the modified OPW method, and generalized to include ℓ-dependent potentials, overlapping orbitals, and relativistic effects. Attention is given to the problem of choosing suitable LCAO's with which to augment the OPW basis set.

The plan of this paper is as follows. In Section 2 we present the relativistic generalization of the OPW method as originally proposed by Herring (1940) and discuss some of the practical problems involved in applications of the method. Previous work (non-relativistic) relating to the modified OPW method is critically reviewed in Section 3. The basic theory necessary for discussing the modified relativistic OPW method is developed in Section 4. As an illustration of the theory the 3d states in copper and germanium are considered in Section 5. A brief summary of the paper and our conclusions are contained in the last section. Some results relating to the Dirac equations are given in the Appendix for easy reference.

Fig. 1. Band-Structure Methods.

2. THE RELATIVISTIC OPW METHOD

In a classic paper (Herring (1940)), Herring proposed the non-relativistic orthogonalized plane-wave (OPW) method for calculating electronic band structures. Although the general theory presented in that paper is complicated, a considerable simplificiation is achieved if it is assumed that the exact crystal core states are known and are non-overlapping, i.e., core states on neighboring lattice sites do not overlap. Hereon we shall refer to the latter approximate theory as the simplified OPW method. In practice, approximate crystal core states are invariably used but the resulting corrections to the simplified OPW method are relatively easy to calculate. Corrections for the non-zero overlap of core states are, however, difficult to calculate since they involve the evaluation of multi-center integrals. At the present time no OPW band-structure calculation has included overlap corrections even though, as argued below, they must be significant in many materials. Nevertheless, even ignoring overlap corrections, the OPW method has been used successfully by many workers (for references see Herman et al. (1969)) for calculating the band structures of a wide range of semiconductors.

The essence of the OPW method is to assume that solutions of (1) may be divided into two classes, i.e., those which describe electrons tightly bound to the nuclei of the crystal (core states) and those which relate to electrons propagating throughout the crystal (valence or conduction states). Whether a state is to be classed as a core or valence state is usually quite clear, but in certain cases (e.g., 3d states in transition metals) states may possess the characteristics of both core and valence states. In the latter case the OPW method ceases to be useful and must be modified (next section). Here we derive a relativistic version of Herring's general theory but allowing for different atomic species in the crystal and a state-dependent (e.g., ℓ-dependent) potential. (A discussion of the relativistic OPW method (simplified theory) has been published elsewhere, i.e., Soven (1965), Herman et al. (1968), Dalton (1970).)

Consider a complex crystal containing N unit cells at the lattice positions $\underline{R}$. Within each unit cell we assume that there are n_a atomic species at the positions $\underline{R} + \underline{\tau}$ characterized by atomic potentials $U_{\underline{\tau}}(\underline{r})$. Furthermore, it is assumed that $U_{\underline{\tau}}(\underline{r})$ may closely approximate the crystal potential $V_{cr}(\underline{r})$ in the region of the atom at $\underline{R} + \underline{\tau}$ but is not necessarily identical with $V_{cr}(\underline{r})$, i.e.,

$$V_{cr}(\underline{r}) \simeq U_{\underline{\tau}}(\underline{r}) \qquad |\underline{r}-\underline{R}-\underline{\tau}| < \frac{1}{2} d_{nn} , \tag{3}$$

where d_{nn} is the nearest neighbor distance. The atomic bound-state solutions of the Dirac equation in the presence of $U_{\underline{\tau}}(\underline{r})$ may be written as follows,

$$\tilde{\phi}^c_{\underline{\tau}1},\ \tilde{\phi}^c_{\underline{\tau}2},\ \ldots\ \tilde{\phi}^c_{\underline{\tau}n},\ \tilde{\phi}^v_{\underline{\tau},n+1},\ \ldots\ \tilde{\phi}^v_{\underline{\tau},n+m}\quad , \tag{4}$$

where we have divided the n+m solutions into those which have an amplitude at $\frac{1}{2}d_{nn}$ of less than δ (the first n, say) and those which have an amplitude greater than δ. Wave functions in the first set are the 'core states' (denoted by the superscript c) and those in the second set are the valence (denoted by the superscript v) or 'outer-core' states. Insofar as the choice of δ is arbitrary, the above division of the wave functions into core and valence states is arbitrary. In a similar manner we may classify the corresponding crystal wave functions into two sets, i.e.,

$$\tilde{\Phi}^c_{\underline{k\tau}1},\ \tilde{\Phi}^c_{\underline{k\tau}2},\ \ldots\ \tilde{\Phi}^c_{\underline{k\tau}n},\ \tilde{\Phi}^v_{\underline{k\tau},n+1},\ \ldots\quad , \tag{5}$$

where the first n states are the crystal core states and the remainder are the valence states. (There are of course an infinite (i.e., continuum) number of crystal valence states as indicated in (5).) To a good approximation the crystal core states are related to the atomic core states by a Bloch sum, i.e.,

$$\tilde{\Phi}^c_{\underline{k\tau}}(\underline{r}) = \frac{1}{\sqrt{N}} \sum_{\underline{R}} \tilde{\phi}_{\underline{\tau}}(\underline{r}-\underline{R}-\underline{\tau})e^{i\underline{k}\cdot\underline{R}}\quad . \tag{6}$$

Due to the non-localized nature of the crystal valence states, it is necessary to expand these in terms of a suitable set of basis functions, e.g.,

$$\tilde{\Phi}^v_{\underline{k}}(\underline{r}) = \sum_{j,s} b_{\underline{k}j,s}\tilde{O}^s_{\underline{k}j}(\underline{r})\quad , \tag{7}$$

where $s = \pm\frac{1}{2}$ denotes the spin quantum number, $\underline{k}_j = \underline{k} + \underline{K}_j$, and $\underline{K}_j$ is a reciprocal lattice vector. Herring (1940) argued that since the valence states describe the propagation of electrons through the crystal, a reasonable choice for the basis function is a plane wave (relativistic) orthogonalized to the low-lying core states, i.e.,

$$\tilde{O}^s_{\underline{k}j}(\underline{r}) = \frac{1}{\sqrt{Nv}}\tilde{U}^s_{\underline{k}j}e^{i\underline{k}_j\cdot\underline{r}} - \sum_{\underline{\tau}\alpha}^{\text{core}} A^c_{\underline{\tau}\alpha}(s,\underline{k}j)\tilde{\Phi}^c_{\underline{k\tau}\alpha}(\underline{r}) \tag{8}$$

where

$$A^c_{\underline{\tau}\alpha}(s,\underline{k}j) = \int \tilde{\Phi}^{c\dagger}_{\underline{k\tau}\alpha}(\underline{r})\frac{1}{\sqrt{Nv}}\tilde{U}^s_{\underline{k}j}e^{i\underline{k}_j\cdot\underline{r}}\,d^3r\quad , \tag{9}$$

α stands for the relativistic quantum numbers (n,κ,μ), and the remaining quantities are defined in the Appendix. It is shown in the Appendix that the orthogonality integral $A^c_{\underline{\tau}\alpha}(s,\underline{k}_j)$ can be written in the form

$$A^c_{\underline{\tau}\alpha}(s,\underline{k}j) = e^{i\underline{k}_j\cdot\underline{\tau}}\,a^s_{\kappa\mu}(\underline{k}_j)\,\mathcal{L}^c_{\underline{\tau}n\kappa}(k_j)\quad , \tag{10}$$

where $a^{s}_{\kappa\mu}(\underline{k}_j)$ may be regarded as the relativistic analogue of the spherical harmonic, and $\mathcal{I}^{c}_{\tau n\kappa}(k_j)$ is the relativistic generalization of the radial Bessel function integral which occurs in the non-relativistic theory. With this definition for $A^{c}_{\tau\alpha}(s,\underline{k}_j)$, $\tilde{O}^{s}_{\underline{k}j}$ is orthogonal to the core states $\tilde{\Phi}_{k\tau\alpha}(\underline{r})$ only if the core states are non-overlapping. In the latter case, it would be necessary to invert a small core-overlap matrix to obtain coefficients $A^{c}_{\tau\alpha}(s,\underline{k}_j)$ which would ensure the orthogonality of $\tilde{O}^{s}_{\underline{k}j}$ to the core states. If the wave function defined by (7), (8) and (9) is used as a trial wave function in the variational principle, then the secular equation for the energy-band problem assumes the form

$$\det \left| H^{ss'}_{ij}(\underline{k}) - E_{\underline{k}} J^{ss'}_{ij}(\underline{k}) \right| = 0 \quad , \tag{11}$$

where

$$H^{ss'}_{ij}(\underline{k}) = \int \tilde{O}^{s\dagger}_{\underline{k}i} \tilde{H}_D \tilde{O}^{s'}_{\underline{k}j} \, d^3r \quad , \tag{12}$$

$$J^{ss'}_{ij}(\underline{k}) = \int \tilde{O}^{s\dagger}_{\underline{k}i} \tilde{O}^{s'}_{\underline{k}j} \, d^3r \quad , \tag{13}$$

and $\tilde{H}_D$ is the Dirac Hamiltonian given in (1). Since the evaluation of the matrix elements is tedious but straightforward, we record here only the final results. For this purpose it is necessary to define the following quantities: $\tilde{\phi}_{\underline{\tau}\alpha}$ and $E^{at}_{\underline{\tau}\alpha}$ are assumed to be the atomic core solutions of the equation

$$[c\tilde{\alpha}\cdot p + \tilde{\beta}mc^2 + U_{\underline{\tau}\alpha}\tilde{I}]\tilde{\phi}_{\underline{\tau}\alpha} = E^{at}_{\underline{\tau}\alpha}\,\tilde{\phi}_{\underline{\tau}\alpha} \quad , \tag{14}$$

where

$$U_{\underline{\tau}}(\underline{r}) = U_{\underline{\tau}\alpha}(\underline{r}) \tag{15}$$

for the state $\tilde{\phi}_{\underline{\tau}\alpha}$ (assumed normalized). Thus we are admitting the possibility of an angular momentum dependent (ℓ-dependent) potential as discussed by Callaway (1955a). If we denote the difference between the atomic potential $U_{\underline{\tau}\alpha}(\underline{r}-\underline{R}-\underline{\tau})$ and the crystal potential in the region of the lattice position $\underline{R}+\underline{\tau}$ by $\delta V_{\underline{\tau}\alpha}(\underline{r}-\underline{R}-\underline{\tau})$, i.e.,

$$\delta V_{\underline{\tau}\alpha}(\underline{r}-\underline{R}-\underline{\tau}) = V_{cr}(\underline{r}) - U_{\underline{\tau}\alpha}(\underline{r}-\underline{R}-\underline{\tau}) \tag{16}$$

$$\simeq \overline{\delta V}_{\underline{\tau}\alpha}(|\underline{r}-\underline{R}-\underline{\tau}|) \quad , \tag{17}$$

where $\overline{\delta V}_{\underline{\tau}\alpha}(|\underline{r}-\underline{R}-\underline{\tau}|)$ is the spherical average of $\delta V_{\underline{\tau}\alpha}(\underline{r}-\underline{R}-\underline{\tau})$; then we require the following integrals involving atomic states:

$$I^{c\ c}_{\underline{\tau}\alpha,\underline{\tau}'\alpha'}(\underline{R}) = \int \tilde{\phi}^{c\dagger}_{\underline{\tau}\alpha}(\underline{r}-\underline{\tau})\tilde{\phi}^{c}_{\underline{\tau}'\alpha'}(\underline{r}-\underline{R}-\underline{\tau}')d^3r \quad , \tag{18}$$

$$\nu^{c}_{\underline{\tau}\alpha}(s,\underline{k}_i) = \frac{1}{\sqrt{v}}\int \tilde{\phi}^{c\dagger}_{\underline{\tau}\alpha}(\underline{r}-\underline{\tau})\delta V_{\underline{\tau}\alpha}(\underline{r}-\underline{\tau})\tilde{U}^{s}_{\underline{k}i}e^{i\underline{k}_i\cdot\underline{r}}\, d^3r \quad , \tag{19}$$

$$\omega^{c\ \ c}_{\underline{\tau}\alpha,\underline{\tau}'\alpha'}(\underline{R}) = \int \tilde{\phi}^{c\dagger}_{\underline{\tau}\alpha}(\underline{r}-\underline{\tau})\delta V_{\underline{\tau}'\alpha'}(\underline{r}-\underline{R}-\underline{\tau}')\tilde{\phi}^{c}_{\underline{\tau}'\alpha'}(\underline{r}-\underline{R}-\underline{\tau}')d^3r \quad . \tag{20}$$

It should be pointed out that even if the core states in the overlap integral refer to the same lattice site (i.e., $\underline{\tau}=\underline{\tau}'$, $\underline{R}=0$), $I^{c\ \ c}_{\underline{\tau}\alpha,\underline{\tau}'\alpha'}(\underline{R})$ is not the usual orthogonality integral, i.e., $I^{c\ \ c}_{\underline{\tau}\alpha,\underline{\tau}\alpha'}(0) = \delta_{\alpha\alpha'}$. The latter result is true only if the same potential ($U_{\underline{\tau}\alpha}(\underline{r}) = U_{\underline{\tau}}(\underline{r})$) is used for all states. (N.B. Since the core states are assumed normalized, $I^{c\ \ c}_{\underline{\tau}\alpha,\underline{\tau}\alpha}(0) = 1$.) With these preliminaries the matrix elements may be written as follows:

$$H^{ss'}_{ij}(\underline{k}) = \overline{H}^{ss'}_{ij}(\underline{k}) + \overline{\overline{H}}^{ss'}_{ij}(\underline{k}) \quad , \tag{21}$$

where

$$\begin{aligned}\overline{H}^{ss'}_{ij}(\underline{k}) &= E_0(k_j)\delta_{ij}\delta_{ss'} + V(\underline{k}_i-\underline{k}_j)\tilde{U}^{s\dagger}_{\underline{k}i}\tilde{U}^{s'}_{\underline{k}j} \\ &- 4\pi\sum_{\underline{\tau}n\kappa}^{\text{core}} e^{i(\underline{k}_j-\underline{k}_i)\cdot\underline{\tau}}\,\overline{E}_{\underline{\tau}\alpha}\mathcal{L}^{c}_{\underline{\tau}n\kappa}(k_i)\mathcal{L}^{c}_{\underline{\tau}n\kappa}(k_j)P^{ss'}_{\kappa}(\hat{\underline{k}}_i,\hat{\underline{k}}_j) \\ &+ 4\pi\sum_{\underline{\tau}n\kappa}^{\text{core}}\sum_{n'\neq n} e^{i(\underline{k}_j-\underline{k}_i)\cdot\underline{\tau}}\,[E^{at}_{\underline{\tau}n'\kappa} + \Delta E^{c\ \ c}_{\underline{\tau}n\kappa,\underline{\tau}n'\kappa}(0)] \\ &\times I^{c\ \ c}_{\underline{\tau}n\kappa,\underline{\tau}n'\kappa}\mathcal{L}^{c}_{\underline{\tau}n\kappa}(k_i)\mathcal{L}^{c}_{\underline{\tau}n'\kappa}(k_j)P^{ss'}_{\kappa}(\hat{\underline{k}}_i,\hat{\underline{k}}_j) \quad ,\end{aligned} \tag{22}$$

$E_0(k_j)$ is the relativistic free electron kinetic energy (see Appendix), $V(\underline{k}_i-\underline{k}_j)$ is the Fourier Transform of the crystal potential, i.e.,

$$V(\underline{k}_i-\underline{k}_j) = \frac{1}{Nv}\int V_{cr}(\underline{r})e^{-i(\underline{k}_i-\underline{k}_j)\cdot\underline{r}}\, d^3\underline{r} \quad , \tag{23}$$

v is the volume of the unit cell, $\tilde{U}^{s\dagger}_{\underline{k}i}\tilde{U}^{s'}_{\underline{k}j}$ is a scalar product depending on the spin quantum numbers s,s' and the vectors $\underline{k}_i,\underline{k}_j$, i.e., (see also Appendix)

$$\begin{aligned}\tilde{U}^{s\dagger}_{\underline{k}i}\tilde{U}^{s'}_{\underline{k}j} &= [U_+(k_i)U_+(k_j) + U_-(k_i)U_-(k_j)\frac{\underline{k}_i\cdot\underline{k}_j}{k_ik_j}]\,\delta_{ss'} \\ &+ iU_-(k_i)U_-(k_j)\langle s|\overline{\sigma}|s'\rangle\cdot\left(\frac{\underline{k}_i\times\underline{k}_j}{k_ik_j}\right) \quad ,\end{aligned} \tag{24}$$

and the energy $\overline{E}_{\underline{\tau}\alpha}$ is related to the atomic core-state energy by the equation

$$\bar{E}_{\underline{\tau}\alpha} = E^{at}_{\underline{\tau}\alpha} + \Delta E_{\underline{\tau}\alpha}(s\underline{k}_i) + \Delta E_{\underline{\tau}\alpha}(s'\underline{k}_j) - \Delta E^{c\ \ c}_{\underline{\tau}\alpha,\underline{\tau}\alpha}(0) \quad , \tag{25}$$

where

$$\Delta E_{\underline{\tau}\alpha}(s\underline{k}_i) = \frac{\nu^c_{\underline{\tau}\alpha}(s\underline{k}_i)}{A^c_{\underline{\tau}\alpha}(s\underline{k}_i)} \quad , \tag{26}$$

and

$$\Delta E^{c\ \ c}_{\underline{\tau}\alpha,\underline{\tau}\alpha}(0) = \frac{\omega^{c\ \ c}_{\underline{\tau}\alpha,\underline{\tau}\alpha}(0)}{I^{c\ \ c}_{\underline{\tau}\alpha,\underline{\tau}\alpha}(0)} \quad . \tag{27}$$

The relativistic analogue of the Legendre polynomial is given by the expression

$$P^{ss'}_{\kappa}(\underline{k}_i \hat{,} \underline{k}_j) = \kappa s_{\kappa} P_{\ell\kappa}(\underline{k}_i \hat{,} \underline{k}_j)\delta_{ss'} + i s_{\kappa}\langle s|\bar{\sigma}|s'\rangle \cdot \left(\frac{\underline{k}_i \times \underline{k}_j}{k_i k_j}\right) P'_{\ell\kappa}(\underline{k}_i \hat{,} \underline{k}_j) \quad , \tag{28}$$

where $\underline{k}_i \hat{,} \underline{k}_j$ denotes the cosine of the angle between $\underline{k}_i$ and $\underline{k}_j$, $P_\ell(x)$ is a Legendre polynomial, and

$$P'_\ell(x) = \frac{dP_\ell(x)}{dx} \quad . \tag{29}$$

The integral $I^{c\ \ c}_{\underline{\tau}n\kappa,\underline{\tau}n'\kappa}$ is related to $I^{c\ \ c}_{\underline{\tau}\alpha,\underline{\tau}'\alpha'}(\underline{R})$ (c.f. (18)) by the relation

$$I^{c\ \ c}_{\underline{\tau}n\kappa\mu,\underline{\tau}n'\kappa'\mu'}(0) = \delta_{\mu\mu'}\delta_{\kappa\kappa'} I^{c\ \ c}_{\underline{\tau}n\kappa,\underline{\tau}n'\kappa} \quad . \tag{30}$$

As mentioned above, for a state independent potential (i.e., $U_{\underline{\tau}\alpha}(\underline{r}) = U_{\underline{\tau}}(\underline{r})$), $I^{c\ \ c}_{\underline{\tau}n\kappa,\underline{\tau}n'\kappa} = \delta_{nn'}$ and thus the last term in (22) is zero. Also, for simplicity, we have assumed $\delta V_{\underline{\tau}\alpha}(\underline{r}-\underline{R}-\underline{\tau}) \simeq \overline{\delta V}_{\underline{\tau}\alpha}(|\underline{r}-\underline{R}-\underline{\tau}|)$ (c.f. (17)) in the above derivation. All terms involving overlap effects are included in $\bar{\bar{H}}^{ss'}(\underline{k})$ (c.f. (21)), i.e.,

$$\bar{\bar{H}}^{ss'}_{ij}(\underline{k}) = \sum_{\underline{\tau}\neq\underline{\tau}'}^{core} \sum_{\alpha,\alpha'}^{core} e^{i(\underline{k}_j\cdot\underline{\tau}'-\underline{k}_i\cdot\underline{\tau})} a^{s*}_{\kappa\mu}(\underline{k}_i) a^{s'}_{\kappa'\mu'}(\underline{k}_j)$$

$$\times \mathcal{L}^c_{\underline{\tau}n\kappa}(k_i)\mathcal{L}^c_{\underline{\tau}'n'\kappa'}(k_j) I^{c\ \ c}_{\underline{\tau}\alpha,\underline{\tau}'\alpha'}(0)[E^{at}_{\underline{\tau}'\alpha'} + \Delta E^{c\ \ c}_{\underline{\tau}\alpha,\underline{\tau}'\alpha'}(0)] \ +$$

$$+ \sum_{\underline{R}\neq 0} \sum_{\underline{\tau}\alpha} \sum_{\underline{\tau}'\alpha'} e^{i(\underline{k}_j\cdot\underline{\tau}'-\underline{k}_i\cdot\underline{\tau})} a^{s*}_{\kappa\mu}(\underline{k}_i) a^{s'}_{\kappa'\mu'}(\underline{k}_j)$$

$$\times \mathcal{L}^c_{\underline{\tau}n\kappa}(k_i)\mathcal{L}^c_{\underline{\tau}'n'\kappa'}(k_j) I^{c\ \ c}_{\underline{\tau}\alpha,\underline{\tau}'\alpha'}(\underline{R})$$

$$\times [E^{at}_{\underline{\tau}'\alpha'} + \Delta E^{c\ \ c}_{\underline{\tau}\alpha,\underline{\tau}'\alpha'}(\underline{R})] e^{i\underline{k}\cdot\underline{R}} \quad . \tag{31}$$

The expressions for $J^{ss'}_{ij}$ are similar to those given above for $H^{ss'}_{ij}$, i.e.,

$$J^{ss'}_{ij}(\underline{k}) = \bar{J}^{ss'}_{ij}(\underline{k}) + \bar{\bar{J}}^{ss'}_{ij}(\underline{k}) \quad , \tag{32}$$

where

$$\bar{J}^{ss'}_{ij}(\underline{k}) = \delta_{ij}\delta_{ss'} - 4\pi \sum^{\text{core}}_{\underline{\tau}n\kappa} e^{i(\underline{k}_j-\underline{k}_i)\cdot\underline{\tau}} \mathcal{L}^c_{\underline{\tau}n\kappa}(k_i)\mathcal{L}^c_{\underline{\tau}n\kappa}(k_j) P^{ss'}_{\kappa}(\underline{k}_i,\hat{\underline{k}}_j)$$

$$+ 4\pi \sum^{\text{core}}_{\underline{\tau}n\kappa} \sum^{\text{core}}_{n'\neq n} e^{i(\underline{k}_j-\underline{k}_i)\cdot\underline{\tau}} I^{c\ \ c}_{\underline{\tau}n\kappa,\underline{\tau}n'\kappa} \mathcal{L}^c_{\underline{\tau}n\kappa}(k_i)\mathcal{L}^c_{\underline{\tau}n'\kappa}(k_j) P^{ss'}_{\kappa}(\underline{k}_i,\hat{\underline{k}}_j) \quad , \tag{33}$$

and the overlap contribution is given by

$$\bar{\bar{J}}^{ss'}_{ij}(\underline{k}) = \sum^{\text{core}}_{\underline{\tau}\neq\underline{\tau}'} \sum^{\text{core}}_{\alpha,\alpha'} e^{i(\underline{k}_j\cdot\underline{\tau}'-\underline{k}_i\cdot\underline{\tau})} a^{s*}_{\kappa\mu}(\underline{k}_i) a^{s'}_{\kappa'\mu'}(\underline{k}_j)$$

$$\times \mathcal{L}^c_{\underline{\tau}n\kappa}(k_i)\mathcal{L}^c_{\underline{\tau}'n'\kappa'}(k_j) I^{c\ \ c}_{\underline{\tau}\alpha,\underline{\tau}'\alpha'}(0)$$

$$+ \sum^{\text{core}}_{\underline{R}\neq 0} \sum^{\text{core}}_{\underline{\tau}\alpha} \sum^{\text{core}}_{\underline{\tau}'\alpha'} e^{i(\underline{k}_j\cdot\underline{\tau}'-\underline{k}_i\cdot\underline{\tau})} a^{s*}_{\kappa\mu}(\underline{k}_i) a^{s'}_{\kappa'\mu'}(\underline{k}_j)$$

$$\times \mathcal{L}^c_{\underline{\tau}n\kappa}(k_i)\mathcal{L}^c_{\underline{\tau}'n'\kappa'}(k_j) I^{c\ \ c}_{\underline{\tau}\alpha,\underline{\tau}'\alpha'}(\underline{R}) e^{i\underline{k}\cdot\underline{R}} \quad . \tag{34}$$

The above formulae for the matrix elements contain the relativistic generalization of Herring's work and incorporate the later extensions of Callaway (1955a). The simplified relativistic theory considered previously by Soven (1965), Herman et al. (1968) and Dalton (1970) is recovered by setting $\bar{\bar{H}}^{ss'}_{ij}(\underline{k}) = \bar{\bar{J}}^{ss'}_{ij} = 0$ and assuming that the potentials $U_{\underline{\tau}\alpha}$ are the same for all states and equal to the crystal potential ($\delta V_{\underline{\tau}\alpha}(\underline{r}-\underline{R}-\underline{\tau}) = 0$), i.e.,

$$H_{ij}^{ss'}(\underline{k}) = E_0(k_j)\delta_{ij}\delta_{ss'} + V(\underline{k}_i-\underline{k}_j)\tilde{U}_{\underline{k}i}^{s\dagger}\tilde{U}_{\underline{k}j}^{s'}$$

$$- 4\pi \sum_{\underline{\tau}n\kappa}^{core} e^{i(\underline{k}_j-\underline{k}_i)\cdot\underline{\tau}} E_{\underline{\tau}\alpha}^{cr}\mathcal{L}_{\underline{\tau}n\kappa}^{c}(k_i)\mathcal{L}_{\underline{\tau}n\kappa}^{c}(k_j)P_{\kappa}^{ss'}(\underline{k}_i,\hat{\underline{k}}_j) \quad , \quad (35)$$

$$J_{ij}^{ss'}(\underline{k}) = \delta_{ij}\delta_{ss'} - 4\pi \sum_{\underline{\tau}n\kappa} e^{i(\underline{k}_j-\underline{k}_i)\cdot\underline{\tau}} \mathcal{L}_{\underline{\tau}n\kappa}^{c}(k_i)\mathcal{L}_{\underline{\tau}n\kappa}^{c}(k_j)P_{\kappa}^{ss'}(\underline{k}_i,\hat{\underline{k}}_j) \quad (36)$$

where $\bar{E}_{\tau\alpha} = E_{\tau\alpha}^{cr}$ is now the exact crystal core eigenvalue. As can be seen from (25) corrections to the atomic eigenvalues $E_{\tau\alpha}^{at}$ are easily calculated, if it is assumed that $\delta V_{\tau\alpha}$ is spherically symmetric, since they involve only one-dimensional radial integrals (c.f. (18),(19),(20),(26),(27)). Attempts at calculating the overlap contributions $\bar{\bar{H}}_{ij}^{ss'}(\underline{k})$, $\bar{\bar{J}}_{ij}^{ss'}(\underline{k})$ have not been made at the present time although one might expect them to be significant in many cases. To gain some idea of the size of these contributions it is instructive to examine the size of the core wave functions at half the nearest-neighbor distance. As an example, these quantities are listed in Table I for the core states in copper and germanium (taken from the tables in Herman-Skillman (1963)). An indication of the significance of the numbers listed in Table I can be obtained by noting that the entry for the 3d state in copper (i.e., 0.2190) is large since this state is regarded as a valence state in OPW calculations. On the other hand, the corresponding state in germanium, which is still approximately half the size of the 3d state in copper, is treated as a core state in germanium. In Fig. 2 we have placed 3d states (radial contribution) on nearest-neighbor lattice sites in copper and germanium to show the relative sizes of the overlap effects in these materials. These observations suggest that the overlap terms in an OPW calculation of the band structure for germanium

Table I. Sizes of Atomic Wave Functions at Half the Nearest Neighbor Distance. *

State	ls	2s	2p	3s	3p	3d	4s
Copper	0.0000	0.0000	0.0000	0.0209	0.0413	0.2190	0.5914
Germanium	0.0000	0.0000	0.0000	0.0077	0.0202	0.0922	0.6550

* [N.B. The quantities tabulated here and plotted in Figs. 2 and 3 refer to r × radial wave functions.]

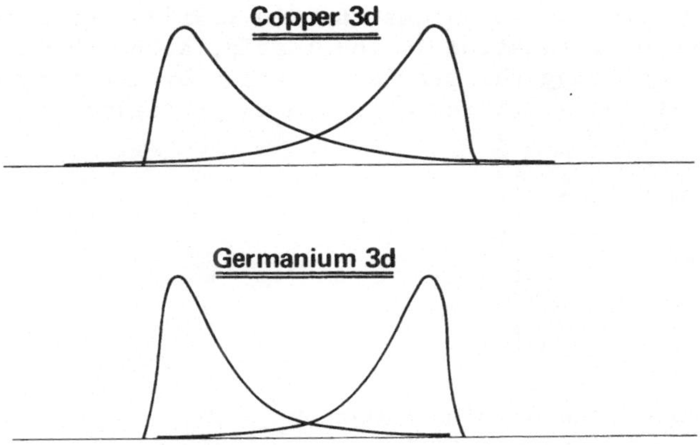

Fig. 2. 3d States in Copper and Germanium

may be significant and should be included. Similarly, but to a lesser extent, one might expect the overlap terms arising from the 3p states in copper to be non-negligible. By the same arguments, the 3d states in zinc and gallium will make larger contributions to the overlap terms $\bar{\bar{H}}_{ij}^{ss'}(\underline{k})$, $\bar{J}_{ij}^{ss'}(\underline{k})$ than in the case of germanium and should be included in an OPW calculation. However, since the overlap terms $\bar{\bar{H}}_{ij}^{\underline{s}\underline{s}'}(\underline{k})$, $\bar{J}_{ij}^{\underline{s}\underline{s}'}(\underline{k})$ will in general be small in absolute magnitude compared to the non-overlap terms $\bar{H}_{ij}^{\underline{s}\underline{s}'}(\underline{k})$, $\bar{J}_{ij}^{\underline{s}\underline{s}'}(\underline{k})$ (c.f. (21) and (32)), the errors in the energy levels resulting from the neglect of overlap should be small, possibly a few thousandth's of a rydberg at most. (Further discussion of overlap effects is contained in Section 5 in connection with the modified OPW method.)

To conclude this section we briefly consider here one of the main disadvantages of the OPW method, i.e., the necessity of using large basis sets (several hundreds or even thousands of OPW's) to obtain adequate convergence. For example, several thousand OPW's are needed to obtain energy levels converged to 0.001 ryds in carbon and zinc oxide, whereas hundreds of thousands would be needed to obtain similar convergence in one of the transition metals. Even in the successful applications of the OPW method (e.g., silicon and germanium) several hundred OPW's are required. In all cases one is faced with the problem of diagonalizing large matrices. In practice two methods have been used to handle such matrices: (a) symmetrization and (b) partitioning of matrices. The first method invokes group theory to block diagonalize the matrix thus considerably reducing the actual matrix size which must be diagonalized. However, this method is restricted to symmetry points in the zone. For calculations of optical properties (e.g., dielectric function) it is necessary to consider large numbers of $\underline{k}$ points (typically a few hundred), most of which do not lie along symmetry axes. In such cases the partitioning technique (Lowdin (1951)) for reducing the size of the secular equation is invaluable, although it has the disadvantage of replacing the secular equation by a determinantal equation. Thus, if the hermitian matrix $\overline{M}$ is partitioned in the following manner,

$$\overline{M} = \left|\begin{array}{c|c} \overline{M}_A & \overline{h} \\ \hline \overline{h}^{\dagger} & \overline{M}_B \end{array}\right| , \tag{37}$$

then the roots of the secular equation

$$\det |\overline{M} - \varepsilon\overline{I}| = 0 \tag{38}$$

are given by the roots of the determinantal equation

$$\det |\overline{M}_A + \overline{h}(\overline{M}_B - \varepsilon\overline{I}_B)^{-1}\overline{h}^{\dagger} - \varepsilon\overline{I}_A| = 0 , \tag{39}$$

where $\overline{I}_A$, $\overline{I}_B$ are unit matrices having the dimensions of $\overline{M}_A$ and $\overline{M}_B$, respectively. This method has been used extensively by Herman and coworkers (Herman et al. (1969)) in applications of the OPW method. Typically, $\overline{M}$ may be of dimension 500 x 500 and $\overline{M}_A$ of dimension 100 x 100. The major difficulty lies in the evaluation of the inverse energy-dependent matrix (c.f. (39)). It has been found in practice that retaining only the diagonal part of $\overline{M}_B$ and replacing the energy in the denominator by a suitably chosen energy $\overline{\varepsilon}$ works well and gives eigenvalues, in the region of $\overline{\varepsilon}$, accurate to about 0.001 ryds compared with treating the determinantal equation exactly. However, now that accurate experimental measurements (for photoemission work see the article by D. Eastman in this volume) are being made over an energy range of a rydberg or so, the approximate procedure described above should be modified, possibly replacing the single energy $\overline{\varepsilon}$ by two energies, ε_1 and ε_2, in order to increase the range of energy over which the approximation is valid. More accurately, the eigenvalues $\lambda(\varepsilon)$ of the energy-dependent matrix could be obtained for particular values of the energy ε, and interpolation techniques used to find the root of the equation $\lambda(\varepsilon) = \varepsilon$. The latter procedure for locating accurate roots of the determinantal equation has recently been implemented by W. E. Rudge and R. K. Nesbet (private communication).

3. THE MODIFIED OPW METHOD (NON-RELATIVISTIC)

One method for avoiding large basis sets in the OPW method is to add auxiliary functions designed to replace the large wave-vector OPW's. Since the latter essentially describe the behavior of the crystal wave function in the region of the nuclei (i.e., $\tilde{\Psi}_{\underline{k}}^{<}$ in (2)), the auxiliary functions are expected to have the character of tight-binding functions (i.e., LCAO's). This proposal was originally put forward by Herring (1940) and is supported by recent work connected with the derivation of 'first-principles' model Hamiltonians from the KKR method. The central problem is to find auxiliary functions which will significantly reduce the number of OPW's needed to obtain a specified degree of convergence (say, 0.005 ryds). Since an unfortunate choice of auxiliary functions may well result in having to use more OPW's (for a given convergence criterion) than if no auxiliary functions were used, it is instructive to examine previous attempts at finding suitable functions. (The formal mathematical theory required for the modified relativistic method is developed in Section 4.)

For definiteness we write the trial wave function for the modified OPW method as follows (the relativistic notation is used here for convenience):

$$\tilde{\Phi}_{\underline{k}}^{v}(\underline{r}) = \underbrace{\sum_{\underline{\tau\alpha}} c_{\underline{\tau\alpha}} \tilde{\Xi}_{\underline{k\tau\alpha}}^{v}(\underline{r})}_{\text{near nuclei}} + \underbrace{\sum_{js} b_{\underline{k}j,s} \tilde{O}_{\underline{k}j}^{s}(\underline{r})}_{\text{interstitial}} , \qquad (40)$$

where $\tilde{O}_{\underline{k}j}^{s}$ is an OPW basis function as defined in Section 2 (c.f. (8)) and $\tilde{\Xi}_{\underline{k\tau\alpha}}^{v}(\underline{r})$ represents a Bloch sum of 'suitable' atomic orbitals, $\tilde{\chi}_{\underline{\tau\alpha}}^{v}(\underline{r})$, orthogonalized to the same set of inner core states which enter the definition of the OPW's. The discussion below will be concerned with the choice of orbitals $\tilde{\chi}_{\underline{\tau\alpha}}^{v}(\underline{r})$. (N.B. The superscript v denotes a valence or outer-core orbital.)

In the region of a nucleus the crystal potential is approximately atomic, i.e., $V_{cr}(\underline{r}) \simeq U_{\underline{\tau\alpha}}(\underline{r})$ (c.f. (3) and (15)) so that, to a good approximation,

$$\tilde{\chi}_{\underline{\tau\alpha}}^{v}(\underline{r}) \simeq \tilde{\chi}_{\underline{\tau\alpha}}^{v}(r,\varepsilon_{\underline{\tau\alpha}}^{v}) , \qquad (41)$$

where $\tilde{\chi}_{\underline{\tau\alpha}}^{v}(r,\varepsilon_{\underline{\tau\alpha}}^{v})$ is a solution of the Dirac (or Schroedinger) equation in the presence of $\tilde{U}_{\underline{\tau\alpha}}(\underline{r})$ at the energy $\varepsilon_{\underline{\tau\alpha}}^{v}$. The latter energy need not correspond to the energy of a bound state in the potential $U_{\underline{\tau\alpha}}(\underline{r})$. In the inner-core region (i.e., $r \ll \frac{1}{2}d_{nn}$), $\tilde{\chi}_{\underline{\tau\alpha}}^{v}(r,\varepsilon_{\underline{\tau\alpha}}^{v})$ is insensitive to small variations in the value of $\varepsilon_{\underline{\tau\alpha}}^{v}$, but in the outer-core (or institial) region the reverse is true. Thus, the choice of $\varepsilon_{\underline{\tau\alpha}}^{v}$ is not critical if $\tilde{\chi}_{\underline{\tau\alpha}}^{v}(\frac{1}{2}d_{nn},\varepsilon_{\underline{\tau\alpha}}^{v})$ is small, but may be important if $\tilde{\chi}_{\underline{\tau\alpha}}^{v}(\frac{1}{2}d_{nn},\varepsilon_{\underline{\tau\alpha}}^{v})$ is large. However, the latter situation is generally the case of interest since we are mainly concerned with finding good approximations to the crystal valence states. It is usually argued that as long as the auxiliary functions accurately describe the behavior of the crystal wave function in the inner-core region, the precise outer-core behavior of the auxiliary functions is not important since the OPW's are available to describe the crystal wave function in the interstitial region. In general this is true but the number of OPW's required to meet a particular convergence criterion will depend upon how well the outer-core behavior of the auxiliary functions describes the crystal wave function in this region. Similarly, the fact that the OPW's will also contribute to the inner-core region cannot be ignored. Thus if a small number (~50-100) of OPW's are to be used, it is essential to choose orbitals, $\tilde{\chi}_{\underline{\tau}\alpha}(r)$, which approximate closely the crystal wave functions in <u>both the inner-core and outer-core</u> (or interstitial) regions. Furthermore, it is suggested here that auxiliary functions constructed from the orbitals defined by (41), with a suitably chosen $\varepsilon^{v}_{\underline{\tau}\alpha}$, should serve as useful basis functions. For example as a starting point, one may choose $\varepsilon^{v}_{\underline{\tau}\alpha} = \varepsilon^{at}_{\underline{\tau}\alpha}$, where $\varepsilon^{at}_{\underline{\tau}\alpha}$ is a bound-state solution of the Dirac equation. In semiconductors it may be preferable to use $\varepsilon^{v}_{\underline{\tau}\alpha} = \varepsilon_{gap}$ where ε_{gap} is an energy in the band gap, if this is the region of interest. In the case of transition metals the choice $\varepsilon^{v}_{\underline{\tau}\alpha} = \varepsilon_0$ where ε_0 is the resonant energy (see review Dalton (1971)) may be appropriate.

One of the disadvantages of using orbitals of the above sort to calculate matrix elements is the necessity of calculating multi-center overlap integrals. This is to be expected since the modified OPW method is essentially a combination of the OPW and LCAO methods (see Fig. 1). To avoid this problem many workers have attempted to use <u>nearest-neighbor cut-off</u> functions of the following sort,

$$\tilde{\chi}^{v}_{\underline{\tau}\alpha}(\underline{r}) = f_{\underline{\tau}\alpha}(r)\tilde{\chi}^{v}_{\underline{\tau}\alpha}(r,\varepsilon^{v}_{\underline{\tau}\alpha}) \quad , \tag{42}$$

where

$$\begin{aligned} \underline{f}_{\underline{\tau}\alpha}(r) &\simeq 1 \quad , \qquad r \ll \tfrac{1}{2}d_{nn} \\ &\simeq 0 \quad , \qquad r \geqslant \tfrac{1}{2}d_{nn} \quad . \end{aligned} \tag{43}$$

Such functions clearly preserve the inner-core behavior of $\tilde{\chi}^{v}_{\underline{\tau}\alpha}(r,\varepsilon^{v}_{\underline{\tau}\alpha})$ but completely distort the outer-core behavior. If $\tilde{\chi}^{v}_{\underline{\tau}\alpha}(r,\varepsilon^{v}_{\underline{\tau}\alpha})$ is large in the region where $f_{\underline{\tau}\alpha}(r)$ differs significantly from unity (i.e., large distortion of $\tilde{\chi}^{v}_{\underline{\tau}\alpha}(r,\varepsilon^{v}_{\underline{\tau}\alpha})$), one would expect large numbers of OPW's to be required to correct for this distortion. In such cases one would also expect that for a given number (say 100) of OPW's the energy levels resulting from the use of such functions would be dependent on the precise shape of $f_{\underline{\tau}\alpha}(r)$. Thus the use of cut-off functions to avoid evaluating multi-center integrals introduces further problems relating to the optimum choice of $f_{\underline{\tau}\alpha}(r)$. Furthermore, in cases where the distortion of $\tilde{\chi}^{v}_{\underline{\tau}\alpha}(r,\varepsilon^{v}_{\underline{\tau}\alpha})$ is large they are inappropriate and may result in having to use <u>more</u> OPW's than if no cut-off orbitals are used at all. Thus in general, multi-center integrals arising from the overlap of atomic orbitals must be evaluated if the modified OPW method is to be successful. With these observation we briefly summarize previous attempts at constructing modified OPW schemes.

Nearest-neighbor cut-off functions have been used widely in view of their simplicity. Callaway (1955b) calculated the band structure of body-centered cubic iron using a single 3d cut-off function plus 78 OPW's. Lithium was studied by Brown and Krumhansl (1958) using a single 2s cut-off function and apparently obtained convergence to about 0.01 ryds with 55 plane waves. Their scheme involved plane waves (PW) rather than OPW's, but, although inherently simpler, has the disadvantage of employing larger secular equations than in the OPW method since they explicitly calculate the inner core states. This may be a significant drawback to the use of their method for the heavier atoms. Cut-off functions obtained by numerical integration of the Schroedinger equation at an energy ε' chosen so that $\tilde{\chi}^{v}_{\underline{\tau}\alpha}(\frac{1}{2}d_{nn},\varepsilon') = 0$ ($f_{\underline{\tau}\alpha}(r) = 1$ all r) have been used by Gray and Brown (1967) for Cu_3Au, and Gray and Karpien (1971) for Cu. Butler et al. (1969) also considered Cu but used a 3d atomic function arbitrarily modified (with a fitted cubic polynomial) to go to zero at $r=\frac{1}{2}d_{nn}$. In the papers so far mentioned no convergence studies were presented or was any attempt made to examine the sensitivity of the 'converged' energy levels to variations in the shape of $f_{\underline{\tau}\alpha}(r)$. The first serious attempt at assessing the convergence properties of the modified OPW method was made by Deegan and Twose (1967) in their study of Nb. In particular, they showed (with their particular choice of $f_{\underline{\tau}\alpha}(r)$) that although certain levels (e.g., Γ_1, H_{15}) were well converged (0.005 ryds) at about 100 OPW's, other levels (e.g., $\Gamma'_{25'}$) did not appear to be converged to the same degree even with 181 OPW's. Although the sensitivity of the levels to variations in $f_{\underline{\tau}\alpha}(r)$ was not discussed in that paper, Deegan commented (private communication) that several levels were significantly changed (0.005 ryds) at 100 OPW's when different cut-off functions were tried. This is supported by the later work of Kleinman and Shurtleff (1969) on Ni who found a change of 0.007 ryds in the Γ_{12} level on varying their cut-off parameter α from 2.04 to 4.18 even when 2083 plane waves were used. Similar problems arising from the use of cut-off functions are clearly evident in the recent work of Euwema on carbon and zinc oxide (1970) and niobium (1971). In the present author's view many of the convergence problems experienced by the above authors result from the use of cut-off functions and would have been significantly reduced if the relevant overlap integrals had been included in their calculations. This view is supported by the work of Kunz (1969) on the alkali halides. On the basis of a detailed study of LiI using a modified PW scheme (not OPW), he concludes "...in order to fully utilise the advantages of mixed basis sets, it is necessary to include overlapping orbitals in the basis set." Studies of model Hamiltonians for transition metals again emphasize the importance of overlapping basis functions. (N.B. In connection with overlapping 3d states and optimum orbitals see also the recent work of Lipari and Deegan (1971).)

4. THE MODIFIED RELATIVISTIC OPW METHOD (THEORY)

The trial wave function for the modified relativistic OPW method consists of an expansion in terms of OPW's and auxiliary LCAO's (for convenience we repeat (40)),

$$\tilde{\Phi}^{v}_{\underline{k}}(\underline{r}) = \sum_{\underline{\tau}\alpha} c_{\underline{\tau}\alpha} \tilde{\Xi}^{v}_{\underline{k}\underline{\tau}\alpha}(\underline{r}) + \sum_{js} b_{\underline{k}j,s} \tilde{O}^{s}_{\underline{k}j}(\underline{r}) \quad , \tag{40}$$

where $\tilde{O}^{s}_{\underline{k}j}$ is a relativistic OPW (c.f. (8)) and $\tilde{\Xi}^{v}_{\underline{k}\underline{\tau}\alpha}(\underline{r})$ is a relativistic Bloch sum of suitable atomic orbitals, $\tilde{\chi}^{v}_{\underline{\tau}\alpha}(\underline{r})$, orthogonalized to the inner-core states, $\tilde{\Phi}^{c}_{\underline{k}\underline{\tau}'\alpha'}(\underline{r})$, i.e.,

$$\tilde{\Xi}^{v}_{\underline{k}\underline{\tau}\alpha}(\underline{r}) = \tilde{X}^{v}_{\underline{k}\underline{\tau}\alpha}(\underline{r}) - \sum_{\underline{\tau}'\alpha'}^{\text{core}} B^{v\ c}_{\underline{\tau}\alpha,\underline{\tau}'\alpha'} \tilde{\Phi}^{c}_{\underline{k}\underline{\tau}'\alpha'}(\underline{r}) \quad , \tag{44}$$

where

$$\tilde{X}^{v}_{\underline{k}\underline{\tau}\alpha}(\underline{r}) = \frac{1}{\sqrt{N}} \sum_{\underline{R}} \tilde{\chi}^{v}_{\underline{\tau}\alpha}(\underline{r}-\underline{R}-\underline{\tau}) e^{i\underline{k}\cdot\underline{R}} \quad , \tag{45}$$

$$B^{v\ c}_{\underline{\tau}\alpha,\underline{\tau}'\alpha'} = \int \tilde{X}^{v}_{\underline{k}\underline{\tau}\alpha}(\underline{r}) \tilde{\Phi}^{c}_{\underline{k}\underline{\tau}'\alpha'}(\underline{r}) d^3r \quad ,$$

$$= \sum_{\underline{R}} I^{v\ c}_{\underline{\tau}\alpha,\underline{\tau}'\alpha'}(\underline{R}) e^{i\underline{k}\cdot\underline{R}} \quad , \tag{46}$$

and the valence-core overlap integral, $I^{v\ c}_{\underline{\tau}\alpha,\underline{\tau}'\alpha'}(\underline{R})$, is defined by (18) with the replacement of one of the atomic core states by an atomic valence state. Application of the variational principle with the trial wave function defined by (40) leads to a secular equation with the following structure (c.f. (11)),

$$\det |\overline{D} - E_{\underline{k}}\overline{K}| = 0 \quad , \tag{47}$$

where

$$\overline{D} = \left| \begin{array}{c|c} T^{v\ v}_{\underline{\tau}\alpha,\underline{\tau}'\alpha'} & h^{v}_{\underline{\tau}\alpha,s'j} \\ \hline h^{v\ \dagger}_{\underline{\tau}\alpha,s'j} & H^{ss'}_{ij} \end{array} \right| \quad , \tag{48}$$

$$T^{v\ v}_{\underline{\tau}\alpha,\underline{\tau}'\alpha'} = \int \tilde{\Xi}^{v\dagger}_{\underline{k}\tau\alpha} \tilde{H}_D \tilde{\Xi}^{v'}_{\underline{k}\tau'\alpha'} d^3r \quad , \tag{49}$$

$$h^{v}_{\underline{\tau}\alpha,s'j} = \int \tilde{\Xi}^{v\dagger}_{\underline{k}\tau\alpha} \tilde{H}_D \tilde{O}^{s'}_{\underline{k}j} d^3r \quad , \tag{50}$$

and $H^{ss'}_{ij}$ is defined by (12) and (21). The matrix elements of $\tilde{K}$ corresponding to $T^{v\ v}_{\underline{\tau}\alpha,\underline{\tau}'\alpha'}$, $h^{v}_{\underline{\tau}\alpha,s'j}$ and $H^{ss'}_{ij}$ are

$$S^{v\ v}_{\underline{\tau}\alpha,\underline{\tau}'\alpha'} = \int \tilde{\Xi}^{v\dagger}_{\underline{k}\tau\alpha} \tilde{\Xi}^{v}_{\underline{k}\tau'\alpha'} d^3\underline{r} \quad , \tag{51}$$

$$g^{v}_{\underline{\tau}\alpha,s'j} = \int \tilde{\Xi}^{v\dagger}_{\underline{k}\tau\alpha} \tilde{O}^{s'}_{\underline{k}j} d^3r \quad , \tag{52}$$

and $J^{ss'}_{ij}$ (c.f. (13) and (32), respectively). Expressions for the OPW matrix elements $H^{ss'}_{ij}$, $J^{ss'}_{ij}$ were given in Section 2. The evaluation of the remaining matrix elements is tedious but straightforward. Thus the final result for $T^{v\ v}_{\underline{\tau}\alpha,\underline{\tau}\alpha}$ may be written as follows:*

*(NB. $\chi^{v}_{\underline{\tau}\alpha}(\underline{r})$ is here assumed to be a bound-state solution similar to the core-states; more generally other terms must be added to the expressions for the matrix elements).

$$T^{v\ v}_{\underline{\tau}\alpha,\underline{\tau}'\alpha'} = t^{v\ v}_{\underline{\tau}\alpha,\underline{\tau}'\alpha'} + y^{v\ v}_{\underline{\tau}\alpha,\underline{\tau}'\alpha'} \quad , \tag{53}$$

where

$$t^{v\ v}_{\underline{\tau}\alpha,\underline{\tau}'\alpha'} = \varepsilon^{v}_{\underline{\tau}\alpha} B^{v\ v}_{\underline{\tau}\alpha,\underline{\tau}'\alpha'} - \sum_{\underline{\tau}''\alpha''}^{\text{core}} \left(\varepsilon^{v}_{\underline{\tau}\alpha} B^{v\ c}_{\underline{\tau}\alpha,\underline{\tau}''\alpha''} B^{c\ \ v*}_{\underline{\tau}''\alpha'',\underline{\tau}'\alpha'} + \varepsilon^{v}_{\underline{\tau}'\alpha'} B^{v\ c*}_{\underline{\tau}\alpha,\underline{\tau}''\alpha''} B^{c\ \ v}_{\underline{\tau}''\alpha'',\underline{\tau}'\alpha'} \right)$$
$$+ \sum_{\underline{\tau}''\alpha''}^{\text{core}} \sum_{\underline{\tau}'''\alpha'''}^{\text{core}} E^{at}_{\underline{\tau}'''\alpha'''} B^{v\ c*}_{\underline{\tau}\alpha,\underline{\tau}''\alpha''} B^{c\ \ c}_{\underline{\tau}''\alpha'',\underline{\tau}'''\alpha'''} B^{c\ \ v*}_{\underline{\tau}'''\alpha''',\underline{\tau}'\alpha'} \quad , \tag{54}$$

$$y^{v\ v}_{\underline{\tau}\alpha,\underline{\tau}'\alpha'} = W^{v\ v}_{\underline{\tau}\alpha,\underline{\tau}'\alpha'} - \sum_{\underline{\tau}''\alpha''}^{\text{core}} \left(W^{v\ c}_{\underline{\tau}\alpha,\underline{\tau}''\alpha''} B^{c\ \ v*}_{\underline{\tau}''\alpha'',\underline{\tau}'\alpha'} + B^{v\ c*}_{\underline{\tau}\alpha,\underline{\tau}''\alpha''} W^{c\ \ v}_{\underline{\tau}''\alpha'',\underline{\tau}'\alpha'} \right)$$
$$+ \sum_{\underline{\tau}''\alpha''}^{\text{core}} \sum_{\underline{\tau}'''\alpha'''}^{\text{core}} B^{v\ c*}_{\underline{\tau}\alpha,\underline{\tau}''\alpha''} W^{c\ \ c}_{\underline{\tau}''\alpha'',\underline{\tau}'''\alpha'''} B^{c\ \ v*}_{\underline{\tau}'''\alpha''',\underline{\tau}'\alpha'} \quad , \tag{55}$$

and, for example,

$$W^{v\ v}_{\underline{\tau}\alpha,\underline{\tau}'\alpha'} = \sum_{\underline{R}} \omega^{v\ v}_{\underline{\tau}\alpha,\underline{\tau}'\alpha'}(\underline{R})e^{i\underline{k}\cdot\underline{R}} \quad . \tag{56}$$

The valence-valence ($B^{v\ v}_{\underline{\tau}\alpha,\underline{\tau}'\alpha'}$) and core-core ($B^{c\ c}_{\underline{\tau}''\alpha'',\underline{\tau}'''\alpha'''}$) integrals are defined by (46) with the correct insertion of the relevant core or valence wave function. The energy integrals $\omega^{v\ v}_{\underline{\tau}\alpha,\underline{\tau}'\alpha'}(\underline{R})$, $\omega^{v\ c}_{\underline{\tau}\alpha,\underline{\tau}''\alpha''}(\underline{R})$ are defined by (20) with the appropriate choice of wave functions. Conventional (relativistic) tight-binding theory for the valence states is obtained by setting all the valence-core integrals ($B^{v\ c}_{\underline{\tau}\alpha,\underline{\tau}''\alpha''}$) equal to zero. Thus much of the (non-relativistic) work of Slater and Koster (1954) (for cubic systems) and Miasek (1957) (for hexagonal systems) can be used here to simplify the evaluation of integrals of the form $B^{v\ v}_{\underline{\tau}\alpha,\underline{\tau}'\alpha'}$ and $W^{v\ v}_{\underline{\tau}\alpha,\underline{\tau}'\alpha'}$.

As expected, the expression for $S^{v\ v}_{\underline{\tau}\alpha,\underline{\tau}'\alpha'}$ is almost identical with that for $t^{v\ v}_{\underline{\tau}\alpha,\underline{\tau}'\alpha'}$, i.e.,

$$S^{v\ v}_{\underline{\tau}\alpha,\underline{\tau}'\alpha'} = B^{v\ v}_{\underline{\tau}\alpha,\underline{\tau}'\alpha'} - \sum_{\underline{\tau}''\alpha''}^{\text{core}} \Big(B^{v\ c}_{\underline{\tau}\alpha,\underline{\tau}''\alpha''} B^{c\ v*}_{\underline{\tau}''\alpha'',\underline{\tau}'\alpha'} + B^{v\ c*}_{\underline{\tau}\alpha,\underline{\tau}''\alpha''} B^{c\ v}_{\underline{\tau}''\alpha'',\underline{\tau}'\alpha'} \Big)$$
$$+ \sum_{\underline{\tau}''\alpha''}^{\text{core}} \sum_{\underline{\tau}'''\alpha'''}^{\text{core}} B^{v\ c*}_{\underline{\tau}\alpha,\underline{\tau}''\alpha''} B^{c\ c}_{\underline{\tau}''\alpha'',\underline{\tau}'''\alpha'''} B^{c\ v*}_{\underline{\tau}'''\alpha''',\underline{\tau}'\alpha'} \quad . \tag{57}$$

The hybridization integrals $h^{v}_{\underline{\tau}\alpha,s'j}$, $g^{v}_{\underline{\tau}\alpha,s'j}$ are given by analogous expressions which now include also the plane-wave orthogonalization integrals $A^{c}_{\underline{\tau}\alpha}(s\underline{k}_j)$ (c.f. (9) and (10)), i.e.,

$$h^{v}_{\underline{\tau}\alpha,s'j} = \overline{h}^{v}_{\underline{\tau}\alpha,s'j} + \delta\overline{h}^{v}_{\underline{\tau}\alpha,s'j} \quad , \tag{58}$$

where

$$\overline{h}^{v}_{\underline{\tau}\alpha,s'j} = \varepsilon^{v}_{\underline{\tau}\alpha} A^{v}_{\underline{\tau}\alpha}(s'\underline{k}) - \sum_{\underline{\tau}''\alpha''}^{\text{core}} E^{at}_{\underline{\tau}''\alpha''} \Big(B^{v\ c}_{\underline{\tau}\alpha,\underline{\tau}''\alpha''} A^{c}_{\underline{\tau}''\alpha''}(s'\underline{k}_j) + B^{v\ c*}_{\underline{\tau}\alpha,\underline{\tau}''\alpha''} A^{c}_{\underline{\tau}''\alpha''}(s'\underline{k}_j) \Big)$$
$$+ \sum_{\underline{\tau}''\alpha''}^{\text{core}} \sum_{\underline{\tau}'''\alpha'''}^{\text{core}} E^{at}_{\underline{\tau}'''\alpha'''} B^{v\ c*}_{\underline{\tau}\alpha,\underline{\tau}''\alpha''} B^{c\ c}_{\underline{\tau}''\alpha'',\underline{\tau}'''\alpha'''} A^{c}_{\underline{\tau}'''\alpha'''}(s'\underline{k}_j) \quad , \tag{59}$$

$$\delta \bar{h}^{v}_{\underline{\tau}\alpha,s'j} = \nu^{v}_{\underline{\tau}\alpha}(s'\underline{k}_j) - \sum_{\underline{\tau}''\alpha''}^{\text{core}} \Big(B^{v\ c}_{\underline{\tau}\alpha,\underline{\tau}''\alpha''} \nu^{c}_{\underline{\tau}''\alpha''}(s'\underline{k}_j) + W^{v\ c}_{\underline{\tau}\alpha,\underline{\tau}''\alpha''} A^{c}_{\underline{\tau}''\alpha''}(s'\underline{k}_j)\Big)$$

$$+ \sum_{\underline{\tau}''\alpha''}^{\text{core}} \sum_{\underline{\tau}'''\alpha'''}^{\text{core}} B^{v\ c*}_{\underline{\tau}\alpha,\underline{\tau}''\alpha''} W^{c\ c}_{\underline{\tau}''\alpha'',\underline{\tau}'''\alpha'''} A^{c}_{\underline{\tau}'''\alpha'''}(s'\underline{k}_j) \quad , \qquad (60)$$

$$g^{v}_{\underline{\tau}\alpha,s'j} = A^{v}_{\underline{\tau}\alpha}(s'\underline{k}_j) - \sum_{\underline{\tau}''\alpha''}^{\text{core}} \Big(B^{v\ c}_{\underline{\tau}\alpha,\underline{\tau}''\alpha''} A^{c}_{\underline{\tau}''\alpha''}(s'\underline{k}_j) + B^{v\ c*}_{\underline{\tau}\alpha,\underline{\tau}''\alpha''} A^{c}_{\underline{\tau}''\alpha''}(s'\underline{k}_j)\Big)$$

$$+ \sum_{\underline{\tau}''\alpha''}^{\text{core}} \sum_{\underline{\tau}'''\alpha'''}^{\text{core}} B^{v\ c*}_{\underline{\tau}\alpha,\underline{\tau}''\alpha''} B^{c\ c}_{\underline{\tau}''\alpha'',\underline{\tau}'''\alpha'''} A^{c}_{\underline{\tau}'''\alpha'''}(s'\underline{k}_j) \quad , \qquad (61)$$

and the integrals $\nu^{c}_{\underline{\tau}\alpha}(s'\underline{k}_j)$, $\nu^{v}_{\underline{\tau}\alpha}(s'\underline{k}_j)$ are defined by (19).

The general expressions derived above together with the formulae given in Section 2 are new and incorporate both the relativistic OPW and LCAO methods as special cases. The success of the modified OPW method clearly depends upon the ease with which the relevant multi-center integrals (in particular two-center and three-center) can be evaluated. Methods for evaluating such integrals have been developed by quantum chemists in connection with molecular calculations. It is therefore likely that the same methods, suitably modified, can be used here to calculate integrals of the form $I^{c\ c}_{\underline{\tau}\alpha,\underline{\tau}'\alpha'}(\underline{R})$ and $\omega^{c\ c}_{\underline{\tau}\alpha,\underline{\tau}'\alpha'}(\underline{R})$ (c.f. (18) and (20)).

5. 3d STATES IN COPPER AND GERMANIUM

At the time of writing, a computer program has been developed which incorporates the theory of Sections 2 and 4 with the exception of the overlap terms. The starting point for this program was a non-self-consistent relativistic (simplified theory) OPW program. (The latter program was originally provided by Van Dyke but was subsequently modified by Rudge.) Routines to calculate multi-center integrals, and hence the overlap terms are to be developed and should soon be available. Nevertheless, the present program can be used to illustrate the main points of this paper, i.e., the use of auxiliary functions to increase the convergence of the OPW method and the importance of including overlap terms. Since our ultimate aim is to develop a program which will deal equally well with metals and semiconductors, we consider here in parallel both copper and germanium. The bands of interest in copper arise directly from the 3d states whereas in germanium the higher lying states (e.g., 4s, 4p) are of particular interest. However, for simplicity and purposes of comparison, we shall restrict our attention to the 3d states in both materials, and in particular the lowest 3d state (i.e., Γ_8^+). Since the standard OPW method is incapable of calculating states with d symmetry (recall that Kleinman and Shurtleff (1969) used 4279 plane waves in an expansion of the Γ_{12} level of Ni without any sign of convergence), this example is particularly instructive.

As a first attempt at calculating the 3d bands in copper and germanium a standard OPW was used with the following states treated as core states, i.e., $1s_{\frac{1}{2}}$, $2s_{\frac{1}{2}}$, $2p_{\frac{1}{2}}$, $2p_{\frac{3}{2}}$ $3s_{\frac{1}{2}}$, $3p_{\frac{1}{2}}$, $3p_{\frac{3}{2}}$ The latter were obtained from a self-consistent relativistic atomic calculation with Kohn-Sham exchange based on Liberman's program (Liberman _et al._ (1965)). In the present work the crystal potential, in both copper and germanium, was constructed from a sum of overlapping self-consistent relativistic atomic potentials with Kohn-Sham exchange. The calculations were then repeated but with the OPW basis set augmented by suitable 3d atomic orbitals. For definiteness the latter were taken to be self-consistent relativistic atomic $3d_{\frac{3}{2}}$, $3d_{\frac{5}{2}}$ states. (In future studies of copper it would probably be more appropriate to use the 3d resonant orbital discussed by Hubbard (1967) and Heine (1967). See also the recent work by Anderson and Kasowski (1971).) The results of these calculations are recorded in Table II for the level Γ_8^+.

Table II. Convergence of Γ_8^+ in Copper and Germanium

COPPER

OPW's	STANDARD RELOPW	RELOPW + 3d-ATOMIC STATES
9	−2.939 11	−4.361 38
27	−3.005 02	−4.363 99
51	−3.060 59	−4.364 73
59	−3.112 31	−4.365 19
65	−3.112 31	−4.365 20
89	−3.152 25	−4.371 68
113	−3.178 41	−4.367 21

$\Gamma_6^+ = -5.042\ 60$ ryds.

GERMANIUM

OPW's	STANDARD RELOPW	RELOPW + 3d-ATOMIC STATES
9	–	−4.683 14
27	−2.281 72	−4.685 78
51	−2.394 89	−4.685 86
59	−2.394 89	−4.685 90
65	−2.394 89	−4.686 02
89	−2.399 95	−4.686 04
113	−2.457 07	−4.686 07

$\Gamma_6^+ = -3..886\ 33$ ryds.

Consider first the results for germanium. At 113 OPW's the standard OPW method is predicting that the lowest 3d level (Γ_8^+) lies approximately 1.4 ryds above the level Γ_6^+ (which corresponds to the 4s atomic level), whereas it is well-known to lie below Γ_6^+. Considering the rate of convergence of the OPW method for this level it is clear that several hundreds or even thousands of OPW's are needed to obtain convergence to 0.001 ryds (say). The dramatic improvement in the rate of convergence resulting from the addition of the 3d auxiliary functions is evident in the last column of Table II. Convergence to 0.001 ryds is apparently obtained at 59 OPW's. The same convergence behavior is seen in the case of copper. At 113 OPW's the separation between Γ_8^+ and Γ_6^+ predicted by the standard OPW method is approximately 2 ryds, whereas one would expect from other band calculations (KKR and APW) for copper a separation close to 0.5 ryds. The addition of the 3d orbitals reduces the separation of 2 ryds to approximately 0.65 ryds even with only 9 OPW's in the basis set. Although these results are encouraging, a close study of the convergence in the case of copper reveals an apparent violation of the variational principle, i.e., levels should converge monotonically. However, the erratic behavior shown in column 3 of Table II can be traced to the neglect of the overlap terms in the tight-binding matrix elements. We first note that the overlap matrix ($\bar{K}$ in (47)) must be positive definite. If errors are made in the calculation of $\bar{K}$, such as the neglect of overlap terms, and the errors are large enough, then $\bar{K}$ will cease to be positive definite. Hence one method of assessing the importance of overlap terms is to examine the sign of the determinant of $\bar{K}$. If the latter becomes negative, this is strong evidence that significant overlap terms have been omitted in the calculation of $\bar{K}$. The signs of the determinant of $\bar{K}$ at the symmetry points Γ, X and L for copper and germanium are recorded in Table III. We note that whereas the determinant remains positive in the case of germanium, it becomes negative at 65 OPW's at L, and at 89 OPW's at X, in the case of copper. The change in the sign of the determinant in the latter case coincides precisely with the onset of the erratic convergence behavior of the level Γ_8^+. Since the only known errors in the calculation of $\bar{K}$ arise from the neglect of the overlap terms, we must conclude that the anomalous convergence behavior found in copper is due to the latter. The apparent lack of a similar behavior in germanium is due to the much smaller contribution of the overlap terms, although one cannot exclude such behavior if more OPW's are used. Problems arising from the neglect of overlap terms are also evident in the recent work of Euwema (1971) relating to niobium at the point Γ. The lower entries in Table III are derived from his work. The parameter τ is a measure of the degree of overlap neglected in his 'non-overlapping' gaussian basis functions. Clearly his present scheme, which uses nearest-neighbor 'cut-off' functions, is inadequate and must be modified to include overlap terms.

Table III. Sign of the Determinant of the Matrix $\overline{K}$

"Atomic 3d - Functions"

OPW's		9	27	51	59	65	89	113
Germanium	Γ	+	+	+	+	+	+	+
	X	+	+	+	+	+	+	+
	L	+	+	+	+	+	+	+
Copper	Γ	+	+	+	+	+	+	+
	X	+	+	+	+	+	—	—
	L	+	+	+	+	—	—	—

"Non - Overlapping" Gaussian Functions (Euwema (1971))

PW's		1	13	19	43	55	79
NIOBIUM at Γ	τ = .005	+	+	+	+	+	—
	τ = .007	+	+	+	+	—	—
	τ = .009	+	+	+	—	—	—

To conclude this section we remark that in many cases it would be desirable to improve the convergence of the higher lying ('excited') levels. In general these will require auxiliary orbitals which are large at half the nearest-neighbor distance. For example, the convergence of the level Γ_6^+ in copper and germanium could be considerably improved by adding 4s atomic-like orbitals to the basis functions. A glance at Fig. 3 (c.f. Fig. 2) shows the large overlap resulting from the use of such orbitals, and in particular indicates the futility of using nearest-neighbor cut-off functions to approximate such states.

SUMMARY AND CONCLUSIONS

In this paper we have generalized the work of Herring (1940) relating to the non-relativistic OPW method for calculating band structures to include the possibility of overlapping core orbitals and ℓ-dependent potentials within a completely relativistic framework (Section 2). Methods for treating large OPW basis sets were discussed (Section 2) and the importance of including overlapping core states in certain materials was stressed. In all current

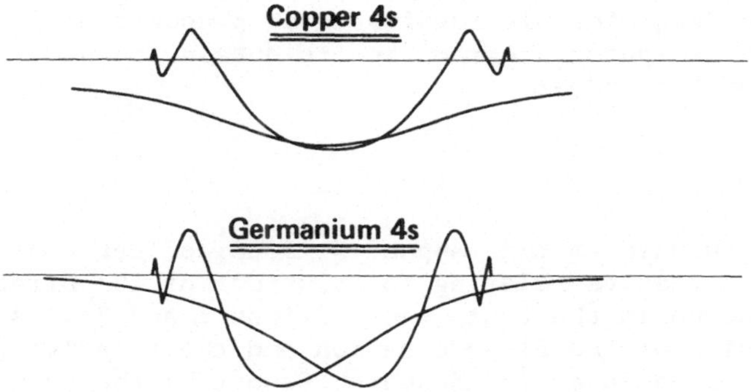

Fig. 3. 4s Atomic States in Copper and Germanium

applications of the OPW method no allowance is made for the possibility of overlapping core states. Previous attempts at increasing the convergence of the OPW method by augmenting the basis states with suitable auxiliary functions (LCAO's) were then reviewed, and it was argued that in most applications nearest-neighbor cut-off functions are not suitable functions (contrary to much recent work) (Section 3). Instead, solutions of the Dirac (or Schroedinger) equation in an appropriate atomic potential, $U_{I\alpha}(r)$ (see text) at some energy $\varepsilon_{I\alpha}$ would appear to be more useful, even though their use would necessitate the evaluation of multi-center integrals. The general theory of the modified relativistic OPW method was presented in Section 4. A simplified version of the theory (i.e., neglecting overlap) was used in Section 5 to study the 3d states in copper (transition metal) and germanium (semi-conductor) to illustrate the importance of including overlap terms.

The scheme proposed here for calculating band structures would seem to be flexible enough to treat virtually all materials, i.e., non-relativistic and relativistic metals and semi-conductors with arbitrary crystal lattice structure. Although various suggestions as to the precise choice of the auxiliary orbitals have been made in the text, only extensive numerical work, aided by further theoretical studies, will be able to decide the optimum choice of orbitals in particular materials.

ACKNOWLEDGEMENTS

I would like to thank Frank Herman for originally stimulating my interest in the OPW method and for several informative discussions relating to the material presented in this paper. My thanks are also due to William Rudge for his invaluable help connected with the development of the computer program used to obtain the numerical results in Section 5 of this paper.

APPENDIX

The main purpose of this appendix is to collect together a few definitions and results relating to solutions of the Dirac equation which were assumed in the text (see Sections 1 and 2). A more detailed discussion of the Dirac equation and the relativistic OPW method is contained in an IBM Research Report by the author (Dalton (1970)). If it is not otherwise stated, the unit of energy is the rydberg (i.e., $me^2/(2\hbar^2) \simeq 13.6$ ev) and the unit of length is the Bohr atomic unit (i.e., $\hbar^2/(me^2) \simeq 0.53$Å) in all subsequent discussion. Furthermore, α denotes the fine-structure constant (i.e., $e^2/c\hbar \simeq 1/137$).

The Dirac equation is generally written as follows,

$$[c\tilde{\alpha}\cdot\bar{p} + \tilde{\beta}mc^2 + V(\underline{r})\tilde{I}_4]\tilde{\psi} = E\tilde{\psi} \quad , \tag{A1}$$

where c is the velocity of light, m is the mass of the electron, $V(\underline{r})$ is a scalar potential, $\tilde{I}_4$ is a 4-dimensional unit operator, $\bar{p}$ is the momentum operator,

$$\tilde{\alpha} = \begin{pmatrix} 0 & \bar{\sigma} \\ \bar{\sigma} & 0 \end{pmatrix} \quad , \quad \tilde{\beta} = \begin{pmatrix} \tilde{I}_2 & 0 \\ 0 & -\tilde{I}_2 \end{pmatrix} \quad , \tag{A2}$$

$\tilde{I}_2$ is the two-dimensional unit operator, and $\bar{\sigma}$ is the Pauli matrix with components

$$\bar{\sigma}_x = \begin{pmatrix} 0 & 1 \\ 1 & 0 \end{pmatrix} \quad , \quad \bar{\sigma}_y = \begin{pmatrix} 0 & -i \\ i & 0 \end{pmatrix} \quad , \quad \bar{\sigma}_z = \begin{pmatrix} 1 & 0 \\ 0 & -1 \end{pmatrix} \quad . \tag{A3}$$

$\tilde{\psi}$ is a four-component spinor (column vector) whose structure depends upon the choice of $V(\underline{r})$. The two cases of interest here are $V(\underline{r})=0$ (i.e., relativistic plane waves) and $V(\underline{r}) = V(r)$ (i.e., spherically symmetric).

Solutions for $V(\underline{r}) = 0$

The relativistic plane-wave solutions ($\tilde{\psi}_{PW}$) of the Dirac equation can be written most conveniently in the form

$$\tilde{\psi}_{PW} = \tilde{U}^s_{\underline{k}} e^{i\underline{k}\cdot\underline{r}} \qquad s = \pm\tfrac{1}{2} \quad , \tag{A4}$$

where $\underline{k}$ is a wave vector with magnitude k,

$$\tilde{U}^s_{\underline{k}} = \begin{pmatrix} U_+(k)\tilde{\chi}^s \\ U_-(k)\frac{\bar{\sigma}\cdot\underline{k}}{k}\tilde{\chi}^s \end{pmatrix} , \tag{A5}$$

$$U_\delta(k) = \left(\frac{E_0(k) + \delta\frac{2}{\alpha^2}}{2E_0(k)} \right)^{\frac{1}{2}} , \qquad \delta = \pm 1 \tag{A6}$$

$$E_0(k) = \frac{2}{\alpha^2}\left(1 + (\alpha k)^2\right)^{\frac{1}{2}} , \tag{A7}$$

and

$$\chi^{\frac{1}{2}} = \begin{pmatrix} 1 \\ 0 \end{pmatrix} , \qquad \chi^{-\frac{1}{2}} = \begin{pmatrix} 0 \\ 1 \end{pmatrix} . \tag{A8}$$

Solutions for $V(\underline{r}) = V(r)$

An electron moving in a central field potential gives rise to various constants of motion. The commutation properties of the operators describing the latter with the Dirac Hamiltonian result in the following relativistic quantum numbers:

$$\begin{aligned} &\text{principal quantum number:} && n = 1, 2, 3, 4, \ldots \\ &\text{spin-orbit quantum number:} && \kappa = \pm 1, \pm 2, \ldots, \pm(n-1), -n, \\ &\text{total angular momentum:} && j = |\kappa| - \tfrac{1}{2} \\ &\text{magnetic quantum number:} && \mu = -j, -j+1, \ldots, j-1, j \\ &\text{orbital angular momentum (upper):} && \ell_A = \kappa \qquad \kappa > 0 \\ &&& \quad\; = -\kappa-1, \quad \kappa < 0 \\ &\text{orbital angular momentum (lower):} && \ell_B = \kappa-1 \qquad \kappa > 0 \\ &&& \quad\; = -\kappa \qquad \kappa < 0 \end{aligned} \tag{A9}$$

With these definitions the four-component spinor solution of the Dirac equation may be written as follows:

$$\tilde{\Psi}_{n\kappa\mu}(\underline{r}) = \begin{pmatrix} \frac{1}{r} G_{n\kappa}(r)\tilde{\chi}^\mu_\kappa(\hat{\underline{r}}) \\ i\frac{1}{r} F_{n\kappa}(r)\tilde{\chi}^\mu_{-\kappa}(\hat{\underline{r}}) \end{pmatrix} , \tag{A10}$$

where $F_{n\kappa}(r)$ and $G_{n\kappa}(r)$ are solutions of the coupled equations

$$\frac{2}{\alpha}\left(\frac{dF_{n\kappa}}{dr} - \frac{\kappa}{r} F_{n\kappa}\right) + \left(E - V(r) - \frac{2}{\alpha^2}\right)G_{n\kappa} = 0 \quad ,$$

$$\frac{2}{\alpha}\left(\frac{dG_{n\kappa}}{dr} + \frac{\kappa}{r} G_{n\kappa}\right) - \left(E - V(r) + \frac{2}{\alpha^2}\right)F_{n\kappa} = 0 \quad . \tag{A11}$$

The central field spinor $\tilde{\chi}^{\mu}_{\kappa}(\hat{\underline{r}})$ is defined by the equation

$$\tilde{\chi}^{\mu}_{\kappa}(\hat{\underline{r}}) = \sum_{s=\pm\frac{1}{2}} C(\ell_{\kappa}\tfrac{1}{2}j;\mu-s,s)Y^{\mu-s}_{\ell_{\kappa}}(\hat{\underline{r}})\tilde{\chi}^{s} \quad , \tag{A12}$$

where $C(\ell_{\kappa}\frac{1}{2}j;\mu-s,s)$ is the spin $\frac{1}{2}$ Clebsh-Gordon coefficient, $Y^{\mu-s}_{\ell_{\kappa}}(\hat{\underline{r}})$ is the standard Spherical Harmonic, $\hat{\underline{r}}$ is a unit vector in the direction of $\underline{r}$, and ℓ_{κ} is the orbital angular momentum quantum which may equal either ℓ_A or ℓ_B. If the Spherical Harmonics are orthonormalized, then,

$$\int \tilde{\chi}^{\mu\dagger}_{\kappa}\tilde{\chi}^{\mu'}_{\kappa'}d^3\underline{r} = \delta_{\mu\mu'}\delta_{\kappa\kappa'} \quad . \tag{A13}$$

The evaluation of the matrix elements in Section 2 is simplified if the following relation is used,

$$\tilde{U}^{s}_{\underline{k}}e^{i\underline{k}\cdot\underline{r}} = \sum_{\kappa'\mu'} a^{s}_{\kappa'\mu'}(\underline{k})\begin{pmatrix} U_{+}(k)j_{\ell_{\kappa'}}(kr)\tilde{\chi}^{\mu'}_{\kappa'}(\hat{\underline{r}}) \\ \\ iU_{-}(k)s_{\kappa'}j_{\ell_{-\kappa'}}(kr)\tilde{\chi}^{\mu'}_{-\kappa'}(\hat{r}) \end{pmatrix} \quad , \tag{A14}$$

where

$$a^{s}_{\kappa'\mu'}(\underline{k}) = 4\pi i^{\ell_{\kappa}}C(\ell_{\kappa}\tfrac{1}{2}j;\mu-s,s)Y^{\mu-s*}_{\ell_{\kappa}}(\underline{k}) \quad , \tag{A15}$$

and $j_{\ell}(x)$ is the spherical Bessel function. The following identity is required for performing summations over the magnetic quantum number μ, i.e.,

$$\sum_{\mu} a^{s\dagger}_{\kappa\mu}(\underline{k}_i)a^{s'}_{\kappa\mu}(\underline{k}_j) = 4\pi\tilde{\chi}^{s\dagger}\overline{M}_{\kappa}(\underline{k}_i,\underline{k}_j)\tilde{\chi}^{s'} \quad , \tag{A16}$$

where

$$\overline{M}_{\kappa}(\underline{k}_i,\underline{k}_j) = s_{\kappa}[\kappa I_2 - \frac{1}{\hbar}\,\overline{\sigma}\cdot\overline{L}]P_{\ell_{\kappa}}(\underline{k}_i\hat{,}\underline{k}_j) \quad , \tag{A17}$$

s_κ is the sign of κ (= ±1), $\overline{L}$ is the orbital angular momentum operator, and $P_\ell(x)$ is the Legendre polynomial.

To illustrate the methods used to derive the expressions in Section 2, we consider here the evaluation of $A^c_{\underline{\tau}\alpha}(s\underline{k}_j)$ as defined by (9) and (10), i.e.,

$$A^c_{\underline{\tau}\alpha}(s\underline{k}_j) = \int \tilde{\Phi}^{c\dagger}_{\underline{k}\tau\alpha} \frac{1}{\sqrt{Nv}} \tilde{U}^s_{\underline{k}j} e^{i\underline{k}_j\cdot\underline{r}} d^3r \quad ,$$

i.e.,

$$A^c_{\underline{\tau}\alpha}(s\underline{k}_j) = e^{i\underline{k}_j\cdot\underline{\tau}} \frac{1}{\sqrt{v}} \int \tilde{\phi}^{c\dagger}_{\underline{\tau}\alpha}(\underline{r})\tilde{U}^s_{\underline{k}j} e^{i\underline{k}_j\cdot\underline{r}} d^3r \quad , \text{(using (6))}$$

i.e.,

$$\underline{A^c_{\underline{\tau}\alpha}(s\underline{k}_j) = e^{i\underline{k}_j\cdot\underline{\tau}} a^s_{\kappa\mu}(\underline{k}_j)\mathcal{L}^c_{\tau n\kappa}(k_j)} \quad , \tag{A18}$$

where

$$\mathcal{L}^c_{\tau n\kappa}(k_j) = \frac{1}{\sqrt{v}} \int_0^\infty r\tilde{R}^{c\dagger}_{\alpha\tau}\tilde{J}_\kappa(k_j)dr \tag{A19}$$

and

$$\tilde{R}^c_{\alpha\tau} = \begin{pmatrix} G^c_{n\kappa}(r) \\ iF^c_{n\kappa}(r) \end{pmatrix} , \quad \tilde{J}_\kappa(k_j) = \begin{pmatrix} U_+(k_j)j_{\ell_\kappa}(k_jr) \\ iU_-(k_j)s_\kappa j_{\ell_{-\kappa}}(k_jr) \end{pmatrix} . \tag{A20}$$

In the last step above we have used (A10) and (A14) for the core state and plane wave, respectively, and performed the angular integrations by virtue of the orthonormality of the $\tilde{\chi}^\mu_\kappa$'s (c.f. (A13)). The integral $\nu^c_{\underline{\tau}\alpha}(s,\underline{k}_i)$ (c.f. (19)) may be evaluated in a similar manner.

REFERENCES

Anderson, O. K. and Kasowski, R. V. (1971) "Electronic States as Linear Combinations of Muffin-tin Orbitals" Phys. Rev. B. 4 1064.

Brown, E. and Krumhansl, J. A. (1958) "Energy Band Structure of Lithium by a Modified Plane-Wave Method" Phys. Rev. 109 30

Butler, F. A., Bloom, F. K. and Brown, E. (1969) "Modification of the Orthogonalised Plane-Wave Method Applied to Copper" Phys. Rev. 180 744.

Callaway, J. (1955a) "Orthogonalised Plane-Wave Method" Phys. Rev. 97 933.

Callaway, J. (1955b) "Electronic Energy Bands in Iron" Phys. Rev. 99 500.

Dalton, N. W. (1970) "Notes on the Dirac Equation and the Relativistic OPW Method" IBM Research Report RJ 785 (unpublished).

Dalton, N. W. (1971) "Approximate KKR Band-Structure Schemes for Transition Metals" Computational Methods in Band Theory (Plenum Press) p. 225.

Euwema, R. N. (1971) "Rapid Convergence of Crystalline Energy Bands by Use of a Plane-Wave-Gaussian Mixed Basis Set" Intern. J. Quantum Chem. 5, 61.

Euwema, R. N. (1971) "Plane'Wave-Gaussian Energy-Band Study of Nb" Phys. Rev. B 4 4332.

Gray, D. and Brown, E. (1967) "Electron Energy Levels in Cu_3Au" Phys. Rev. 160 567.

Gray, D. and Karpien, R. J. (1971) "Some Notes on a Modified OPW Method" Computational Methods in Band Theory (Plenum Press) p. 144.

Heine, V. (1967) "s-d Interaction in Transition Metals" Phys. Rev. 153 673.

Herman, F.,Kortum, R. L., Ortenburger, I. B. and Van Dyke, J. P. (1968) "Relativistic Band Structure of GeTe, SnTe, PbSe and PbS" J. de Phys. C4, sup 11-12, 62.

Herman, F., Kortum, R. L., Ortenburger, I. B. and Van Dyke, J. P. (1969)" Electronic Structure and Optical Spectrum of Semi-Conductors" ARL Technical Report (Aerospace Research Laboratòries, ARL 69-0080).

Herman, F. and Skillman, S. (1963) "Atomic Structure Calculations" (Prentice-Hall, Englewood Cliffs, New Jersey).

Herring, C. (1940) "A New Method for Calculating Wave-Functions in Crystals" Phys. Rev. 57 1169.

Hubbard, J. (1967) "The Approximate Calculation of Electronic Band-Structures" Proc. Phys. Soc. 92 921.

Kleinman, L. and Shurtleff, R. (1969) "Modified Augmented-Plane-Wave Method for Calculating Energy Bands" Phys. Rev. 188 1111.

Kohn, W. and Rostkoker, N. (1954) "Solution of the Schroedinger Equation in Periodic Lattices with an Application to Metallic Lithium" Phys. Rev. 94 1111.

Korringa, J. (1947) "On the Calculation of a Bloch-Wave in a Metal" Physica 13 392.

Kunz, A. B. "Combined Plane-Wave Tight-Binding Method for Energy-Band Calculations with Application to Sodium Iodide and Lithium Iodide" Phys. Rev. 180 934.

Liberman, D., Waber, J. T. and Cromer, D. T. (1965) "Self-Consistent-Field Dirac-Slater Wave-Functions for Atoms and Ions. I. Comparison with Previous Calculations" Phys. Rev. 137 A27.

Lipari, N. O. and Deegan, R. A. (1971) "Wave-Functions and Energy-Bands for Narrow Band Materials: A Modified Tight-Binding Calculation for the d-bands in Cu" (preprint).

Lowdin, P. (1951) "A Note on the Quantum-Mechanical Perturbation Theory" J. Chem. Phys. 19 1396.

Miasek, M. (1957) "Tight-Binding Method for Hexagonal Close-Packed Structure" Phys. Rev. 107 92.

Phillips, J. C. and Kleinman, L. (1959) "New Method for Calculating Wave-Functions in Crystals and Molecules" Phys. Rev. 116 287.

Phillips, J. C. (1968) "Significance of Model Hamiltonians in Energy-Band Theory" Adv. in Phys. 65 79.

Slater, J. C. (1937) "Wave-Functions in a Periodic Potential" Phys. Rev. 51 846.

Slater, J. C. and Koster, G. F. (1954) "Simplified LCAO Method for the Periodic Potential Problem" Phys. Rev. 94 498.

Soven, P. (1965) "Relativistic Band-Structure and Fermi Surface of Thallium" Phys. Rev. A137 1706.

Wigner, E. and Seitz, F. (1933) "On the Constitution of Metallic Sodium. I" Phys. Rev. 43 804.

Wigner, E. and Seitz, F. (1934) "On the Constitution of Metallic Sodium. II" Phys. Rev. 46 509.

MAPW-CALCULATIONS WITH SCREENED EXCHANGE.

APPLICATION TO COPPER [x]

H. Bross and H. Stöhr

Sektion Physik der Universität München

1. INTRODUCTION

The recent calculations of energy bands from first principles point of view (1,2,3) give quite different results when different methods of approximating the Hartree-Fock exchange are used. In addition these Hartree-Fock results may be different from those one obtains when model potentials are used. Therefore, it is desirable to incorporate exchange and correlation effects within the framework of the single particle model as well as possible so that one can be confident that the results obtained reflect the quasi-particle excitations in a solid. On the other hand, as the amount of numerical work becomes extremely large, it is necessary to make simplifications the extent of which is prescribed by the computational capacity being at disposal. This paper reports results of band calculations for Cu performed by the MAPW-method (4) and taking into account many-body effects by means of a non-local potential.

2. THE ONE-PARTICLE EFFECTIVE HAMILTONIAN

We start from the homogeneous equation (5,6,7)

$$\left[E_{n\vec{k}} - \frac{p^2}{2m} - V(\vec{r})\right]\chi_{n\vec{k}}(\vec{r}) - \int \Sigma(\vec{r},\vec{r}\,',E_{n\vec{k}})\,\chi_{n\vec{k}}(\vec{r}\,')\,d^3r' = 0 \qquad (2.1)$$

which characterizes quasi-particle states with wave-vector $\vec{k}$ and band index n. $\frac{p^2}{2m}$ is the kinetic energy and $V(\vec{r})$ consists of the bare nuclear potential together with the Hartree field of the electrons. $\Sigma(\vec{r},\vec{r}\,',\omega)$ is the (complex) self-energy operator which includes all exchange and correlation effects. The eigenvalue $E_{n\vec{k}}$ corresponds to a single particle excitation of the system, which may be observed by suitably chosen experiments. Since $\Sigma(\vec{r},\vec{r}',\omega)$ is not necessarily Hermitian, in general the eigenvalue spectrum $E_{n\vec{k}}$ will be complex and the eigenfunctions $\chi_{n\vec{k}}(\vec{r})$ will not be orthogonal. It is obviously impossible to treat $\Sigma(\vec{r},\vec{r}',\omega)$ in the case of noble metals exactly; however, one can approximate it. In the following we shall describe the different approaches only briefly, since other authors (8,5,7,1,2,3) have proceeded in a similar way.

i) Neglecting vertex corrections the self-energy operator $\Sigma(\vec{r},\vec{r}',\omega)$ may be approximated by

$$\Sigma(\vec{r},\vec{r}',\omega) = i\int G(\vec{r},\vec{r}',\omega-\omega')\, V_{scr}(\vec{r},\vec{r}',\omega')\, d\omega', \qquad (2.2)$$

where $G(\vec{r},\vec{r}',\omega)$ is the exact one-particle Green function and $V_{scr}(\vec{r},\vec{r}',\omega)$ is the effective interaction between the electrons. The definitions and the basic properties of these magnitudes may be found in any monograph treating many-body problems in solids (9). V_{scr} is related to the unscreened two-body interaction $v(\vec{r},\vec{r}') = |\vec{r}-\vec{r}'|^{-1}$ by

$$V_{scr}(\vec{r},\vec{r}',\omega) = \int v(\vec{r},\vec{r}'')\, \epsilon^{-1}(\vec{r}'',\vec{r}',\omega)\, d^3r''. \qquad (2.3)$$

$\epsilon^{-1}(\vec{r},\vec{r}',\omega)$ denotes the time-ordered inverse microscopic dielectric function of the medium (10,11).

ii) In order to get an Hermitian eigenvalue problem we insert the static value for the inverse dielectric function with the result that $V_{scr}(\vec{r},\vec{r}',\omega)$ does not depend on the frequency any longer and that the effective interaction between the electrons becomes instantaneous. The integration over the frequency ω' on the right-hand side of Eq. (2.2) yields the self-energy

$$\Sigma(\vec{r},\vec{r}',\omega) = i\, G(\vec{r},\vec{r}',t=0^-)\, V_{scr}(\vec{r},\vec{r}') \qquad (2.4)$$

which is not only independent of frequency but is Hermitian as well. The eigenvalue spectrum $\epsilon_{n\vec{k}}$ corresponding to Eq. (2.1) will then be real and the set of eigenfunctions $\varphi_{n\vec{k}}$ is both complete and orthonormal. It is evident that these $\varphi_{n\vec{k}}$'s are a well adapted basis to determine matrix elements which are necessary for a further many-body analysis as well as for other theoretical investigations of physical properties. However, on the other hand, due to the assumption made above, additional plasmaron peaks (12) will not appear in the eigenvalue spectrum $\epsilon_{n\vec{k}}$.

iii) If we represent the exact Green function $G(\vec{r},\vec{r}',\omega)$ using these optimally chosen basis functions, by the relation

$$G(\vec{r},\vec{r}',\omega) = \sum_{n,n',\vec{k}} G_{nn'}(\vec{k},\omega)\, \varphi_{n\vec{k}}(\vec{r})\, \varphi_{n'\vec{k}}(\vec{r}')^*, \qquad (2.5)$$

it is to be expected that the Green function $G_{nn'}(\vec{k},\omega)$ does not differ strongly from the zero-order approximation

$$G^{(o)}_{nn'}(\vec{k},\omega) = \frac{\delta_{nn'}}{\omega-\epsilon_{n\vec{k}}+i\,\delta_{\epsilon_{n\vec{k}}}} \qquad (2.6)$$

where $\delta_{\epsilon_{n\vec{k}}}$ is a infinitesimally small magnitude being positive for energies $\epsilon_{n\vec{k}}$ larger than the Fermi energy and negative for energies $\epsilon_{n\vec{k}}$ lower than the Fermi energy. With this assumption the Green function $G(\vec{r},\vec{r}',t=0^-)$ reduces to the usual exchange density up to a minus sign,

$$G(\vec{r},\vec{r}',t=0^-) = -\sum_{n,\vec{k}} \varphi_{n\vec{k}}(\vec{r})\,\varphi^*_{n\vec{k}}(\vec{r}') = -\rho_{ex}(\vec{r},\vec{r}'). \qquad (2.7)$$

Here the sum extends over the occupied states of the solid only.

iv) In order to find an expression for the effective interaction well suited for numerical calculations, we note that in the Fourier expansion

$$\epsilon^{-1}(\vec{r},\vec{r}',\omega) = \frac{1}{(2\pi)^3}\sum_j \int \epsilon^{-1}(\vec{p},\vec{K}_j,\omega)\, e^{i\vec{K}_j\cdot\vec{r}}\, e^{i\vec{p}(\vec{r}-\vec{r}')}d^3p \qquad (2.8)$$

the sum extending over the reciprocal lattice vectors $K_j \neq 0$ gives rise to the so-called local field corrections whereas the term $K_j = 0$ determines the macroscopic behaviour of the solids. In the case of Na and possibly in the cases of the other alkali metals, too, the local field corrections turn out to be very small (13). As no details of $\epsilon^{-1}(\vec{p},\vec{K}_j,\omega)$ are known up to now it seems reasonable to neglect all terms with $K_j \neq 0$ and to approximate $\epsilon^{-1}(\vec{p},0,\omega)$ by the reciprocal value of the dielectric constant $\epsilon(\vec{p},0,\omega)$. Of course, $\epsilon(\vec{p},0,\omega)$ may be determined from the proper polarization diagrams in the usual way (9). In the case of metals the following behaviour can be derived in the long wave limit without detailed knowledge of the wave-functions by summing over the most divergent graphs,

$$\epsilon(\vec{p},0,0) = 1 + \left(\frac{\kappa}{p}\right)^2. \qquad (2.9)$$

The constant κ essentially depends on the density of states at the Fermi level and reduces to the Thomas-Fermi screening constant in the free-electron case. Assuming this asymptotic relation to be valid for any value of $\vec{p}$ leads to a screened potential

$$V_{scr}(\vec{r},\vec{r}') = \frac{1}{|\vec{r}-\vec{r}'|}\, e^{-\kappa|\vec{r}-\vec{r}'|} \qquad (2.10)$$

which has the correct behaviour at small and very large values of $|\vec{r}-\vec{r}'|$ and which is well suited for numerical calculations. The same result was derived by directly solving the Dyson equation for the effective potential in the limit $p\to 0$ and $\omega\to 0$ (14).

With regard to the different assumptions made in deriving the self-energy it is not reasonable to determine the screening constant from the density of states. However, we may get it by fitting calculated energy values to experimental results. By this appropriate choice we hope to fold the dominant many-body effects caused by the electron-electron interaction into the formalism to the greatest extent.

All effects due to the electron-electron interaction not taken into account and due to the electron-phonon interaction may be evaluated by a many-body analysis using the $\varphi_{n\vec{k}}$'s as basis set. Since the $\varphi_{n\vec{k}}$'s are a complete set for each fixed value of $\vec{k}$, the function $\chi_{n\vec{k}}$ characterizing a quasi-particle state of the wave-vector $\vec{k}$ may be represented by the series

$$\chi_{n\vec{k}} = \sum_{n'} S_{n'n} \varphi_{n'\vec{k}} \tag{2.11}$$

where the matrix-elements $S_{n'n}$ are determined by Eq. (2.1). This will lead to the linear homogeneous equation

$$(E_{n\vec{k}} - \epsilon_{n\vec{k}}) S_{n'n} + \sum_{n''} \langle n'k | \sum(\vec{r},\vec{r}\,',E_{nk}) - \frac{e^{-\kappa|\vec{r}-\vec{r}\,'|}}{|\vec{r}-\vec{r}\,'|} | n''\vec{k}\rangle S_{n''n} = 0 \tag{2.12}$$

The excitation energies $E_{n\vec{k}}$ of the quasi-particles are defined by the roots of the corresponding secular equations, and are not given by the expectation value of the mass-operator, as generally is assumed (5,6).

3. THE MAPW-FORMALISM TAKING INTO ACCOUNT MANY-BODY EFFECTS

The approximations of Sec. 2 lead to the integro-differential equation

$$\left[\epsilon_{n\vec{k}} - \frac{p^2}{2m} - V(\vec{r})\right] \varphi_{n\vec{k}}(\vec{r}) - \int \sum\nolimits_o(\vec{r},\vec{r}\,')\, \varphi_{n\vec{k}}(\vec{r}\,')\, d^3r' = 0 \tag{3.1}$$

with

$$\sum\nolimits_o(\vec{r},\vec{r}\,') = |\vec{r}-\vec{r}\,'|^{-1}\, e^{-\kappa|\vec{r}-\vec{r}\,'|}\, \rho_{ex}(\vec{r},\vec{r}\,') \tag{3.2}$$

determining the eigenvalue spectrum $\epsilon_{n\vec{k}}$ and the corresponding set of eigenfunctions $\varphi_{n\vec{k}}$. It is non-linear since its kernel $\sum_o(\vec{r},\vec{r}\,')$ depends on the eigenfunctions $\varphi_{n\vec{k}}$ of the occupied states. We linearize it by inserting the wave-functions evaluated by some other procedure or by a lower step of a self-consistent procedure.

For numerical calculations it is reasonable to consider Eq. (3.1) as the Euler-Lagrange equation of a variational principle which is solved by a Rayleigh-Ritz procedure using MAPW-trial functions (4). Details of the most effective version of the MAPW-procedure are published elsewhere (15,16). An essential characteristic of this formalism is that it produces wave-functions which are not only orthogonal to the wave-functions of the core-states but, as well as their first derivatives, are continuous in the whole space. Thus, these functions are well suited for any accurate prediction of experimentally interesting quantities, such as optical properties, X-ray scattering, Knight-shift, or Mößbauer shift.

In addition, similar as in Slater's APW (17) version, the extremal procedure allows for a general potential (18,19,20). This turns out to be not only very important in the case of some insulators where the muffin-tin potential is most likely unable to describe accurately the electronic properties (21) but also in the case of metals. The present investigations lead to the result that the width of the bands corresponding to non-localized electrons is very sensitive to the variations of the potential outside the APW-sphere whereas the non-spherical components of the potentials inside the APW-sphere are of smaller influence. In the case of paramagnetic Ni the first mentioned effect yields a change of bandwidth of the magnitude 0.05 - 0.1 Ry (22). Corrections due to the nonspherical potential within the APW-sphere are in the magnitude of 0.001 Ry in the case of Cu. In any case these deviations from the muffin-tin potential are important for "high-precision" calculations quoting errors of 0.001 Ry or less.

4. RESULTS (23)

a) Lithium

The results of our MAPW-calculation using a non-local potential were already reported (15,16). In preparing the following discussions for the case of Cu we only want to note that the energies of the conduction band do not depend strongly on the special value of κ lying within the range from 0.7 k_{TF} to 0.9 k_{TF} where k_{TF} denotes the Thomas-Fermi-screening constant of Li.

b) Copper

The linearization of both the charge density and the exchange density due to the occupied electron states has been achieved by wave functions determined by the MAPW-procedure using Chodorow's potential (24,25).

Within the limits of accuracy the energy values were found to agree with those evaluated by Burdick (26). Further, the calculated values of the Fourier coefficients of the electron density deviate from the experimental results (27) only slightly.

The Hartree-potential $V(\vec{r})$ evaluated by considering 123 different $\vec{k}$-points in the Brillouin zone is shown in Figure 1 for

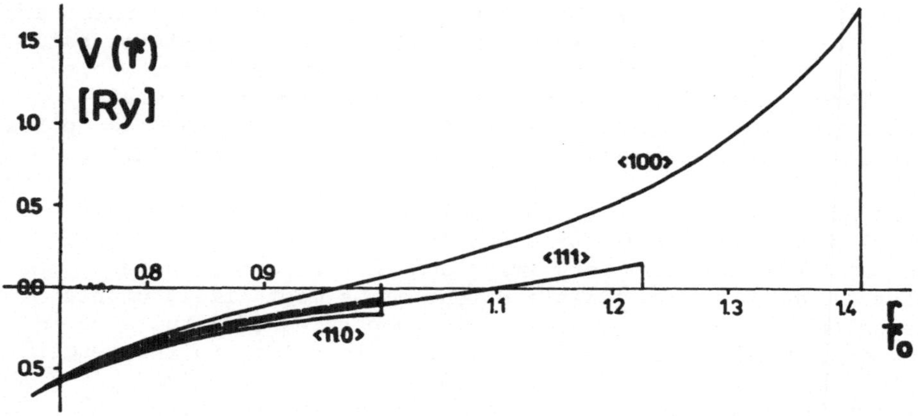

Figure 1. The Coulomb potential of Cu for the directions ⟨100⟩ , ⟨110⟩ , and ⟨111⟩ .

the directions ⟨100⟩ , ⟨110⟩ , and ⟨111⟩ . The dashed line corresponds to the spherical average of the potential. We remark that the potential deviates from the spherical average, already within the radius of the APW-sphere, r_o . However, the corrections of the energies due to these non-spherical contributions are smaller than 10^{-4} Ry for core states and 10^{-3} Ry for valence states. Outside the APW-sphere, the potential turns out to be strongly varying and therefore it cannot be well approximated by a constant value. As mentioned above, the MAPW-procedure allows to take into account potentials of this kind exactly.

The calculations of the contributions to the matrix elements due to the non-local potential is rather time consuming. In order to find a way to reduce this computational work we have calculated the energies corresponding to the end points of the directions ⟨100⟩ and ⟨111⟩ of $\vec{k}$-space, X and L, respectively, by summing over different numbers of terms in the exchange density. From these results we conclude the error to be smaller than 0.003 Ry if only two different points for each of the ⟨001⟩ , ⟨011⟩ , and ⟨111⟩ directions are considered in the exchange density.

The structure of the valence bands of Cu is schematically shown in Figure 2. It consists of five small d bands and a relatively broad s-p band where the latter one is hybridized with one of the d bands. Characteristic energies are the lower and the upper edge of the s-p band with the energies E_{Γ_1} and $E_{X_4'}$, respectively, and the energies $E_{\Gamma_{12}}$ and $E_{\Gamma_{25}'}$ of the d-bands. Notice that in the bandstructure of Cu compatible with the measurement, d bands lie inmidst of the s-p band what means that $E_{\Gamma_1} < E_{\Gamma_{12}} < E_{\Gamma_{25}'}$ $E_{X_4'}$.

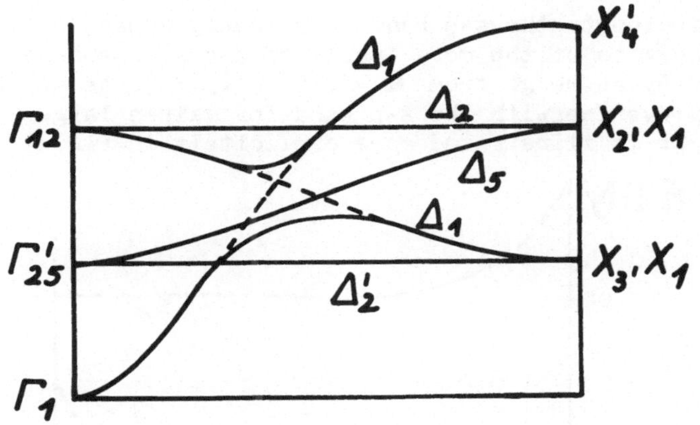

Figure 2. The valence bands of Cu (schematically). The dashed lines correspond to the non-hybridized bands.

The dependence of these energies on the screening constant is shown in Figure 3. κ is measured in atomic units. The value of the Thomas-Fermi-screening constant is in the order of magnitude of 1. Similar as in the case of Li the energies of the electron states which are not localized in the atomic polyhedron and which

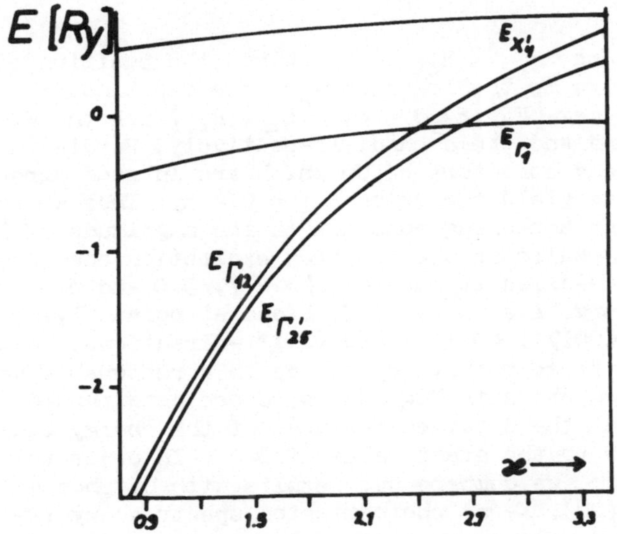

Figure 3. Dependence of the energies E_{Γ_1}, $E_{X_4'}$, $E_{\Gamma_{12}}$, and $E_{\Gamma_{25}'}$, on the parameter κ.

may be attributed to the s-p band only weakly depend on κ . However, the energies of the more localized d-states change from -2.0 Ry to 0 Ry as we go from κ = 0.9 to κ=3.3. As we can see the d bands only overlap with the s-p band for values larger than 2.5. This behaviour is illustrated more explicitely in Fig. 4 where the

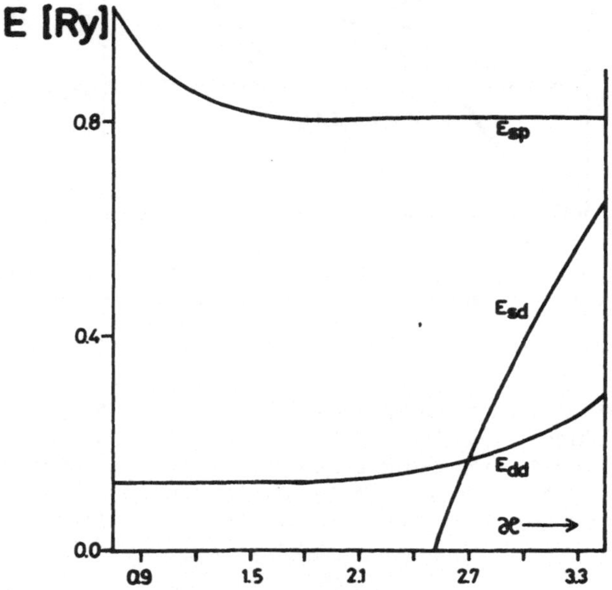

Figure 4. Characteristic energy differences as function of x.

energy difference $E_{sd} = E_{\Gamma_{25}'} - E_{\Gamma_1}$ gives the position of the d bands relative to the lower edge of the s-p band. $E_{sp} = E_{X_4'} - E_{\Gamma_1}$ and $E_{dd} = \max[E_{X_2'}, E_{X_5}] - \min[E_{X_1}, E_{X_3}]$ are the widths of the s-p band and the d band, respectively. Burdick's calculations (26) of the band structure of Cu which are in good agreement with the experiments yield the value E_{sd} = 0.4 Ry. Thus we conclude we must choose the screening constant in the magnitude of $\kappa = 3$ in order to find realistic one-electron excitation energies. Up to now we did MAPW-calculations for κ= 2.7, 3.0 and 3.3 at the points of high symmetry, i.e. Γ , X, K, L and at points lying in the midst of the ⟨001⟩ , ⟨011⟩ , and ⟨111⟩ -directions, respectively. By connecting these results by curves in a reasonable way the band structures shown in Fig. 5a, b, c are constructed. It is surprising that the detailed features of the energy bands are very sensitive to the exact value of κ . In order to find the best value of κ we compare our results with UV-photoemission spectroscopy (28), X-ray photoemission spectroscopy (29) experiments and with measurements of the strain dependence of the optical constants (30). The corresponding values are listed in Table 1 together with results of other APW-calculations using

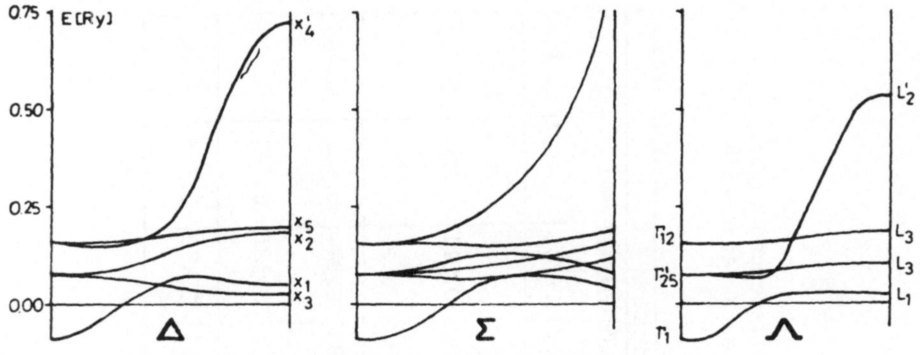

Figure 5a. The valence band of Cu with x = 2.7

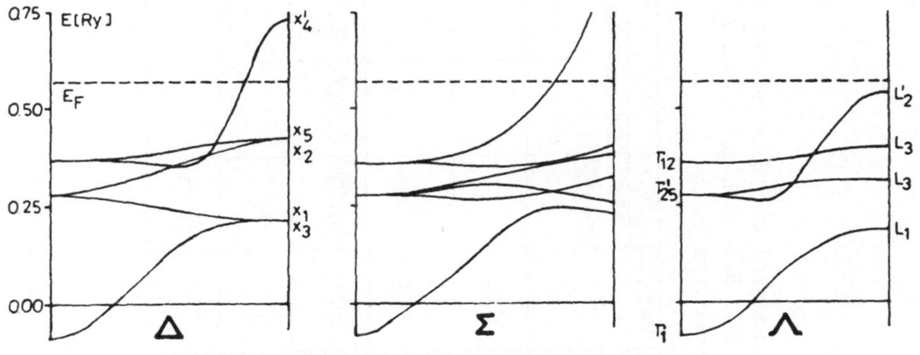

Figure 5b. The valence band of Cu with x = 3.0

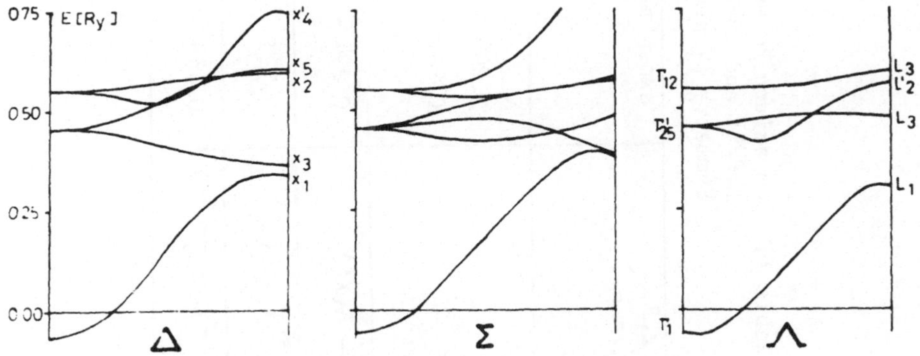

Figure 5c. The valence band of Cu with x = 3.3

Table 1. Comparison Between Experimental and Theoretical Energy Differences in Cu. (L_1^{upper} and L_3^{upper} Mean the Second Eigenvalues of the Valence Bands with Symmetry L_1 and L_3, respectively.)

ΔE (in Ry)	experimental	Meth.	theoretical						
			APW Chod.	APW-selfconsist		MAPW: Present			
				α=5/6	α=1	χ=2.7	χ=3.0	χ=3.3	χ=3.04
E_{dd} = $\max(X_2,X_5)$ - $\min(X_1,X_3)$	0.206 0.221	UPS XPS	0.249	0.224	0.192	0.165	0.211	0.281	0.218
$E(X_5) - E(X_4')$	0.294 ± 0.01	SOS	0.292	0.317	0.407	0.523	0.324	0.141	0.299
$E(L_1^{upper}) - E(L_2')$	0.323	UPS	0.335			0.257	0.321	0.397	0.330
$E(L_1^{upper})-E(L_3^{upper})$	0.459 ± 0.02	SOS	0.444			0.610	0.470	0.373	0.455

Chodorow's potential (26) or using Slater's approximation for the exchange potential with different values of the multiplying factor α (31,32). Again we note the strong dependence of the energy differences considered on κ. A least squares' fit yields the value $\kappa = 3.04$. The energies interpolated for this value lie within the limits of error of the measured values. The APW calculations using Chodorow's potential or Slater's exchange approximations do not yield such a good agreement with the experiments as the MAPW procedure does with a non-local potential.

As mentioned in Sec. 2 the MAPW-procedure is well suited for the calculations of core states, too. It is not surprising that the energies of these localized states are strongly sensitive to the value of κ. In Table 2 the energy differences of the 3 s and 3 p states from the Fermi energy interpolated for the value $\kappa = 3.04$ are listed together with experimental results found by soft X-ray spectroscopy (33).

Table 2. The Energy-Differences Between the Higher Core States and the Fermi-Energy E_F

ΔE (in Ry)	theoretical $\kappa = 3.04$	corrected	experimental
$E_F - M_I$	8.9	9.3	8.8
$E_F - M_{II}$	5.1	5.4	5.7
$E_F - M_{III}$		5.2	5.4

In addition we have entered the corrections of the theoretical values due to relativistic effects taken from literature (34). The theoretical and experimental results satisfactorily agree with one another.

5. CONCLUSION

The preliminary investigations show that in the case of Cu the energies of the quasi-particles lying in the energy range of 10 Ry below the Fermi energy may be obtained using a screened exchange potential with one value of the screening constant. These investigations are further performed in order to find more details of the energy bands especially in the neighbourhood of the Fermi energy.

REFERENCES AND FOOTNOTES

x Research supported by the Deutsche Forschungsgemeinschaft

1) N.O. Lipari and W.B. Fowler, Phys.Rev. B 2, 3354 (1970)

2) N.O. Lipari, phys.stat.sol. 40, 691 (1970)
3) N.O. Lipari and A.B. Kunz, Phys.Rev. B 3, 491 (1971)
4) H. Bross, Phys.kondens.Materie 3, 119 (1964)
5) L.T. Hedin, Quantum Chemistry Group, Uppsala, Technical Report No. 84, (1962) unpublished; Phys.Rev. 139, A 796 (1965); Arkiv Fysik 30, 231 (1965)
6) L. Hedin and S. Lundqvist, in Solid State Physics, edited by F. Seitz, D. Turnbull, and H. Ehrenreich, Academic Press, New York and London 1969, vol. 23, p. 1
7) W. Brinkman and B. Goodman, Phys.Rev. 149, 597 (1966)
8) F. Bassani, J. Robinson, B. Goodman, and J.R. Schrieffer, Phys.Rev. 127, 1969 (1962)
9) See for example the list cited in the monograph (6).
10) D.S. Falk, Phys.Rev. 118, 105 (1960)
11) S.L. Adler, Phys.Rev. 126, 413 (1962)
12) B.I. Lundquist, Phys.kondens.Materie 6, 193 and 206 (1967)
13) A. Rauh and H. Bross, Z.Naturforsch. in press
14) H. Bross, Z.Physik 215, 485 (1968)
15) H. Bross, G. Bohn, G. Meister, W. Schubö, and H. Stöhr, Phys.Rev. B 2, 3098 (1970)
16) H. Bross, G. Bohn, G. Meister, W. Schubö, and H. Stöhr, in P.M. Marcus, J.F. Janak, and A.R. Williams, Plenum Press, New York-London 1971, p. 44
17) J.C. Slater, Phys.Rev. 51, 846 (1937)
18) H. Schlosser and P.M. Marcus, Phys.Rev. 131, 2529 (1963)
19) P.M. Marcus, Intern.J.Quantum Chem. 1S, 567 (1967)
20) W.E. Rudge, Phys.Rev. 181, 1024 (1969)
21) A.B. Kunz, W.B. Fowler, and P.M. Schneider, Phys.Lett. 28A, 553 (1969)
22) G. Meister, Dissertation München 1971
23) All the calculations were done using the TR 4 and TR 440 of the Leibniz-Rechenzentrum of the Bayerische Akademie der Wissenschaften and the IBM 360/91 of the Max-Planck-Institut für Plasmaphysik, Garching.
24) M.I. Chodorow, Dissertation, M.I.T. 1939
25) F.I. Arlinghaus, Phys.Rev. 153, 743 (1967)
26) G.A. Burdick, Phys.Rev. 129, 138 (1963)
27) L.D. Jennings, D.R. Chipman, J.J. DeMarco, Phys.Rev. 135, A 1612 (1964)
28) C. N. Berglund and W.E. Spicer, Phys.Rev. 136, A 1044 (1964)
29) C.S. Fadley and D.A. Shirley, Phys.Rev.Lett. 21, 980 (1968)
30) U. Gerhardt, Phys.Rev. 172, 651 (1968)
31) E.C. Snow, Phys.Rev. 171, 785 (1968)
32) E.C. Snow and J.T. Waber, Phys.Rev. 157, 570 (1967)
33) Y. Cauchois, J.Phys.Rad. 16, 253 (1955)
34) F. Herman and S. Skillman, Atomic Structure Calculations, Prentice-Hall, New Jersey 1963

SOME REMARKS ON CURRENT RAPW CALCULATIONS FOR SILVER, MOLYBDENUM, AND VANADIUM

Niels Egede Christensen

Physics Laboratory I, The Technical University of Denmark, DK-2800 Lyngby, Denmark

I. Silver

The two parameters, width and position of the d-bands of the noble - and transition metals respond sensitively to the choice of potential. In a series of preliminary APW calculations for silver[1] 16 different potentials were constructed, and the corresponding band structures were compared to results of experimental work on optical properties and Fermi surface properties. The potentials examined were based on atomic orbitals obtained from Hartree-Fock, Hartree-Fock-Slater, and Dirac-Slater calculations. The amount of Slater-exchange was varied parametrically. Also Gaspar potentials with various exchange weights were included in this study.

It was found - in this semiempirical fitting scheme - that the best non-relativistic APW calculation of the energy bands of silver was obtained from a potential based on (relativistic) Dirac-Slater atomic orbitals including full Slater exchange in the atomic as well as in the crystal potential. The exchange - and Coulomb contributions were treated separately, within the muffin tin spheres as well as in the region between the spheres.

This method of potential construction has been applied to gold[2,3,4,5] in a RAPW band calculation. This band structure was checked against a large variety of experimental results - static and modulated reflectance, photoemission and Fermi surface data[6], and all experimental results could be quantitatively explained in terms of the model.

Now, the fundamental choice of potential was based on (non-relativistic) APW calculations for silver. It is therefore interesting to return to silver and perform the same detailed calculations of observables as was done for gold. A few of these results will be shown here. Fig. 1 shows the RAPW band structure of silver.

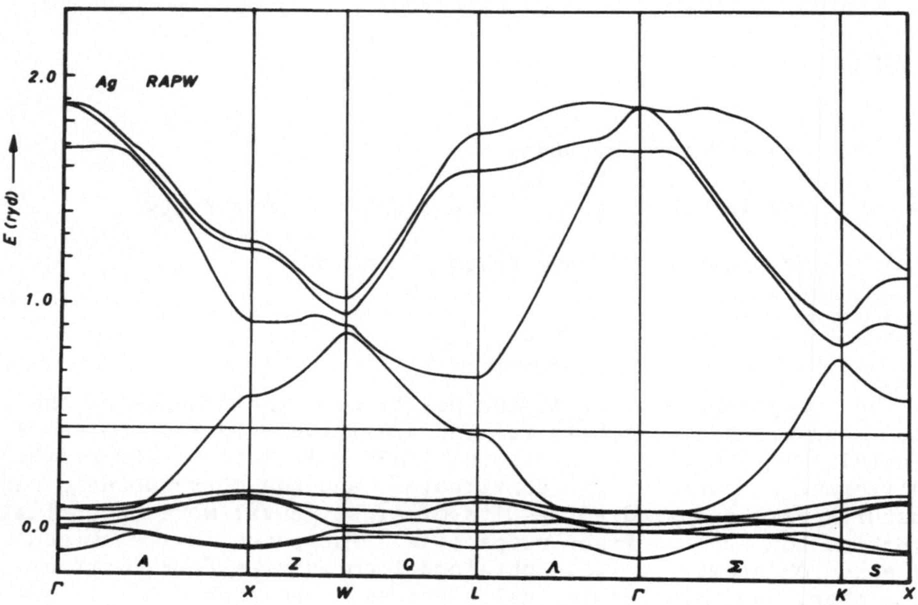

Fig. 1. RAPW energy bands of silver. (Labels of irreducible representations are given in ref. 3 and 4).

In gold we observed[2,3,4] that the L-gap $L_{2'} \rightarrow L_1$ was reduced by 3 eV when relativistic effects were taken into account. The effect in silver is smaller (∿ 1 eV) but still very important - and maybe implying more interesting consequences. The reduction in Ag is namely sufficient to pull the L-gap below the spectral value of the main interband edge. Thus, in silver we do not only have a tail below $\hbar\omega_i$ as in gold but there will be a distinct edge below the main edge. Fig. 2 shows the joint density-of-states function for silver near the main edge $\hbar\omega_i$.

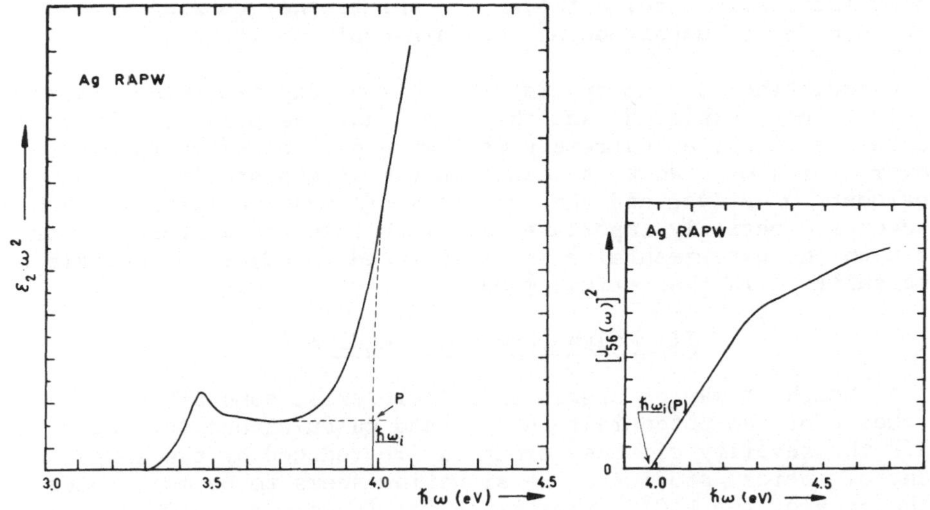

Fig. 2. The calculated function $\varepsilon_2 \cdot \omega^2$ near the interband edge of silver. The edge is $\hbar\omega_i$ = 3.98 eV, and the lower edge - superposing the tail - is at 3.45 eV. The right figure shows how the interband edge ($\hbar\omega_i$) is defined as the parabolic footpoint of the partial joint density-of-states function J_{56}. The indices 5 and 6 indicate that transitions from band 5 to band 6 are responsible for the steep increase of the absorption at $\hbar\omega_i$.

The experimental value for $\hbar\omega_i$ is 3.97 eV[7,8,9], and the present calculation yields $\hbar\omega_i$ = 3.98 eV. Thus, the d-band position is given correctly by the model. The strong absorption is due to transitions between band 5 and 6, and the region in k-space where the transitions occur is probably as in the case of gold[4]. The second edge - the one that superposes the tail below $\hbar\omega_i$ - is situated at 3.45 eV, and these transitions occur in a region around L from band 6 (E_F) to band 7. If the present band calculation is correct, it will thus give a simple interpretation of the structure in ε_2 and $\Delta\varepsilon_2$ observed by Garfinkel et al.[7] at $\hbar\omega$ = 3.4 eV. Also the piezoreflectance traces presented by Nilsson and Sandell[8] contain strong unexplained elements of structure at 3.4 eV. It should be emphasized that the present interpretation is in contrast to the conclusions made by Liljenvall and Matthewson[9]. Their arguments are, however, somewhat weakened by the fact that they relate their experimental results to a non-relativistic model although the shifts they consider are of the same order of magnitude as the spin-orbit splittings.

Calculations of energy distributions of photo-emitted electrons have just been completed, and they show that the present RAPW band structure also agrees extremely well with photoemission results. However, still much work is needed in the interpretation of optical experiments on silver and the present model must be further checked. It becomes especially important that such interpretations are correct when the experimental results are used to adjust potentials or phase shifts[10] in theoretical models.

II. Molybdenum and Vanadium

Although it may be argued that there exist some ambiguities in the choice of the potentials in the band calculations for the noble metals the severity of these troubles are reduced by the large amount of optical studies. The situation seems to be more complex in the case of the B.C.C. transition metals Mo and V. Only very little optical work has been done on these metals.

The difficulties in setting up the crystal potential for Mo or V are not restricted to a choice of how to include correlation, what exchange factor should be used etc., but it is also unclear what atomic configuration should be applied. Mattheiss found[11] that the d-band width of V decreased by a factor of two when the configuration was altered from $3d^4 4s^1$ to $3d^3\ 4s^2$.

In Fig. 3 we make a similar comparison for Mo. The d-band width of Mo calculated from the $4d^4\ 5s^2$- configuration is 0.048 Ry smaller than the value corresponding to the $4d^5 5s^1$ configuration. Thus we find for Mo a much smaller sensitivity in the band structure to changes of configuration than observed for V[11]. In order to examine this in more detail, the V bands were calculated. The $V(3d^4 4s^1)$ calculation agrees well with the one by Mattheiss, but going to $V(3d^3\ 4s^2)$ only reduces the d-band width at H by 0.08 Ry - again a small effect compared to that of ref. 11 (0.20 Ry). However, if the $3d^4\ 4s^1$-orbitals of the atoms are 'frozen' and the new configuration formed by merely changing the occupation numbers, then our result is much closer to that of ref. 11. Thus, it is essential whether the configuration change is done self-consistently or not. Further, it should be mentioned that Mattheiss used Hartree-Fock atomic calculations and therefore his potentials became more configurational dependent than our potentials using Slater exchange.

The Fermi surface of Mo has been calculated, and it agrees well with rf-size effect measurements. Unfortunately the Fermi-level falls just at the center of the d-bands. Therefore it moves with the d-bands when the potential is altered without any big

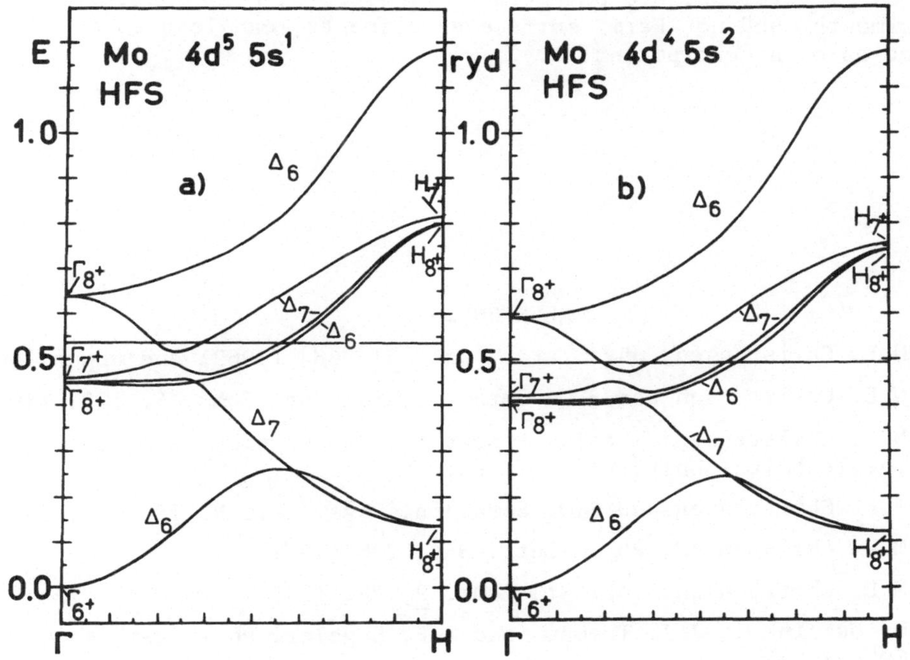

Fig. 3. RAPW band calculations for Mo. The two atomic calculations were both carried through to self-consistency.

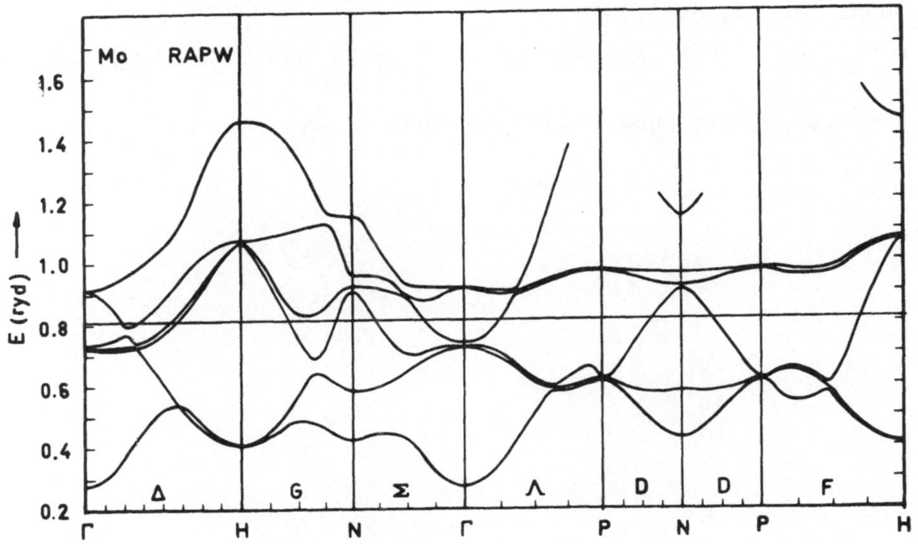

Fig. 4. Energy bands of Mo calculated along symmetry lines. Potential as in fig. 3a.

changes of Fermi surface parameters. Therefore, only optical experiments, and not Fermi-surface experiments, can lead to a selection of a best potential.

REFERENCES

1. N.E. Christensen, phys. stat. sol. 31, 634 (1969) and unpublished.
2. N.E. Christensen and B.O. Seraphin, Sol. St. Comm. 8, 1221 (1970)
3. N.E. Christensen, Thesis, Report No. 75, Physics Laboratory I, The Technical University of Denmark
4. N.E. Christensen and B.O. Seraphin, Phys. Rev. B. 15. Oct. 1971
5. N.E. Christensen, Phys. Lett. 35A, 206 (1971)
6. N.E. Christensen, Sol. St. Comm. 9, 749 (1971)
7. M. Garfinkel, J.J. Tieman, and W.E. Engeler, Phys. Rev. 148, 695 (1966)
8. P.-O. Nilsson and B. Sandell, Sol. St. Comm. 8, 721 (1970)
9. H.G. Liljenvall and A.G. Matthewson, J. Phys. C: Metal Phys. Suppl. No. 3, 341 (1970)
10. B.R. Cooper, E.L. Kreiger, and B. Segall, Phys. Rev. B4,1734 (1971)
11. L.F. Mattheiss, Phys. Rev. 134, A970 (1964)

RECENT KKR BAND CALCULATIONS

Ulrich Rössler

Institut für theoretische Physik II, Marburg

1. INTRODUCTION

Among the computational methods for energy band calculations the KKR method has obtained an important place. The transparent physical concept of scattering on which this method is based has two immediate advantages. 1) In angular momentum representation the lattice geometry and the potential of the ions are separated, thus structure constants which are determined only by the lattice geometry can be calculated once and than be used in all crystals with same symmetry. 2) Energy eigenvalues converge rapidly with increasing angular momentum, thus the dimension of secular determinants is kept small. These advantages together with the simplicity of the muffin-tin approximation for the crystal potential made most recently possible energy band calculations for small gap semiconductors, semiconductors, ionic and van-der Waals crystals on which we report in section 3.

The muffin-tin approximation for the crystal potential is a poor model in covalent semiconductors, where localized bond charges cause a strong deviation of the single site potential from spherical symmetry. For this case the KKR method has been extended to non-muffin-tin potentials. We will discuss this point in section 2 together with considerations on the calculation of KKR wave functions, which play an ever increasing role, as the simple critical point analysis of energy bands in comparison with optical measurements is going to be replaced by optical constants calculations.

The final section is concerned with some problems where KKR energy bands have been used as a starting point: high resolution band calculations, density of states calculations, and the exciton problem.

2. RECENT DEVELOPMENT OF THE KKR METHOD

Recent developments in the KKR theory have been presented in detail last year at the IBM Yorktown Heights Symposium on Computational Methods in Band Theory. Thus, for a complete representation we refer to the proceedings of this symposium[1] and restrict our remarks here to what we regard as the most important contributions.

a) KKR Wave Function in the Muffin-Tin Approximation[2,3]

While all other computational methods for energy band calculations start out from an Ansatz for the wave function in the entire unit cell, the KKR-Ansatz is made only for the scattered wave inside the muffin-tin sphere (radius ρ)

$$\Psi_{\vec{k}}(E,\vec{r}) = \sum_L C_L(\vec{k})\, Y_L(\hat{r})\, R_{\ell E}(r) \quad , \ r<\rho . \tag{1}$$

Here the notation L for angular momentum and $Y_L(\hat{r})$ for sperical harmonics is used. $R_{\ell E}(r)$ is the solution of the radial Schrödinger equation

$$R_{\ell E}(r) = j_\ell(\kappa r) + \kappa\int dr_1 r_1^2\, j_\ell(\kappa r_<)\, n_\ell(\kappa r_>)\, v(r_1)\, R_{\ell E}(r_1), \quad \kappa=\sqrt{E} \tag{2}$$

which consists of the incoming and scattered waves. $r_>$ ($r_<$) being the larger (smaller) of r and r_1, $v(r_1)$ is a single muffin-tin potential.
The wave function inside and outside the muffin-tin spheres is given by the integral equation

$$\Psi_{\vec{k}}(E,\vec{r}) = \int d^3r_1\, \mathcal{G}_{\vec{k}}(\vec{r}-\vec{r}_1; E)\, v(r_1)\, \Psi_{\vec{k}}(E,\vec{r}_1). \tag{3}$$

The structural Green's function with singularities at each lattice site $\vec{R}$

$$\mathcal{G}_{\vec{k}}(\vec{r}-\vec{r}_1; E) = \sum_{\vec{R}\neq 0} e^{i\vec{k}\vec{R}}\, G_{\vec{k}}(\vec{r}-\vec{r}_1-\vec{R}; E) + G_{\vec{k}}^{0}(\vec{r}-\vec{r}_1; E) \tag{4}$$

can be expanded in spherical harmonics

$$G_{\vec{k}} = \sum_{LL'} \{ A_{LL'}(\vec{k},E) j_l(\kappa r) j_{l'}(\kappa r_1) + \kappa \delta_{LL'} j_l(\kappa r_<) n_l(\kappa r_>) \} Y_L(\hat{r}) Y_{L'}(\hat{r}_1). \tag{5}$$

Using (1) and (4) equation (3) can be rewritten for $r<\rho$

$$\sum_L C_L(\vec{k}) Y_L(\hat{r}) R_{lE}(r) = \int d^3r_1 \sum_{LL'} \{ A_{LL'}(\vec{k},E) j_l(\kappa r) j_{l'}(\kappa r_1) + \kappa \delta_{LL'} j_l(\kappa r_<) n_l(\kappa r_>) \} Y_L(\hat{r}) Y_{L'}(\hat{r}_1) v(r_1) \sum_{L''} C_{L''}(\vec{k}) Y_{L''}(\hat{r}_1) R_{l''E}(r_1). \tag{6}$$

If we consider (2), the scattered wave contribution to the left hand side of (6) is equal to the contribution of the Green's function $G^0_k(\vec{r}-\vec{r}_1;E)$ in (4) to the right hand side, i.e. the incoming wave in the muffin-tin sphere at $\vec{R}=0$ cancels the tails of the waves scattered from all potentials $\vec{R}\neq 0$. Therefore, after integrating over $\hat{r}_1$ and making use of (2) we can write instead of (6)

$$\sum_L C_L(\vec{k}) Y_L(\hat{r}) j_l(\kappa r) = \sum_{LL'} A_{LL'} \int dr_1 r_1^2 j_{l'}(\kappa r_1) v(r_1) R_{l'E}(r_1) C_{L'}(\vec{k}) Y_{L'}(\hat{r}) j_{l'}(\kappa r). \tag{7}$$

This leads to a set of linear equations for the expansion coefficients $C_L(\vec{k})$

$$C_L(\vec{k}) = \sum_{L'} C_{L'}(\vec{k}) \langle \kappa l | v | l E \rangle A_{LL'}(\vec{k},E). \tag{8}$$

The integral is the reactance or K-matrix of a single muffin-tin potential, which is determined by the scattering phase shifts δ_l of the potential v(r)

$$\int_0^\rho dr\, r^2 j_l(\kappa r) v(r) R_{lE}(r) = -\frac{1}{\kappa} \tan\delta_l . \tag{9}$$

A simple matrix operation gives the KKR energy secular determinant

$$\| A_{LL'}(\vec{k},E) + \kappa \cot\delta_l \, \delta_{LL'} \| = 0. \tag{10}$$

We now use (3) in the form (6) for $r > \varrho$ to calculate the wave function outside the muffin-tin sphere. Integrating over $\hat{r}_1$ and using (8) and (9) yields

$$\Psi_{\vec{k}}(E,\vec{r}) = \sum_{L} \{ j_l(\kappa r) - \tan\delta_l \, n_l(\kappa r) \} Y_L(\hat{r}) C_L(\vec{k}). \tag{11}$$

Equation (2) in connection with (9) tells us that the radial solution for $r > \varrho$ is

$$R_{lE}(r) = j_l(\kappa r) - \tan\delta_l \, n_l(\kappa r) \tag{12}$$

i.e. by comparing (12) and (11) with (1) we find that the expansion (1) which originally was restricted for $r < \varrho$ is valid also outside the muffin-tin spheres. For simple lattices with only one atom per unit cell the expansion (1), which according to the above considerations can be used throughout the Wigner-Seitz cell, is the same as in the cellular method. The calculation of dipole matrix-elements should be no problem in this case.

Moreover, as has been shown by Andersen[3], the cellular expansion can be used throughout the unit cell also in crystals with more than one atom per unit cell. This is due to the fact, that the cancellation in a muffin-tin sphere of the incoming wave by the tails of the scattered waves of all the other muffin-tin spheres, which was mentioned in connection with (7), is a general aspect and does not depend on the lattice symmetry.

b) KKR Theory for Non-Muffin-Tin Potentials[4]

The simplicity in formulating the energy band problem and the expressions for the KKR wave functions in the muffin-tin approximation is caused by the constant potential in the interspace between muffin-tin spheres. The propagation of electrons between scattering events takes place with kinetic energy $E=\kappa^2$, i.e. the electrons are scattered elastically at the ionic potentials. If we allow the potentials to overlap, there will be no region with constant potential energy. The propagation of the electron between scattering events takes place with energies which depend on the overlap of potential tails at the place of the electron. Therefore, the energy of the electron before and after scattering must not be the same, which is typical for inelastic scattering. Writing the Green's function in plane wave representation

$$\mathcal{G}_{\vec{k}}(\vec{r}-\vec{r}_1;E) = G^o_{\vec{k}}(\vec{r}-\vec{r}_1;E) + \frac{1}{(2\pi)^3}\sum_{\vec{R}\neq 0} e^{i\vec{k}\vec{R}} \int d^3q \, \frac{e^{i\vec{q}(\vec{r}-\vec{r}_1-\vec{R})}}{E-q^2} \tag{13}$$

one finds, that the elastic scattering is described by the residues of the integral at $q=\pm\sqrt{E}$. The detailed study of $\mathcal{G}_{\vec{k}}(\vec{r}-\vec{r}_1;E)$ by Williams[4] shows, that the remainder of the integral in (13) vanishes for $|\vec{r}-\vec{r}_1| < R$, which in the sense of (6) is synonymous with the muffin-tin approximation. Thus the structural Green's function (4) can be split into $\mathcal{G}^e_{\vec{k}}$ for the elastic and $\mathcal{G}^i_{\vec{k}}$ for the inelastic scattering

$$\mathcal{G}_{\vec{k}}(\vec{r}-\vec{r}_1;E) = G^o_{\vec{k}}(\vec{r}-\vec{r}_1;E) + \mathcal{G}^e_{\vec{k}}(\vec{r}-\vec{r}_1;E) + \mathcal{G}^i_{\vec{k}}(\vec{r}-\vec{r}_1;E).$$

When the inelastic Green's function is considered in a perturbation expansion, the energy secular determinant for the overlapping potential problem differs from the muffin-tin secular equation (10) only in that the K-matrix (9) of a single ionic potential has to be replaced by the K-matrix of the total potential in a given unit cell.

In an actual application of this method[2,4] $\mathcal{G}^i_{\vec{k}}$ was considered only to first order in the perturbation expansion. In this case the inelastic correction is related to an overlap integral of single site potentials. The results for Si and Ge are in very good agreement with experimental data. Unfortunately, since in these calculations a l-dependent screened ion potential (Heine Abarenkov potential) was used, a comparison of the results with KKR-muffin-tin energy bands from l-independent atomic potentials (Herman-Skillman) is not possible, as the effect of overlapping potentials can not be separated from the effect which is introduced by the l-dependence.

The relativistic formulation of the KKR method for non-muffin-tin potentials has been accomplished most recently by Inoue and Okazaki[5].

3. RECENTLY CALCULATED KKR ENERGY BANDS

As was mentioned in the introduction the KKR method with the muffin-tin approximation has found a number of applications to various materials during the last year.

Since always the energy bands for a group of similar solids have been calculated and a detailed discussion of the results was given in the original papers, we briefly present here only one band model out of each group and mention the main points of interest.

a) Solid Rare Gases[6]

As an example the energy bands of Kr are shown in Fig.1. The tightly bound 4p valence electrons of Kr result in an extremely flat valence band (band width ~0.5eV) and a large gap energy (11.4eV). The conduction bands correspond to excited atomic 5s and 4d levels. There has been some interest in the energy band problem of solid rare gases in the last two years. Unfortunately the different band models do not agree and in particular the calculated separations of s- and d-like conduction bands deviate considerably from each other[7]. Since because of strong excitonic effects on the optical properties of solid rare gases a critical point analysis is not possible to check the band models with experimental data, this

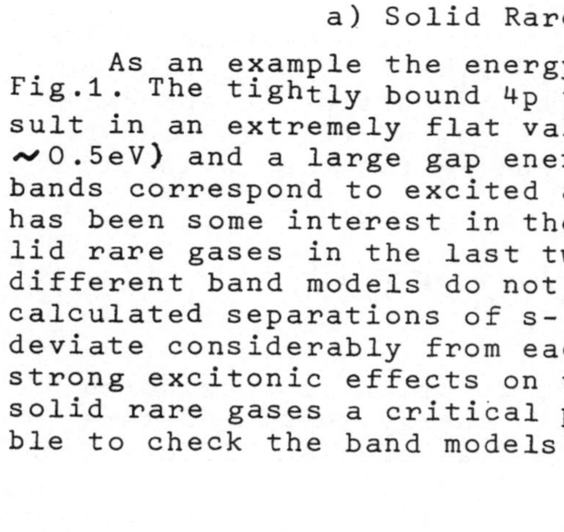

Fig. 1 :

Energy bands of solid Kr

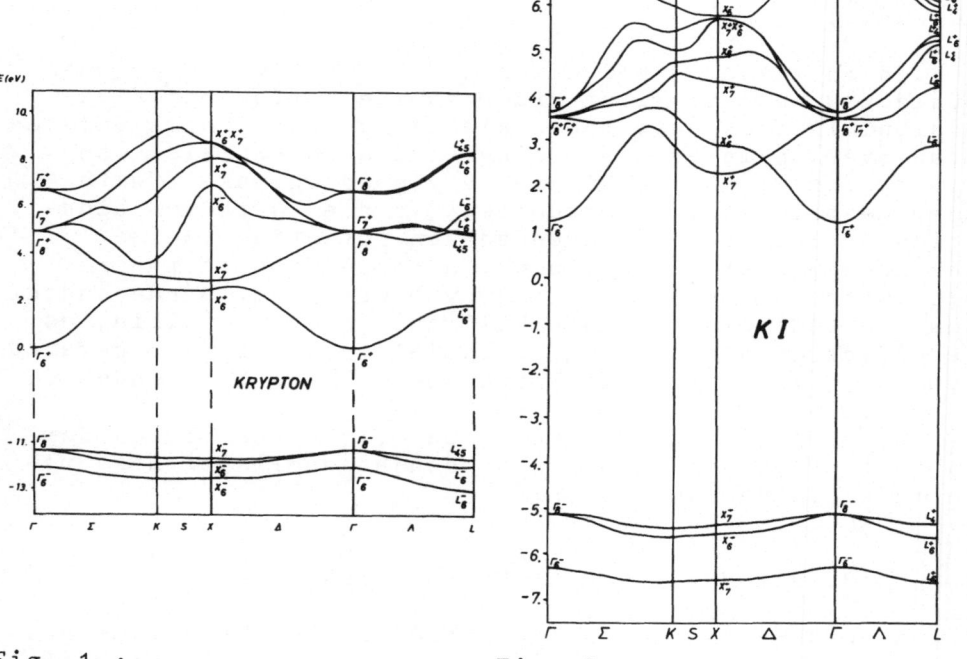

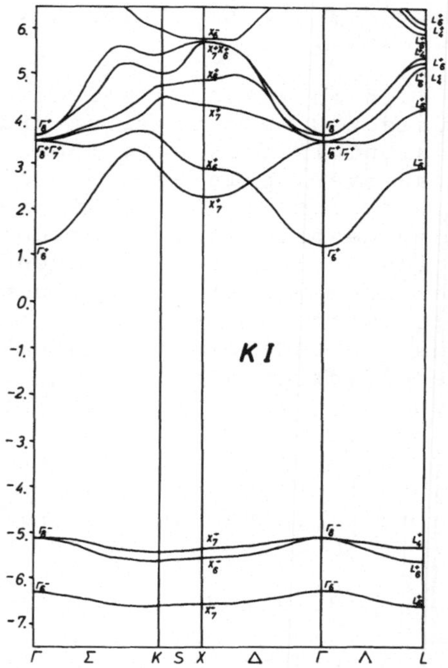

Fig. 2 ;

Energy bands of KI

controversy of s-d separation can be solved only by density of states and exciton calculations. This problem will be discussed in the final section.

b) Potassium Halides[8]

The general features of the potassium halide energy bands are similar to those of the solid rare gases. This can be understood by the isoelectronic neighborhood of the alkali halides and the solid rare gases. By comparing the band model of KI in Fig.2 with the Kr energy bands of Fig.1 one observes the crossing of the two lowest conduction bands on the Δ-axis in KI. This causes a second conduction band minimum at the X point and, therefore, two sets of excitons, namely at the Γ and X point can occur. Experimental data (valence absorption by one and two photon processes, and core absorption from K-3p) are interpreted assuming the same binding energies for both, Γ and X excitons. The agreement found under this assumption between the calculated $X_7^+ - \Gamma_6^+$ separation and experimental data is reasonable, but a more conclusive discussion is possible only with a calculated optical constant which includes the electron-hole Coulomb interaction.

c) Thallous Halides[9]

The thallous halides have found considerable interest in the last time because of their rare selection of physical properties. Although these materials are predominantly ionic, the two Tl-6s electrons exceeding the closed shell result in a large electronic dielectric constant, and contribute to the uppermost valence band with maximum in point X (Fig.3), the lowest conduction band with a large spin-orbit splitting (Δ_{so}=2.4eV at Γ) is due to atomic Tl-6p states and has minima at R and X. The direct gap at point X allows for exciton intervalley scattering, an effect which for the first

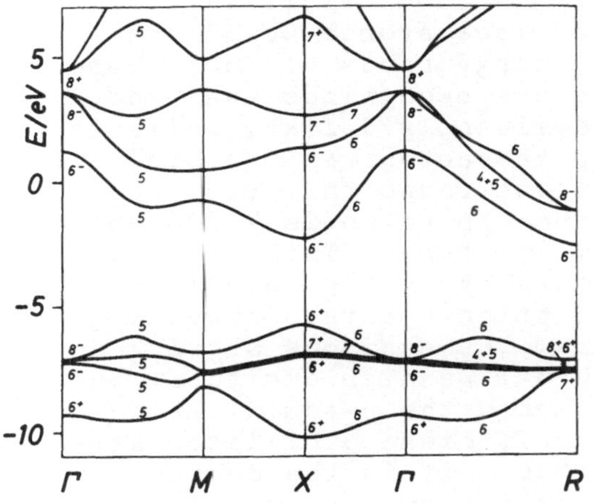

Fig. 3 :
Energy bands of TlCl

time has been found in these materials[10].

The simple muffin-tin approximation of the crystal potential, which was used in all band calculations mentioned in this section, contains some ambiguity concerning the choice of the potential inside the muffin-tin spheres. This became apparent in the application of the KKR method to potassium and Tl halides discussed before. While the use of atomic and ionic potentials with the same exchange (Slater) led to almost identical results for KI, the atomic potentials had to be ruled out in favor of the ionic potentials for the Tl halides, since the s-like valence bands were too deep below the halide-p bands, when atomic potentials were used.. So far there exists no rigorous concept as to which potentials are most appropriate in a KKR calculation. The ambiguity in the choice of the potentials is increased by different possibilities for the approximation of the exchange potential. More detailed studies of these problems would be of interest.

d) Cu Halides[11]

Yet unpublished band calculations for Cu halides are of particular interest because of the mixing of Cu-3d states and halide p-valence electrons, i.e. p-d hybridisation. From a negative spin-orbit splitting in the upper valence band which is indicated by a 1:2 ratio of the excitons in CuCl, it was concluded[12] that the uppermost valence band is almost d-like. The same result is suggested by the relative position of free atomic Cu-3d and Cl-3p levels. Fig.4 shows the energy bands of CuCl. Here from the symmetry of the bands one recognizes that the uppermost valence band is predominantly d-like, while the lower valence band group has p-character. From a study of the partial wave amplitudes it was found that for the upper Γ_7, Γ_8 valence bands the d/p ratio is 70/30 while being 30/70 for the lower Γ_8, Γ_7 terms. This relatively large mixing shows up experimentally by the coincidence of one[12] and two[13] photon absorption spectra concerning the spin-orbit split Γ excitons. The excitons showing up in these measurements are both n=1 Wannier excitons, but while in the one photon experiments the p-admixture to the hole wave function causes sufficient oscillator strength to exhibit the excitonic structure, it is the d-component of the valence band which allows the two quantum process.

Comparing the higher interband transitions in CuCl some problems arise with the relative position of p and d bands. When adjusting the direct gap energy between the

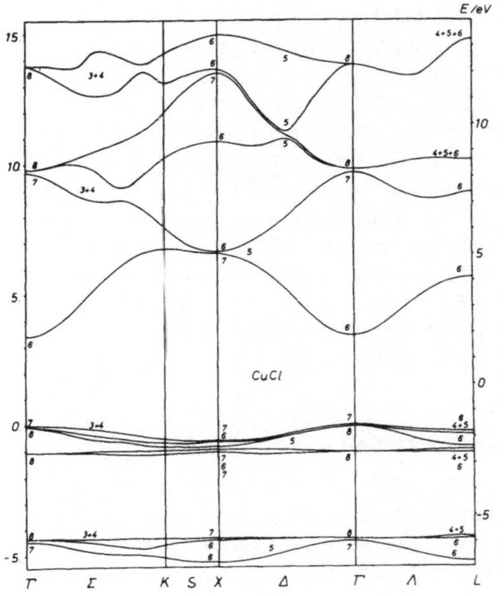

Fig. 4 :
Energy bands of CuCl

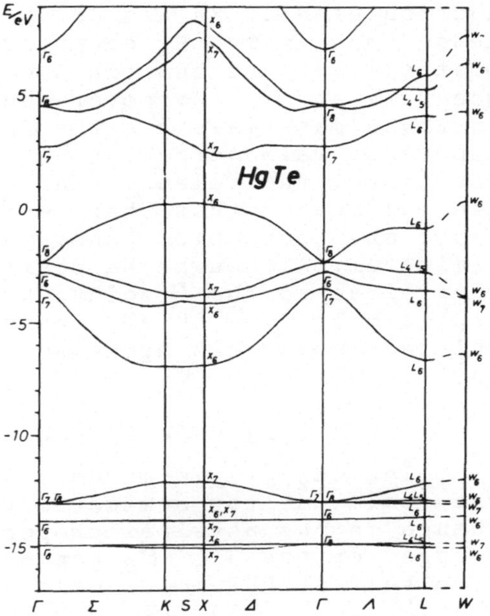

Fig. 5 :
Energy bands of HgTe

more d-like valence band and the conduction band minimum then the transitions from the lower valence bands are in error of 1eV with the experimental data[12].

The question of the relative position of d-bands with respect to p- and s-like bands in KKR band models has been raised also recently by photoemission experiments for ZnO[14]. The d-bands are sensitive to a change in the potential, i.e. when taking nonsuperimposed potentials inside the muffin-tin spheres instead of superimposed potentials the d-bands are shifted by 2-3eV to lower energies, while the rest of the energy bands remains unchanged. Thus, because of the ambiguity in the choice of the potential KKR results for the relative position of d-valence bands with respect to s- and p-bands can be in error by several eV.

e) Hg Chalcogenides[15]

The band gap in II-VI compounds (Γ_6- Γ_8) decreases with increasing atomic number. According to this tendency in Hg chalcogenides the gap is negative. The s-like lowest conduction band of Zn and Cd chalcogenides appears as a valence band in the Hg compounds (Fig.5) where the uppermost filled band and the lowest conduction band are degenerate at Γ_8. Moreover the lack of inversion symmetry in these materials has the important consequence that the uppermost valance band (Γ_8) has nonzero slope at Γ and, therefore, the valence band maxima are at $\vec{k}\neq 0$. Both facts are the reason that there exists a small overlap of valence and conduction bands which is characteristic for a semimetal. Although the position of the band maxima is very close to the Γ point ($k\sim 2\pi/a)0.01$) and the overlap is only 0.3-0.4meV, these quantities have been calculated and are in agreement with existing experimental data.

f) CdTe-HgTe Mixed Crystals[16]

The $Cd_xHg_{1-x}Te$ mixed crystal system is an example for the semiconductor-semimetal transition. The corresponding change in the physical properties is consistent with a change for the Γ_6-Γ_8 gap energy from positive to negative values. KKR model calculations were based on shifting the Γ_6 term relative to Γ_8 by varying the constant potential value only for s-like bands. In Fig.6 the semiconductor-semimetal transition is demonstrated by three band models. The left model corresponds to HgTe, Γ_6 is below Γ_8, the lowest conduction band is a subband compatible with Γ_8. The right model is equivalent to CdTe with a positive band gap, the lowest conduction band is now s-like. The third model shows the situation where Γ_6 and Γ_8 are nearly degenerate and the lowest conduction band changes from being a subband of the p-like band to a s-conduction band.

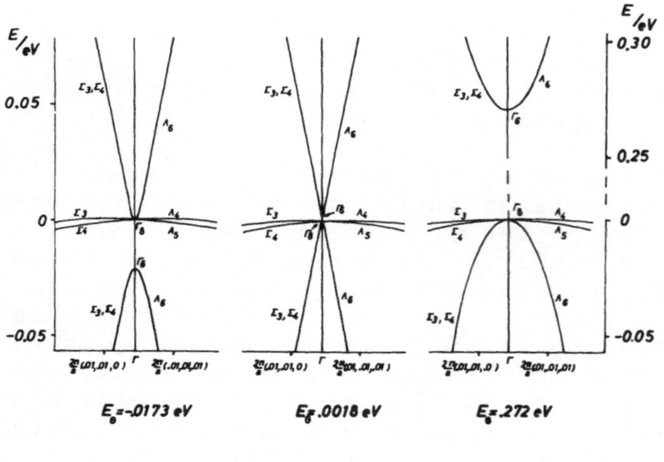

Fig. 6 :

Band model for CdTe-HgTe mixed crystals

4. GOING BEYOND BAND CALCULATIONS

As mentioned in the previous sections, KKR energy bands on symmetry axes in the Brillouin zone do not provide encugh information on the electronic structure for a complete understanding of experimental results. Therefore, attempts have been made to go beyond simple band calculations. We mention here problems for which KKR energy bands have been used as a starting point.

a) High Resolution Band Calculations[16]

The results of KKR energy bands for Hg compounds and HgTe-CdTe mixed crystals emphasized the aptitude of the method to detect even very small details of the energy bands. A more comprehensive representation of the $Cd_xHg_{1-x}Te$ energy bands was possible by using $\vec{k}\vec{p}$ formulas to compute constant energy contours around the Γ point. The parameters entering the $\vec{k}\vec{p}$ expressions were determined by a fit to the detailed KKR bands along the symmetry axes.

b) Density of States Calculations[17]

For an interpretation of core absorption spectra the imaginary part of the dielectric constant can be calculated in the simplified form

$$\epsilon_2(E) \sim \frac{1}{E^2} \sum_{f,i} n_{fi}(E) \tag{14}$$

by assuming constant transition matrix elements. Since the initial core states (i) do not depend on $\vec{k}$ the joint density of states $n_{fi}(E)$ is a replica of the conduction band density of states. Calculations of $\epsilon_2(E)$ in this approximation have been performed for solid rare gases. The results (Fig.7a) for 3d-absorption in Kr show an almost complete one-to-one correspondence between density of states structures and experimental peaks[18] with an overall agreement of about 0.2eV.

Since this model permits a comparison of an energy band model with experimental data which goes beyond the critical point analysis, it can be used to solve the controversy of the calculated s-d conduction band separation mentioned in 3a). For this purpose density of states calculations for Lipari's[7] Kr energy bands have been performed. The results deviate strongly from the experimental absorption curve (Fig.7b). This is certainly caused by the too large separation of s- and d-like conduction bands in this model.

c) The Exciton Problem[19]

Valence absorption spectra of solid rare gases and alkali halides are strongly influenced by excitonic effects. A direct comparison of these experimental data with theoretical band models is, therefore, in general not possible. Theoretical efforts must aim at optical constant calculations which include the electron-hole Coulomb interaction. This is possible along the lines of Hermanson's model[20] which considers the Coulomb potential as a perturbation of $\epsilon_2(E)$ in (14) by using a Green's function formalism. The density of pair states replacing the $\epsilon_2(E)$ of (14) can be expressed as

$$n^{pair}(E) = -\frac{1}{\pi}\,\mathrm{Im}\,\langle 0|G|0\rangle$$

$$= -\frac{1}{\pi}\,\mathrm{Im}\sum_S \langle 0|G_0|S\rangle\langle S|\frac{1}{1-VG_0}|0\rangle \qquad (15)$$

in terms of the two-particle Green's function G or the one-electron Green's function G_0 and the Coulomb potential V. S denotes shells of lattice vectors with equal length. The Green's function G_0 of the band problem can be calculated, once the energy bands throughout the Brillouin zone are known. From reasons of computational effort so far only a two band model was used. In this case the Coulomb potential matrix V is given by Hermanson[20] for Kr and Xe. In Fig.8 we compare the results of (15) with the absorption measurements of Baldini[21]. Exciton states in front of the absorption edge and metastable resonances, i.e. maxima in the density of pair states continuum correspond closely

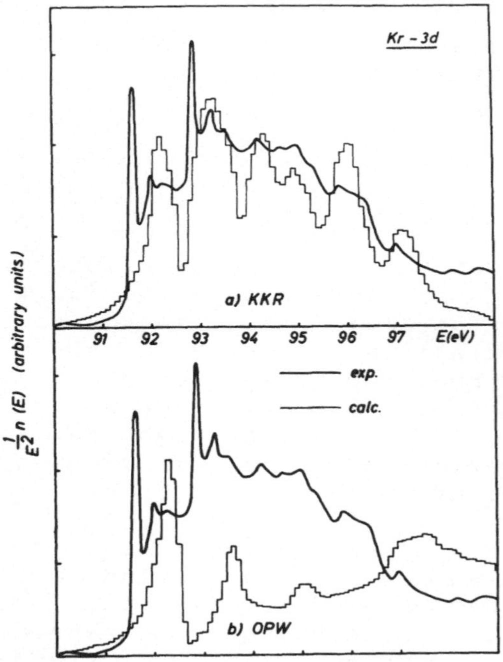

Fig. 7 :

Kr-3d absorption and $\epsilon_2(E)$ calculated from KKR (a) and OPW (b) energy bands.

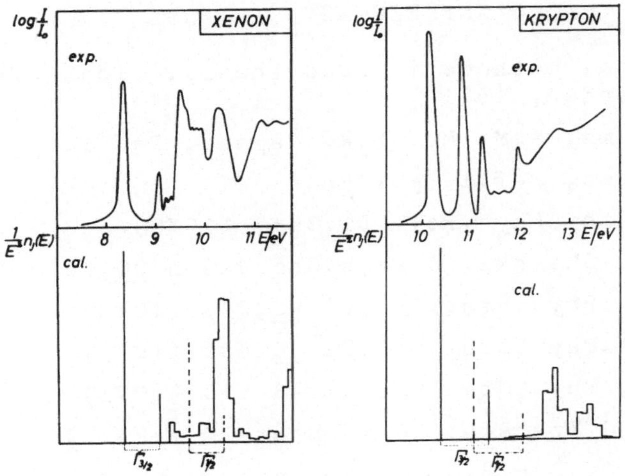

Fig. 8 : Valence absorption in solid Xe and Kr.

to experimental peaks.

The outline given in this paper on the recent development of the KKR method and on its applications to a variety of solids confirmes the important role of this method in the energy band problem. Morevoer it is indicated that KKR energy bands are reliable to serve as a starting point for problems which go beyond the energy band problem. One can expect, therefore, that this method will turn out to be a useful concept in efforts at forthcoming problems of solid state physics.

REFERENCES

1. Computational methods in band theory, Plenum Press New York,London, 1971
2. A.R. Williams, S.M. Hu, D.W. Jepsen, Ref.1, p.157
3. O.K. Andersen, Ref.1, p.178
4. A.R. Williams, Phys.Rev. B1,3417 (1970)
5. M. Inoue,M. Okazaki, J.Phys.Soc.Japan 30,1575 (1971)
6. U. Rössler, Phys.Stat.Solidi 42,345 (1970)
7. N.O. Lipari,Phys.Stat.Solidi 40,691 (1970)
8. H. Overhof, Phys.Stat.Solidi 43,575 (1971)
9. H. Overhof,J.Treusch, Solid State Commun. 9,53 (1971)
10. E. Mohler, G. Schlögl, J. Treusch, Phys.Rev.Letters 27,424 (1971)
11. J. Rathje, Diplomarbeit, Marburg, 1971
12. M. Cardona, Phys.Rev. 129,69 (1963)
13. D. Fröhlich, B. Staginnus, E. Schönherr, Phys.Rev. Letters 19,1032 (1967)
14. R.A. Powell, W.E. Spicer, J.C. McMenamin, Phys.Rev. Letters 27,97 (1971)
15. H. Overhof, Phys.Stat.Solidi 43,221 (1971)
16. H. Overhof, Phys.Stat.Solidi 45, 315 (1971)
17. U. Rössler, Phys.Stat.Solidi 45,483 (1971)
18. R. Haensel, G. Keitel, P. Schreiber, Phys.Rev. 188, 1375 (1969)
19. O. Schütz, U. Rössler, to be published
20. J. Hermanson, Phys,Rev. 166,893 (1968)
21. O. Baldini, Phys.Rev. 128,1562 (1962)

THE ATOMIC ARRANGEMENTS AND RADIAL DISTRIBUTION FUNCTIONS OF AMORPHOUS SILICON AND GERMANIUM

Douglas Henderson

IBM Research Laboratory, San Jose, California &

Department of Applied Mathematics, University of Waterloo

Waterloo, Ontario, Canada

Experimental scattering measurements of the radial distribution function (RDF), g(R), of amorphous silicon[1,2] and amorphous germanium,[1,3] plotted in Fig. 1, show that these substances preserve the local tetrahedral structure of their crystalline forms with peaks representing four first-neighbors at R_0 and twelve second neighbors at $(8/3)^{1/2}R_0$, where R_0 is the crystalline first-neighbor distance. The short-range crystalline and amorphous RDF are, of course, quantitatively different. The first- and second-neighbor peaks in the amorphous RDF are somewhat broader than the corresponding peaks in the crystalline RDF. Thus, the local tetrahedral structure is somewhat distorted in the amorphous forms. However, it is only at larger values of R that qualitative differences between the crystalline and amorphous RDF become apparent. For example there is no peak in the amorphous RDF at the third-neighbor distance, $(11/3)^{1/2}R_0$, of the crystal.

In addition, the small angle scattering data[1] and the model building of Polk[4] indicate that the densities of fully-annealed amorphous Si and Ge are very nearly equal to that of their crystalline forms. A fully-annealed amorphous solid is, of course, an idealization never fully realized in practice. Actual samples of amorphous Si and Ge are less dense than their crystalline forms because they contain such imperfections as voids or unsatisfied bonds. However, it is a highly useful starting point or canonical model for theoretical studies.

Recently, the author and F. Herman[5] have made a computer simulation of the structure of amorphous Si and Ge by considering a system of 64 atoms in a cubic box with periodic boundary conditions.

The atoms are initially located randomly. Each atom is considered in sequence and its four first-neighbors and twelve second-neighbors are moved radially towards the distances R_0 and $(8/3)^{1/2}R_0$, respectively. The process is repeated until a satisfactory RDF is obtained. This takes only a few seconds. Additional computing shows no appreciable change.

In this way a random tetrahedral network of atoms is constructed. In this network each atom is surrounded by a nearly tetrahedral cluster of four neighbors. Thus, the first- and second-neighbor peaks in the amorphous RDF are preserved. However, neighboring tetrahedral units can be randomly rotated about their common bond. Thus, there is no peak in the amorphous RDF at the third-neighbor distance of the crystal.

The RDF of the simulated structure is plotted in Fig. 1. Comparison with experiment shows good agreement. The RDF of the simulated structure has not only the first- and second-neighbor peaks but also has peaks at about 2.5 R_0, 3.25 R_0, and 4.1 R_0, as does the experimental RDF. The correct location of these latter peaks came as somewhat of a surprise because the use of periodic boundary conditions can lead to distortions at $R > L/2$, where L is the length of the edge of the cubic box. For 64 atoms at the crystal density $L = 8\ R_0/(3)^{1/2} \simeq 4.62\ R_0$.

The peaks in the simulated RDF are perhaps too broad. This is because this method does not lead to a static structure. However, the agreement with experiment is sufficiently good to make the static structure which is closest to this simulation a promising starting point in the study of the electronic properties of amorphous Si and Ge.

Shevchik and Paul[3] have made similar simulation studies. They used a finite cluster of 1000 atoms and built their cluster in a manner closely analogous to the actual deposition of an amorphous film. Thus, their cluster is less dense than the crystal and contains a small number of unsatisfied bonds. Their structure also shows the correct location of the peaks in the RDF.

On the basis of these studies, it may be concluded that the structure of amorphous Si and Ge is that of a random tetrahedral network.

The author is grateful to Dr. N. J. Shevchik for suggesting the examination of the simulation RDF for $R > L/2$.

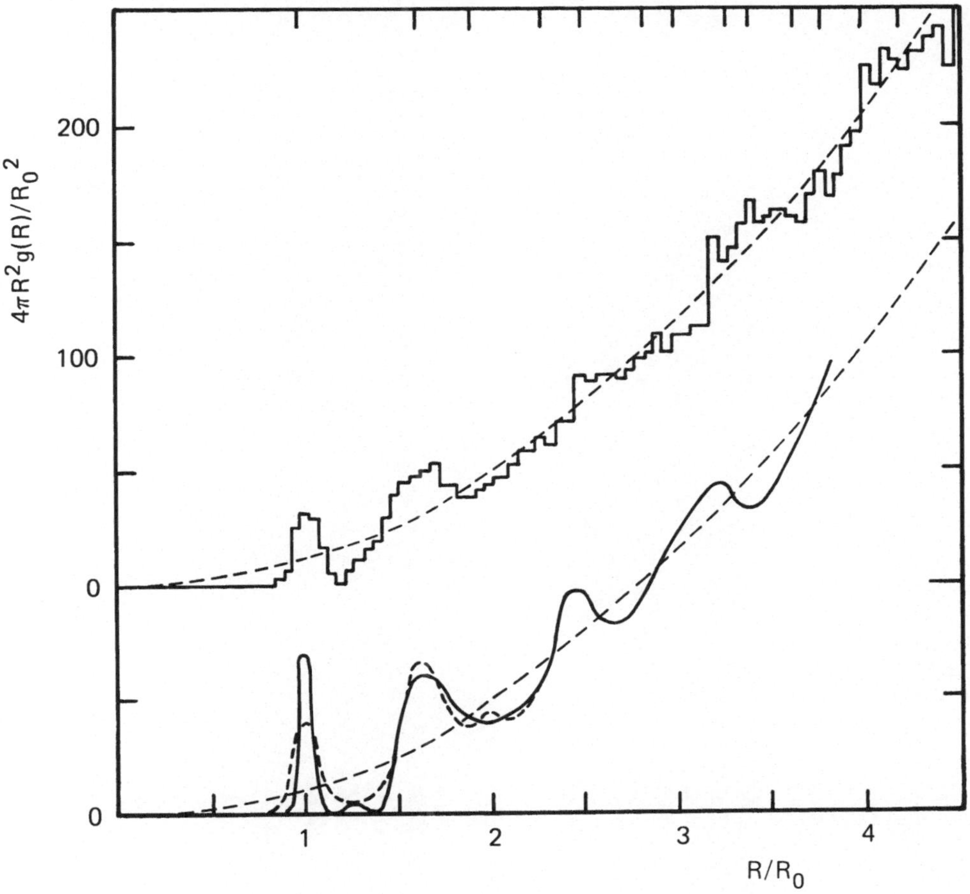

Fig. 1. Radial distribution function of amorphous Si and Ge and of simulated structure. The broken and solid curves, respectively, give the experimental results of Moss and Graczyk (ref. 2) for Si and of Shevchik and Paul (ref. 3) for Ge. The histogram gives the simulation results of Henderson and Herman (ref. 5). The marks at the top of the figure give the location of the neighbors in the crystalline form of these substances.

---

1. V. H. Richter and G. Breiting, Z. Naturforsch 13a, 988 (1958).

2. S. C. Moss and J. F. Graczyk, Proc. Tenth Int. Conf. Phys. Semicond. (USAEC, 1970) p. 658.

3. N. J. Shevchik and W. Paul, J. Non-Crystalline Solids (in press).

4. D. E. Polk, J. Non-Crystalline Solids 5, 365 (1971).

5. D. Henderson and F. Herman, J. Non-Crystalline Solids (in press).

OPTICAL PROPERTIES OF POLYTYPES OF GERMANIUM†

W. E. Rudge and I. B. Ortenburger

IBM San Jose Research Laboratory, San Jose, California,

95114

The atomic arrangement of the ideal amorphous form of germanium is thought to be a tetrahedrally coordinated random network. Bearing in mind that the diamond structure is only one crystalline example of such a network, it is instructive to compare the density of states and the optical spectrum of amorphous Ge not only with these same quantities calculated for cubic Ge, but also with those for such simple polytypes as the wurtzite-like 2H structure with four atoms per unit cell and the 4H structure with eight atoms per unit cell. All three structures have identical first and second nearest neighbors, but differ at the third nearest neighbors and beyond. We can gain insight into both Brillouin zone and short range order effects by studying these three theoretically possible polytypes, assigning to each the same atomic density and nearest neighbor separation.

Our computations indicate that while N(E) for these polytypes becomes progressively more like presently available experimental N(E) for amorphous Ge as one continues to add atoms to the unit cell, the gross features of the optical spectra of these polytypes are all very nearly alike, and are consistently different from the amorphous material.

The band structures for the three polytypes were calculated using the pseudopotential method. The density of states [N(E)], the joint density of states, and the imaginary part of the dielectric constant [$\varepsilon_2(\omega)$] (assuming direct transitions) were obtained for all three structures by the Gilat-Raubenheimer scheme.

The basic structure of three well-defined valence band regions in N(E) is maintained in all three structures. Detailed differences are undoubtedly due in large measure to differences in the Brillouin zone structures for the three polytypes. The maximum of the highest energy peak in the valence band region shifts by 1.0 eV and 0.4 eV to higher energies as one goes from cubic to 2H and to 4H. Moreover, the sharp structure in the conduction band region is washed out as one proceeds from cubic to 2H to 4H. Our results for 4H Ge are more nearly like those obtained experimentally by Spicer and Donovan for amorphous Ge than are our results for 2H Ge or cubic Ge. The implication is that the more complex the unit cell, the more closely N(E) resembles the amorphous N(E).

The calculated $\varepsilon_2(\omega)$ curves for the three polytypes of Ge are compared with each other and with the experimental amorphous $\varepsilon_2(\omega)$ curves in Fig. 1. It is evident that the sharp structure is wiped out as one proceeds from cubic to 2H to 4H, but that the location of the main peak moves only slightly closer to the experimental main peak in this sequence. The principal result here is that the gross features of the optical spectra of the three polytypes are all very nearly the same, the differences being due primarily to Brillouin zone differences.

Calculations based on the non-direct transition model with constant matrix elements were also carried out for the three polytypes. This model cannot account for more than a small fraction of the shift of the main peak to lower energies, as is required by experiment as one goes from crystalline to amorphous Ge. In order to reconcile theory with experiment, it is necessary to introduce some additional factor, such as a suitably chosen energy-dependent oscillator strength. Alternatively or additionally, there may be changes in the energy level structure.

Our calculations suggest that the N(E) of tetrahedrally coordinated polytypes of Ge approach the N(E) obtained experimentally by Donovan and Spicer for amorphous Ge as one proceeds to progressively more complex polytypes; that is to say, polytypes containing progressively more atoms per unit cell. On the other hand, the optical spectra of such polytypes all appear to have the same gross features, and their more or less common main peak does not fall at the location of the main peak for amorphous Ge. Thus, the shift of the main peak from the crystalline to the amorphous location must be attributed to disorder effects explicitly.

†For a more complete report of this work, with appropriate references, see: I. B. Ortenburger, W. E. Rudge and F. Herman, Proceedings of the 4th International Conference of Amorphous and Liquid Semiconductors, Journal of Non-Crystalline Solids, in press.

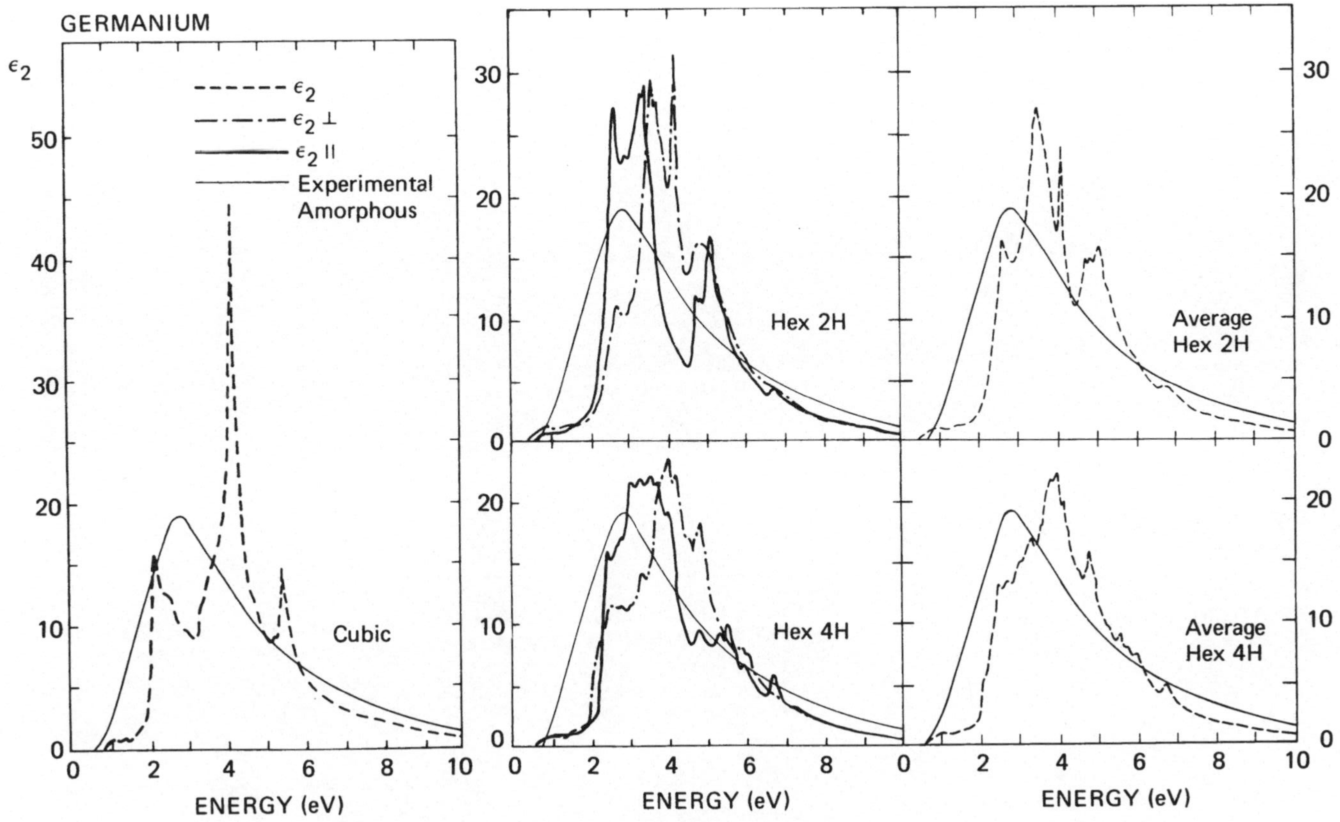

Fig. 1. Calculated $\varepsilon_2(\omega)$ for three polytypes of germanium, all superimposed on the experimental curve for amorphous germanium.

CHARGE DENSITY IN THE GALLIUM ARSENIDE CRYSTAL

N. W. Dalton[†] and D. E. Schreiber

IBM Research Laboratory, San Jose, California 95114

ABSTRACT

Approximating the charge density in crystalline gallium arsenide by a spatial superposition of self-consistent charge densities of gallium and arsenic atoms, we have calculated several contours of constant charge density and examined the three-dimensional character of these contours. A novel feature of the present work is that computer graphics techniques have been used to present the results in the form of three short color movies.

1. INTRODUCTION

A knowledge of the electronic charge distribution in insulators or semiconductors is essential for a complete understanding of the nature of the chemical bond in such materials. Accurate calculations of crystal charge densities are becoming feasible at the present time as a result of faster computers and improved methods for calculating electronic band structures. Significant progress in this direction has recently been made by Walter and Cohen (1971) using pseudopotential methods. In particular they have calculated charge densities for several diamond and zinc-blende type semiconductors using wave functions derived from pseudopotential band-structure calculations. Their principal objective was to determine the critical ionicity which plays an important role in Phillips'

[†] Permanent address: Theoretical Physics Division, A.E.R.E., Harwell, Berkshire, England.

recent theory (Phillips (1970)) of the chemical bond. They found it instructive to present their charge density data in the form of contour maps and charge 'clouds' to illustrate many important qualitative features of their results.

One of the mainstream activities of this laboratory is the development of programs for calculating accurate self-consistent first-principles electronic band structures and related physical properties for a wide range of metals and semiconductors. As a preliminary step in this direction we have performed calculations similar to those of Walter and Cohen (1971). However, the present work differs from that of the previously mentioned authors in two respects, (i) the use of a simplified model to calculate the charge density, and (ii) the presentation of our results in the form of a computer-generated movie. Since our model is not based on a self-consistent band-structure calculation, we are unable to show the build-up of charge density due to self-consistent crystal charge rearrangement which is evident in the work of Walter and Cohen (1971). Future calculations of the charge density will be carried out based upon a self-consistent OPW method so that useful comparisons can be made between our work and that of Walter and Cohen (1971) and experiment. Therefore, the present results should be regarded as exploratory rather than final.

Since the emphasis in this paper is on the use of computer graphics in computational solid state physics rather than on the accurate calculation of crystal charge densities, we have simplified the latter calculation by representing the crystal charge density as follows:

$$\rho^{v,c}_{GaAs}(\underline{r}) = \sum_{\underline{R}} [\rho^{v,c}_{Ga}(\underline{r}+\underline{R}) + \rho^{v,c}_{As}(\underline{r}+\underline{R}+\underline{\tau})] \quad , \qquad (1)$$

where $\rho^{v,c}_{GaAs}(\underline{r})$ denotes the gallium arsenide crystal valence (v) or core (c) charge density at $\underline{r}$, $\rho^{v,c}_{Ga}(\underline{r}+\underline{R})$ denotes the self-consistent atomic valence or core charge density at $\underline{r}+\underline{R}$ due to a gallium atom at the lattice position $\underline{R}$, and $\rho^{v,c}_{As}(\underline{r}+\underline{R}+\underline{\tau})$ denotes the self-consistent valence or core charge density at $\underline{r}+\underline{R}+\underline{\tau}$ due to an arsenic atom at the lattice position $\underline{R}+\underline{\tau}$. The atomic charge densities were calculated using the relativistic program of Liberman et al. (1965) with Kohn-Sham exchange. Core charge densities were constructed from 1s to 3d atomic states whereas 4s and 4p states were used for the valence charge density. The lattice structure for gallium arsenide is sphalerite of which a part is drawn in Figure 1.

In the following section we explain how computer graphics methods can be used to display the charge density and, in particu-

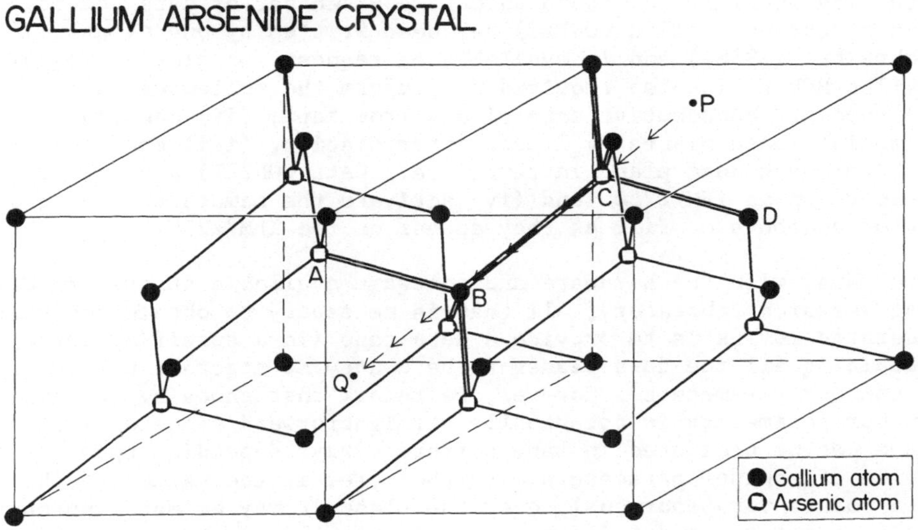

FILM 1. Relates to atoms B and C in the plane ABCD

FILM 2. Relates to the rotation of the plane ABCD about the line BC

FILM 3. Relates to a walk from P to Q through the atoms at C and B respectively.

Figure 1.

lar, how we constructed three computer-generated movies of $\rho^{v,c}_{GaAs}(\underline{r})$. For purposes of illustration we also include four instantaneous snapshots taken from two of the movies.

2. GRAPHICS SUBROUTINES AND RESULTS

Since the crystal charge density is a scalar function of position in three dimensions and since a single contour map can sample only two dimensions, it is clear that several contour maps are required to map out the crystal charge density. This leads quite naturally to the idea of constructing a computer-generated movie to display the charge density. Fortunately, all the necessary hardware and software for making computer-generated movies is available at the IBM San Jose Research Laboratory so that it is a comparatively simple matter to set up computer displays of the charge density. The basic hardware required is an IBM/2250 screen to display pictures of contours and a camera system to record the pictures as they appear on the screen. In addition, a routine is required for constructing specified contours through a given set of data points. Such a routine (called CONPLT) has been written by one of us (Schreiber (1968)) and is available on request. A simple program (called MOVIE) is also required to perform the following operations: (i) read two consecutive data planes from tape, (ii) construct intermediate data planes by linear interpolation, (iii) construct contours of each data plane in turn (i.e., CALL CONPLT) and display contours on an IBM/2250, and (iv) activate the camera to record each set of contours on film as they appear on the IBM/2250.

Thus, with the hardware and software available at the IBM San Jose Research Laboratory, all that is necessary to obtain computer-generated movies is to provide a data tape (in a specified format) containing all the data planes to be contoured together with a set of contour parameters. However, we remark that choosing suitable contour parameters is not usually straightforward since a given data plane can be contoured in many different ways depending upon the choice of contour parameters. Furthermore, if the values of the data points vary enormously over the plane it may be more appropriate to process the data (e.g., logarithmically compress or filter the data) before contouring. Some illustrative contour maps are depicted in Figure 2.

To produce the movies described below we used Eq. (1) to calculate separately the crystal core and valence charge densities on a 60 x 60 mesh (i.e., 3600 data points) in the planes of interest. Sufficient terms in the lattice summation were retained to ensure that the crystal charge density was converged to 1 part in 10^5. Six black and white movies were made originally (i.e., three valence and three core). Each movie consisted of approximately 3000 frames although only 50-100 data planes were calculated using Eq. (1). The remainder were obtained by linear interpolation between consecutive planes. Color films (blue for core, yellow for valence) were made from the black and white originals before superimposing the core and valence films. The content of the final three color films may be explained with the aid of Figure 1.

<u>Film 1</u> 'Formation of Gallium Arsenide'

This film shows how the contours of constant charge density (valence and core) evolve as the atoms of gallium and arsenic are brought together to form the crystal. The plane referred to in this film contains atoms A, B, C, and D in Figure 1.

<u>Film 2</u> 'Rotation About the Gallium Arsenide Bond'

Attention is focused on the contours of constant charge density in a plane which passes through the line joining a pair of adjacent gallium and arsenic atoms (i.e., B and

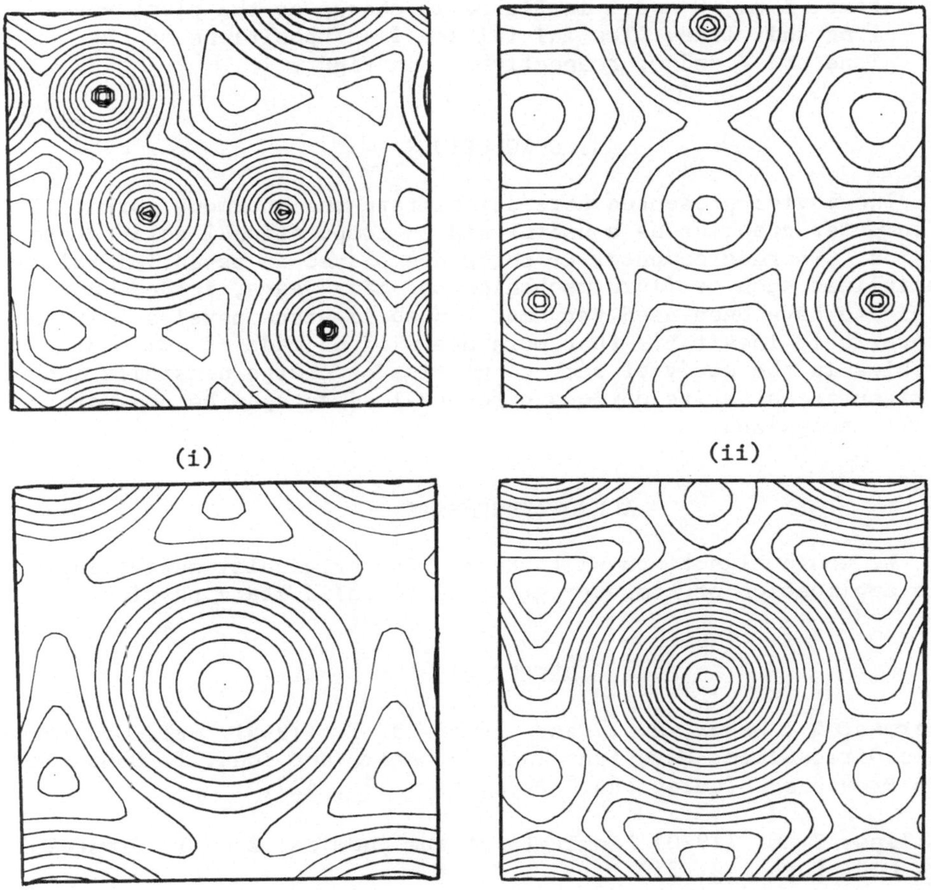

(i) (ii) (iii) (iv)

Figure 2. Representative contour maps from color movies. (i) Valence charge density in the plane ABCD (see Figure 1) (films 1 and 2). (ii),(iii),(iv) Valence charge density in successive planes perpendicular to the axis CB as one moves from P to C (film 3).

C in Figure 1). This film shows how these contours change as this plane is rotated about the bond BC.

<u>Film 3</u> 'A Walk Along the Gallium Arsenide Bond

This film shows how the contours in a plane perpendicular to the line joining a pair of adjacent gallium and arsenic atoms (i.e., C and B in Figure 1) change as the plane moves along this line. The initial and final positions of the plane are P and Q, respectively, in Figure 1.

3. CONCLUSIONS

The movies presented here demonstrate the enormous potential of computer graphics as a useful aid in computational solid state physics. Whereas computer graphics has played an essential role in disciplines such as fluid mechanics for many years, it does not appear to have been used very much in solid state physics. In the future it is inevitable that more use will be made of computer graphics in the study of such quantities as charge densities and fermi surfaces as its obvious perceptual advantages become more widely recognized.

ACKNOWLEDGEMENTS

We would like to thank F. Herman for originally suggesting this application of computer graphics to solid state physics.

REFERENCES

Liberman, D., Waber, J. T. and Cromer, D. T. (1965) "Self-Consistent Field Dirac-Slater Wave Functions for Atoms and Ions. I. Comparison with Previous Calculations." Phys. Rev. 137, A27.

Phillips, J. C. (1970) "Ionicity of the Chemical Bond in Crystals" Rev. Mod. Phys. 42, 317.

Schreiber, D. E. (1968) "A Generalized Equipotential Plotting Routine for Calculating a Scalar Function of Two Variables" IBM Research Report RJ-499.

Walter, J. P. and Cohen, M. L. (1971) "Electronic Charge Densities in Semiconductors" Phys. Rev. Letters 26, 17.

III

Treatment of Exchange and Correlation Effects in Crystals

INTRODUCTORY REMARKS

Per-Olov Löwdin

Dept. of Quantum Chemistry, Uppsala Univ. Sweden

Quantum Theory Project, Univ. of Florida, Gainesville

The organizers have kindly asked me to give a brief introduction to this session. The exchange and correlation effects in solids are closely connected with the existence of the Coulomb repulsion e^2/r_{12} between the electrons which tries to keep the electrons apart. The problem involves also the study of the self-interaction of a single electron.

The effects may be seen by considering the second-order density matrix

$$\Gamma(x_1 x_2 | x_1' x_2') \qquad (1)$$

where $x = (\vec{r}, \zeta)$ is a space-spin coordinate. Since the quantity Γ is antisymmetric in each pair of indices, one has $\Gamma = 0$ whenever $x_1 = x_2$ or $x_1' = x_2'$, i.e.

$$\Gamma(x_1 x_2 | x_1 x_2) = 0, \qquad (2)$$

when $r_{12} = 0$ and $\zeta_1 = \zeta_2$. This is the well-known "Fermi hole" for electrons having parallel spins, which shows that the exclusion principle through the antisymmetry condition automatically keeps the electrons apart. For electrons with antiparallel spins, one could expect the existence of a similar "Coulomb hole" as a result of the Coulomb repulsion, and if this effect is not properly treated, the model is affected by a "correlation error".

In the independent-particle-model, each particle is moving in the "average field" of all the other particles, and the resulting correlation error is often measured in terms of the "correlation energy", which is defined as the difference between the exact energy and the energy of the conventional symmetry-adapted Hartree-Fock (HF) scheme:

$$E_{corr} = E_{exact} - E_{HF}. \qquad (3)$$

Since the virial theorem is valid for both the exact solution and the HF-approximation, one has further

$$\langle T\rangle_{corr} = -\frac{1}{2}\langle V\rangle_{corr} = -\frac{1}{2}\langle L + C\rangle_{corr}, \qquad (4)$$

where T is the kinetic energy and V is the coulomb energy, consisting of the nuclear attraction energy L and the coulomb repulsion C. The correlation effects are hence going to show up not only in quantities connected with the two-electron operator C but also associated with the one-electron operators T and L.

The first discussion of the error in the HF-method based on symmetry-adapted molecular orbitals was given by Slater[1] in 1930 in his famous paper on the cohesive properties of the alkali metals. He chose the H_2-molecules as an example. If a and b are the atomic orbitals of the two atoms involved and $\varphi = a + b$ is the associated molecular orbital, one has

$$\varphi_1\varphi_2 = (a_1 b_2 + b_1 a_2) + (a_1 a_2 + b_1 b_2), \qquad (5)$$

where the first term represents the Heitler-London function for the covalent bond, and the second term represents ionic contributions from the combination ($H^+ + H^-$). For increasing internuclear distance R, the molecule should separate into two neutral atoms, since the coulomb repulsion would prevent the two electrons with antiparallel spins to accumulate on the same proton to form the H^- ion. Since the MO-wave function (5) does not reflect this behaviour for separeted atoms but contains a large ionic contribution of higher energy, the HF-method is affected by a very large "correlation error" for $R = \infty$; see Fig. 1. Slater[2] has emphasized that the existence of this error makes a simple HF-theory of the magnetic properties of solids impossible.

The first estimates of the correlation energy for the alkali metals were carried out by Wigner[3)] in 1933. Dimensional considerations show that the correlation energy may be written in the form

$$E_{corr} = -\frac{e^2}{r_{corr}}, \tag{6}$$

where the "correlation radius" r_{corr} is some form of average distance between two electrons. Wigner introduced the "electronic radius" r_s associated with a density ρ through the relation

$$\frac{4\pi}{3} r_s^3 \rho = 1, \tag{7}$$

which gives $r_s = (3/4\pi\rho)^{1/3}$ and $1/r_s \sim \rho^{1/3}$. The latter relation forms the basis for Slater's treatment of exchange and correlation, to which we will return below. In a first approximation, the quantities r_s and ρ are considered as characteristic constants in which all energy contributions may be expressed. For the free-electron gas, one obtains the well-known exchange energy per electron[4)]:

$$E_{exch} = -\frac{e^2}{r_{exch}} = -0.458\frac{e^2}{r_s}, \tag{8}$$

which gives the average "exchange radius" $r_{exch} = 2.18\, r_s$. For the correlation energy, Wigner obtained the formula

$$E_{corr} = -0.288\frac{e^2}{r_s + 5.1\, a_H}, \tag{9}$$

where a_H is the Bohr radius; this gives $r_{corr} = 3.47\, r_s + 17.7\, a_H$, i.e. the correlation radius is considered as a linear function in r_s. For the self-energy of an electron, an elementary calculation gives finally

$$E_{self} = e^2 \iint \frac{\rho(1)\rho(2)}{r_{12}} dv_1 dv_2 = 1.2\frac{e^2}{r_s}. \tag{10}$$

Wigner used his results in calculating the cohesive energies of the alkali halides.

Atomic correlation energies. The author[5)] became interested in the correlation problem in 1951 in connection with a calculation of the cohesive energy of the sodium metal based on atomic Hartree-Fock functions, which gave a good result without including any correlation effects. It was clear that there was some cancellation of errors which led to this most unexpected result, and a series of studies of electronic correlation were started. Except for Wigner's formula, hardly any data were available. Calculations of the correlation energies for the He-like ions revealed that they were remarkably constant[6]:

H^-	He	Li^+	Be^{2+}	B^{3+}	C^{4+}	
-1.08	-1.14	-1.18	-1.20	-1.22	-1.23 eV	(11)

One would guess that, in the power series $E_{corr} = a_2 + a_3/Z + a_4/Z^2 + \cdots$, the constant a_2 would be the dominating term. Actually this is not the case, but a Chebyskev transformation of the series shows that E_{corr} is very slowly varying for $(1/Z) < 1$.

These calculations were continued by Fröman[7] who obtained Eexact by subtracting the relativistic corrections from the experimental value:

$$E_{exact} = E_{exp} - E_{rel}. \tag{12}$$

He found that the correlation energy for the Be-atom was -2.37 eV, and that the correlation energies for the Ne-like ions were around -11 eV. At a time when many were inclined to consider the correlation energies for an isoelectronic series as approximately constant, Linderberg and Shull[8] could show that, for the Li-like and Be-like ions, the correlation energies contain a term linear in Z:

$$E_{corr} = a_1 Z + a_2 + a_3/Z + a_4/Z^2 + \cdots, \tag{13}$$

which is associated with the $2s$-$2p$ degeneracy for $Z = \infty$. For a more detailed survey of the development in this period, we will refer elsewhere.[9]

During the 1960's, the evaluation of the atomic and molecular correlation energies has been continued by Clementi[10]. For many years, there seemed to be a gap between the atomic data and Wigner's formula (9), but Clementi has finally shown that the two aspects may be joined together[11].

<u>Perturbation aspects on correlation problem</u>. It was generally assumed that the Hartree-Fock model was successful for atoms, molecules, and solids simply because the

interaction was weak. It was hence greatly surprising that, in the 1950's, the same scheme in the form of the shell-model started to work successfully also for the atomic nuclei, where the interaction is definitely very strong. The study of the nuclear correlation problem by Brueckner[12] and his collaborators have helped also the understanding of electronic correlation in solids.

During the last two decades, the mathematical tools for solving the Schrödinger equation $\mathcal{H}\Psi = E\Psi$ have been completely changed. The Hartree-Fock method was based on the variation principle[13], whereas the new approaches are to a large extent based on "transition formulas" of the type

$$E = \frac{\langle \Phi | \mathcal{H} | \Psi \rangle}{\langle \Phi | \Psi \rangle}, \tag{14}$$

where Φ is an arbitrary function at our disposal. The quantity (14) for approximate eigenfunctions Ψ is unbounded but, by introducing appropriate constraints, one may get both upper and lower bounds to the true energy E. In a perturbation problem with $\mathcal{H} = \mathcal{H}_0 + V$, one may choose $\Phi = \varphi_0$, $\langle \varphi_0 | \varphi_0 \rangle = 1$, and introduce the intermediate normalization

$$\langle \varphi_0 | \Psi \rangle = 1 . \tag{15}$$

Using the relation $\mathcal{H}_0 \varphi_0 = E_0 \varphi_0$, one obtains

$$\begin{aligned} E &= \langle \varphi_0 | \mathcal{H}_0 + V | \Psi \rangle = \\ &= E_0 + \langle \varphi_0 | V | \Psi \rangle . \end{aligned} \tag{16}$$

The idea is to go back to an expectation value formula by making the substitutions

$$\Psi = W \varphi_0 , \qquad t = VW , \tag{17}$$

which gives

$$\begin{aligned} E &= E_0 + \langle \varphi_0 | t | \varphi_0 \rangle = \\ &= \langle \varphi_0 | \mathcal{H}_0 + t | \varphi_0 \rangle , \end{aligned} \tag{18}$$

where W is usually referred to as the "wave operator" and

$\mathcal{t} = VW$ as the "scattering operator" or "reaction operator". In the perturbation approach to the correlation problem, the total Hamiltonian is written in the form

$$\begin{aligned} \mathcal{H} &= \sum_i \mathcal{H}_i + \sum_{i<j} \mathcal{H}_{ij} = \\ &= \sum_i (\mathcal{H}_i + \mathcal{U}_i) - \sum_i \mathcal{U}_i + \sum_{i<j} \mathcal{H}_{ij} = \\ &= \mathcal{H}_0 + V \quad , \end{aligned} \tag{19}$$

where the quantities $\mathcal{U}_i$ are potentials at our disposal. Brueckner noticed that, if φ_0 is chosen as a single Slater determinant, the Hartree-Fock energy has the form

$$E_{HF} = \langle \varphi_0 | \mathcal{H}_0 + V | \varphi_0 \rangle \ , \tag{20}$$

whereas the correct energy has the form (18). One can hence obtain the correct result in a SCF-scheme using the "self-consistent-field potentials" $\mathcal{U}_i$, if one everywhere in the Hartree-Fock scheme replaces the perturbation V by the reaction, $\mathcal{t}$, which leads to the exact SCF-scheme[14]. Using (3), (18) and (20), one obtains further

$$E_{corr} = \langle \varphi_0 | \mathcal{t} - V | \varphi_0 \rangle \ , \tag{21}$$

and it may hence seem natural to refer to ($\mathcal{t} - V$) as the "correlation operator". It should be observed, however, that the wave operator W is anything but unique. It may even be chosen as a "correlation factor" $W = \Psi/\varphi_0$, in which case ($\mathcal{t} - V$) is a local potential.

The quantities W and $\mathcal{t} = VW$ have been studied in connection with perturbation theories of Brillouin-type and Schrödinger-type by Schwinger, Brueckner, and their collaborators. By using partitioning technique[15], one may obtained the following closed formulas

$$W = (1 - T_0 V)^{-1} \ , \qquad \mathcal{t}^{-1} = V^{-1} - T_0 \ , \tag{22}$$

where $T_0 = (E - \mathcal{H}_0)^{-1} P$ and $P = 1 - |\varphi_0\rangle\langle\varphi_0|$. Using inner projections, one may instead of the conventional power series derive expressions for $\mathcal{t}$ which are quotients between polynomials[16] and Padé approximants[17] to the conventional series. In his nuclear work, Brueckner has essentially limited himself to the study of the two-body part of the reaction operator, whereas many-body reac-

tions have to be included in the study of electronic correlation due to the long-range character of the Coulomb
repulsion. Work along these lines is still in progress.
For a more detailed study of Brueckner's work, we will
refer elsewhere[18].

   Pair correlation and pair functions. A special method for treating the correlation problem closely related to the perturbation approach and the use of the transition formula (16) has been developed by Nesbet and Sinanoglu in slightly different forms. One expands the e-
xact solution Ψ in determinants built up from the Hartree-Fock functions and one-electron functions out of their
orthogonal complement:

$$\Psi = \varphi_0 + \varphi_{s.e.} + \varphi_{d.e.} + \varphi_{t.e.} + \cdots , \qquad (23)$$

where φ_0 is the Hartree-Fock solution, $\varphi_{s.e.}$ is the sum
of all singly excited determinants with proper coeffici-
ents, $\varphi_{d.e.}$ is the sum of all doubly excited determinants,
etc. Using the Brillouin theorem $\langle \varphi_0 | \mathcal{H} | \varphi_{s.e.} \rangle = 0$ and the
orthogonality properties, one obtains from (16):

$$E = \langle \varphi_0 | \mathcal{H} | \Psi \rangle =$$
$$= E_{HF} + \langle \varphi_0 | \mathcal{H} | \varphi_{d.e.} \rangle =$$
$$= E_{HF} + \langle \varphi_0 | C | \varphi_{d.e.} \rangle , \qquad (24)$$

where C is the operator for the Coulomb repulsion between the electrons, and the last term represents the
correlation energy. This quantity may hence be described
in terms of the relation between φ_0 and $\varphi_{d.e.}$, i.e. in
terms of pair excitations or a finite number of "pair functions". In their approximate treatment of these functions,
the two authors have developed different techniques: Nesbet is using an extension of the Bethe-Godstone method,
whereas Sinanoğlu has analyzed the pair functions using
essentially variational methods. For a more detailed study, we will refer to some fairly recent reviews.[19, 20]

   It has sometimes been said that formula (24) gives
the best physical description of the correlation energy,
since it involves only the Coulomb repulsion C . It
should be remembered, however, that transition formulas
of type (14) and (24) do not have the same simple physical interpretation as expectation value formulas, where

$\Phi = \mathcal{W}\Psi$. For an Hamiltonian of the type $\mathcal{H} = T + L + C$, one should have

$$E_{corr} = \langle T \rangle_{corr} + \langle L \rangle_{corr} + \langle C \rangle_{corr}, \qquad (25)$$

where the contributions fulfil the virial theorem (4). One has hence to distinguish carefully between a strong mathematical tool which renders a correct number and the physical interpretations. A recent study[2)] has revealed that also the nuclear attraction L gives an interesting contribution $\langle L \rangle_{corr}$ to the correlation energy.

Correlation in solids. Many authors have treated the wave operator W as a "correlation factor" of a more or less complicated form. In solid-state theory, Bohm and Pines[22)] have studied a plasma model and deduced a correlation factor of the form

$$W = \exp\left\{-\sum_{k<k_c}\sum_{ij}{}' \frac{2\pi e^2}{k^2}\,\frac{e^{i\vec{k}\cdot\vec{r}_{ij}}}{\hbar\,\omega_p}\right\}, \qquad (26)$$

which describes the "collective motions" of the electrons; here k_c is a cut-off vector for the plasma oscillations and ω_p is the plasma frequency. The rather wide gap between this approach and Wigner's method has been at least partly bridged by Krisement[23)]. Additional aspects on the correlation problem for the electron gas have been given by Brueckner.[18)]

Important contributions to the understanding of the correlation effects in solids using many-body theory have been given by Hubbard, Hedin, and Lundqvist, who are all here - we are looking forward to their lectures today.

It is clear that all energy contributions may be expressed in terms of the electron radius r_s given by (7) or the quantity $\rho^{1/3}$. If the density $\rho = \rho(P)$ is known not only in the point P but also in its neighbourhood, one knows also its first derivatives, the second derivatives, etc. Part of the discussion has been dealing with the problem to what an extent it would be necessary to introduce gradρ (and the higher derivatives) explicitly in the formula for the correlation energy.

In his treatment of exchange, Slater[24)] has proposed the use of a local potential which is proportional to $\rho^{1/3}$, and this approach has later been generalized to the X_α-method. In recent applications, Slater has found that this method gives excellent results in applications

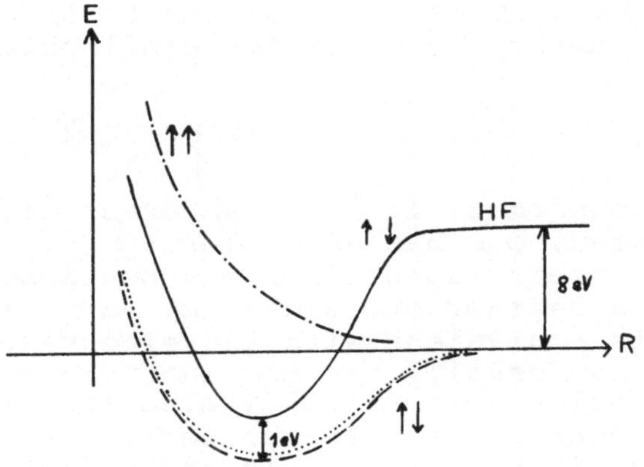

Fig. 1. Energy as a function of internuclear distance: the full line refers to a single determinant, the lower curve (- - -) represents the true behavior, the upper curve (-·-·) gives the energy of a state with parallel spins. The dotted line(···) represents the results of Slater's X_α-method. The numbers are taken from ref. 9.

to atoms, molecules, and solids at equilibrium distance and that it has correct asymptotic behaviour for separated atoms; see Fig. 1. Slater has used the last feature to conclude that the X_α-method treats not only exchange but also correlation.

For the moment, the X_α -method seems to be the simplest method for treating exchange and correlation in atoms, molecules, and solids. In addition to Slater, several authors have contributed to the development of the method, and here I would only like to mention Frank Herman's excellent work explicitly. In the $X_{\alpha\beta}$-method, one has included also certain effects associated with grad ρ , and Karlheinz Schwarz is going to present some very interesting applications from his work with Frank Herman at San José using fixed values of α and β .

Hedin and Lundqvist are going to attack the same problem using many-body theory, and they obtain results which are strikingly similar to the X_α -method with the difference that α is no longer a constant but a slowly varying function.

Symmetry and correlation. Before concluding this introduction, I would like to say a few words about the importance of the symmetry properties in treating the correlation problem. If Λ is a normal constant of motion or $G=\{g\}$ is a symmetry group of the Hamiltonian, one has

$$\mathcal{H}\Lambda = \Lambda\mathcal{H}, \qquad \mathcal{H}g = g\mathcal{H}. \tag{27}$$

It is easily shown that, if Ψ is an exact eigenfunction to the Hamiltonian, one has also $\Lambda\Psi=\lambda\Psi$, i.e. Ψ is automatically symmetry-adapted or may be chosen that way in the case of a degeneracy. Since the variation principle $\delta\langle\mathcal{H}\rangle=0$ is equivalent with the eigenvalue problem $\mathcal{H}\Psi=E\Psi$, one has usually assumed that the symmetry properties would follow automatically also from the variation principle even in the case of restricted variations. However, if the eigenfunction Ψ is approximated by a single Slater determinant D, there is no obvious reason why the relation

$$\Lambda D = \lambda D \tag{28}$$

should follow from the variation principle $\delta\langle D|\mathcal{H}|D\rangle=0$, and the relation (28) should hence be considered as a <u>constraint</u>. It has been shown[25] that such a constraint is "self-consistent", and hence there exist solutions to the Hartree-Fock problem which are symmetry-adapted, but it has become more and more probable that these solutions correspond only to relative minima of $\langle\mathcal{H}\rangle$.

Slater discovered the symmetry difficulties associated with the HF-method already in his famous 1930 paper[1] in connection with the MO-treatment of the H_2-molecule for separated atoms; see Fig. 1. He emphasized it again in 1951 in connection with the "spin-polarization" problem[26] of the Li-atom, which led to the development of the unrestricted Hartree-Fock (UHF)-method to which Nesbet, Watson, and Freeman made important contributions.

The spin-polarization is associated with open-shell systems, but there seemed to be similar difficulties also for closed-shell systems. It seemed as if the entire Hartree-Fock method based on a single Slater determinant would be affected by a <u>symmetry dilemma</u>:[27]

> the assumption that D should be symmetry adapted or an eigenfunction to Λ leads to an energy high above the absolute minimum, whereas the absolute minimum $\langle\mathcal{H}\rangle$ is associated with a determinant D which is a "mixture" of components of various symmetry types.

Since the correlation energy defined by (3) refers

to the error in the conventional symmetry-adapted HF--scheme, it is clear that a large part of this error may be associated with the constraint (28) and may be diminished by removing the constraint. In such a case, the symmetry properties are lost, and the determinant becomes a mixture of different symmetry types:

$$D = \sum_k D_k , \tag{29}$$

where each one of the components is uniquely defined[28] throught the relation $D_k = O_k D$, where O_k is a projector satisfying the relations $O_k^2 = O_k$, $O_k^\dagger = O_k$, $O_k O_l = 0$, and

$$1 = \sum_k O_k . \tag{30}$$

It is easily seen that one of the components D_k may have an even lower energy, which may be further improved by optimizing D for the new situation. In this Projected Hartree-Fock (PHF)-scheme[29], the total wave function Ψ is hence approximated by a projection of a single Slater determinant D :

$$\Psi \approx D_k = O_k D , \tag{31}$$

which is then determined by minimizing the energy:

$$\langle \mathcal{H} \rangle_k = \frac{\langle D_k | \mathcal{H} | D_k \rangle}{\langle D_k | D_k \rangle} = \frac{\langle D | \mathcal{H} O_k | D \rangle}{\langle D | O_k | D \rangle} , \tag{32}$$

which gives

$$\langle \delta D | (\mathcal{H} - \bar{\mathcal{H}}) O_k | D \rangle = 0 . \tag{33}$$

Since the correlation error is essentially associated with electrons having antiparallel spins, one could expect that one could lower the correlation energy by letting these electron avoid each other, e.g. by introducing "different orbitals for different spins" (DODS). This idea forms the basis for the alternant molecular--orbital (AMO) method[30], which has recently been applied also to the alkali metals by Calais and Sperber[31]. I understand that Calais is going to report the results of some calculations carried out by Sperber in San José during the summer in the discussions.

It is clear that there are many aspects on the correlation problem, and that the problem is particularly important from the point of view that it forces us to construct better and better wave functions. It is difficult to find even approximate solutions to the Schrodinger equation, and many new ideas may be needed. In this connection, it is highly important to have large-scale computers available for testing the new approaches, and we thank IBM for its development work.

References

1. J. C. Slater, Phys. Rev. 35, 509 (1930).
2. J. C. Slater, Phys. Rev. 82, 538 (1951); Rev. Mod. Phys. 25, 199 (1953).
3. E. Wigner, Phys. Rev. 46, 1002 (1934); Trans. Faraday Soc. 34, 678 (1938).
4. F. Bloch, Z. Phys. 52, 555; 57, 545 (1929).
5. P. O. Löwdin, J. Chem. Phys. 19, 1570, 1579 (1951).
6. P. O. Löwdin, Texas J. Science 8, 163 (1956).
7. A. Fröman, Phys. Rev. 112, 870 (1958).
8. J. Linderberg and H. Shull, J. Mol. Spec. 5, 1 (1960).
9. P. O. Löwdin, Adv. Chem. Phys. 2, 207 (Interscience, Ed. l. Prigogine, 1959).
10. E. Clementi, J. Chem. Phys. 38, 2244 (1963); 39, 175 (1963); 42, 2783 (1965).
E. Clementi and A. Veillard, J. Chem. Phys. 44, 3050 (1966);
A. Veillard and E. Clementi, J. Chem. Phys. 49, 2415 (1968);
E. Chementi, W. Kraemer, and G. Salez, J. Chem. Phys. 53, 125 (1970).
11. E. Clementi, Int. J. Quantum Chem. IIIS, 179 (1969).
12. K. A. Brueckner, C. A. Levinson, and H. M. Mahmoud, Phys. Rev. 95, 217 (1954); K. A. Brueckner, ibid. 96, 508 (1954); 97, 1353 (1955); 100, 36 (1955); K. A. Brueckner and C. A. Levinson, ibid. 97, 1344 (1955); H. A. Bethe, ibid. 103, 1353 (1956); J. Goldstone, Proc. Roy. Soc. (London) A239, 267 (1957); H. A. Bethe and J. Goldstone, ibid. A238, 511 (1957); L. S. Rodberg, Ann. Phys. (N.Y.) 2, 199 (1957); to mention only a selection of the rich literature on this subject.
13. J. C. Slater, Phys. Rev. 35, 210 (1930); V. Fock, Z. Physik 61, 126 (1930).
14. P. O. Löwdin, J. Math. Phys. 3, 1171 (1962).
15. P. O. Löwdin, J. Math. Phys. 6, 1341 (1965); J. Chem. Phys. 43, S175 (1965).
16. P. O. Löwdin, Phys. Rev. 139, A357 (1965), particularly p. A366.
17. O. Goscinski, Int. J. Quant. Chem. 1, 521 (1967); 2, 761 (1968); O. Goscinski and E. Brändas, Chem. Phys. Letters, 2, 299 (1968); Phys. Rev. 182, 43 (1969); Proc. Vilnius Int. Symposium on the Shell Strucutre of Atoms and Molecules (1969) (Editor A. Jucys) p. 358. Publishing House Mintis, Vilnius (1971); E. Brändas and O. Goscinski, J. Chem. Phys. 51,975 (1969); Int. J. Quant. Chem. 3S, 383 (1970); Phys. Rev. 1A, 552 (1970); Int. J. Quant. Chem. 4, 571 (1970);

G. L. Bendazzoli, O. Goscinski and G. Orlandi, Phys. Rev. A2, 2 (1970); O. Goscinski, Acta Univ. Uppsala, 162, (1970); O. Goscinski, Int. J. Quant. Chem. 5, 331 (1971); E. Brändas, Int. J. Quant. Chem. 4S, 285, (1971); E. Brändas and R. J. Bartlett, Chem. Phys. Letters, 8, 153 (1971).
18. K. Brueckner, Adv. Chem. Phys. XIV, 215 (Interscience, Ed. R. Lefebvre and C. Moser, 1969).
19. R. K. Nesbet, Adv. Chem. Phys. XIV, 1 (Interscience, Ed. R. Lefebvre and C. Moser, 1969).
20. O. Sinanoglu, Adv. Chem. Phys. XIV, 237 (Interscience, Ed. R. Lefebvre and C. Moser, 1969).
21. J. Gruninger, Y. Öhrn, and P.O. Löwdin, J. Chem. Phys. 52, 5551 (1970).
22. For a survey, see D. Pines, Solid-State Physics 1, 368 (Academic Press 1955).
23. O. Krisement, Phil.Mag. 2, 245 (1957).
24. J.C. Slater, Phys. Rev. 81, 385 (1951).
25. M. Delbrück, Proc. Roy. Soc. (London) A129, 686 (1930); C.C.J. Roothaan, Rev. Mod. Phys. 32, 179 (1960); P.O. Löwdin, J. Appl. Phys. Suppl 33, 251 (1962).
26. J.C. Slater, Rev. Mod. Phys. 6, 209 (1934); Phys. Rev. 81, 385 (1951); 82, 535 (1951).
27. P.O. Löwdin, Rev. Mod. Phys. 35, 496 (1963), particularly p. 498.
28. P.O. Löwdin, Phys. Rev. 97, 1509 (1955); Rev. Mod. Phys. 34, 80 (1962); 36, 966 (1964); 39, 259 (1967).
29. For a survey, see P.O. Löwdin, in Quantum Theory of Atoms, Molecules, and Solid State, 601 (Academic Press, Slater dedicatory volume, 1966).
30. P.O. Löwdin, Nikko Symp. Mol. Phys. (Symposium on molecular physics held at Nikko on the occasion of the Internat. Conf. on Theor. Physics. Sept. 1953 in Tokyo and Kyoto; Maruzen, Tokyo 1954) p., 13; Phys. Rev. 97, 1509 (1955); Proc. 10th Solvay Conf., 1954 p. 71, (Inst. internat. de physique Solvay, 10e conseil de physique tenu à Bruxelles 1954: Les électrons dans le métaux, Rapports et discussions, Bruxelles 1955); Rev. Mod. Phys. 32, 328 (1960). For a survey, see R. Pauncz, Alternant Molecular Orbital Method (W.B. Saunders, 1967).
31. J.L. Calais, Arkiv Fysik 28, 479, 511, 539 (1965); 29, 255 (1965).

EFFECTS OF ELECTRON CORRELATION WITH PARTICULAR REFERENCE TO CHARGE AND MOMENTUM DENSITIES IN CRYSTALS

N. H. March and J. C. Stoddart

Department of Physics, The University

Sheffield, England

I INTRODUCTION

The charge density $\rho(\underset{\sim}{r})$ and the momentum density $P(\underset{\sim}{p})$ of the electrons in a crystal can be measured by Bragg and Compton scattering of X-rays respectively.

We argue in this paper that the effect of electron interactions can be studied by using the measured density $\rho(\underset{\sim}{r})$ to construct a one-body potential $V_{Bragg}(\underset{\sim}{r}) \equiv V_B(\underset{\sim}{r})$, and a Dirac density matrix $\rho_B(\underset{\sim}{r}\,\underset{\sim}{r}')$ given by

$$\rho_B(\underset{\sim}{r}\,\underset{\sim}{r}') = \sum_i^N \psi_i^*(\underset{\sim}{r})\,\psi_i(\underset{\sim}{r}') \tag{1.1}$$

where the ψ_i's are the eigenfunctions of $V_B(\underset{\sim}{r})$ and N is the number of electrons in the crystal, such that $\rho_B(\underset{\sim}{r}\,\underset{\sim}{r}) = \rho(\underset{\sim}{r})$.

The fundamental connection between $\rho(\underset{\sim}{r})$ and $P(\underset{\sim}{p})$ is via the exact first-order density matrix $\gamma(\underset{\sim}{r}\,\underset{\sim}{r}')$ (see March, Young and Sampanthar (1)). We consider ρ_B given by (1.1) to be a good approximation to γ when we are near the diagonal, but to require refinement far from $\underset{\sim}{r} = \underset{\sim}{r}'$. We propose a local density approximation based on uniform electron gas theory to correct ρ_B. Tests can be made via the inequality $\gamma^2 < \gamma$, and by comparing the momentum density $P(\underset{\sim}{p})$ found from our approximation to γ with the Compton line shape. Using a parallel approximation, we treat the exchange and correlation hole round an electron in the inhomogeneous electron density in the crystal by correcting the single-particle pair function obtained from ρ_B. Normalization of the hole will give again a test of our approximation.

Though the exact calculation of ρ_B can be carried out by standard methods of band theory, to obtain γ, the pair function and the exchange and correlation energy density $\epsilon_{xc}[\rho]$, as functionals of the electron density, must be a major aim of this approach. It is shown that progress can be made using a gradient development of ρ_B, and combining this with dielectric response theory we obtain a consistent non-local approximation to the energy density $\epsilon_{xc}[\rho]$. The same gradient development is used to give a systematic way of constructing the momentum density $P_B(\underset{\sim}{p})$ from the diagonal part of the density matrix

II APPROXIMATION TO EXACT FIRST-ORDER DENSITY MATRIX

The momentum density $P(p)$, in the case of spherical symmetry, is related to the Compton line $J(q)$ with the conventional reduced variables, through

$$J(q) = \frac{1}{2}\int_q^\infty \frac{I(p)}{p}\,dp \tag{2.1}$$

where $I(p)dp$ is the probability of an electron having momentum of magnitude between p and $p + dp$. This is related to the exact momentum density through

$$I(p) = 4\pi p^2 P(p). \tag{2.2}$$

Thus, $P(p)$ is accessible to experiment by measuring the Compton profile $J(q)$.

We know that $P(p)$ can be got, quite generally, from the exact first-order density matrix through the relation

$$P(\underset{\sim}{p}) = \iint d\underset{\sim}{r}\, d\underset{\sim}{x}\; e^{i\underset{\sim}{p}\cdot\mathbf{x}}\, \gamma(\underset{\sim}{r}+\underset{\sim}{x},\underset{\sim}{r}). \tag{2.3}$$

Hence, valuable, if limited, information on γ can be obtained experimentally from Compton scattering; we shall discuss how this can be used to test our approximate methods below.

From (1.1), we see that $\rho_B(\underset{\sim}{r}\,\underset{\sim}{r}')\big|_{\underset{\sim}{r}'=\underset{\sim}{r}}$ is exactly equal to $\gamma(\underset{\sim}{r}\,\underset{\sim}{r}')\big|_{\underset{\sim}{r}'=\underset{\sim}{r}}$ and hence we expect ρ_B to be a good approximation to γ $\underset{\sim}{r}' = \underset{\sim}{r}$ near the diagonal. However, complete inclusion of electron interactions will affect the off-diagonal elements when $\underset{\sim}{r}' > \underset{\sim}{r}$ say and we must refine the approximation.

II.1 Example of Two-Electron Ion with Large Atomic Number

To make these remarks concrete, we shall briefly report results on a two-electron ion with large atomic number Z. The density obtained by Schwartz (2) is just the diagonal element of γ as given by Hall, Jones and Rees (3) which is

$$\gamma(\underset{\sim}{r}\,\underset{\sim}{r}') = \frac{2Z^3}{\pi}\exp\{-Z(r+r')\}\left[1+\frac{R(2Zr)}{2Z}+\frac{R(2Zr')}{2Z}\right] \quad (2.4)$$

where

$$R(x)=\frac{5x}{8}+\frac{1}{4}\left(3\ln 2-\frac{23}{4}\right)-\frac{1}{2}e^{-x}-\frac{3}{4x}\left(e^{-x}-1\right)+\frac{3}{4}\int_0^x dt\,\frac{e^{-t}-1}{t}\,. \quad (2.5)$$

From the density $\rho(\underset{\sim}{r})$ of Schwartz, we obtain the one-electron orbital simply as $\psi_B(\underset{\sim}{r}) = \{\rho(\underset{\sim}{r})/2\}^{\frac{1}{2}}$ giving

$$\rho_B(\underset{\sim}{r}\,\underset{\sim}{r}') = \frac{2Z^3}{\pi}\exp\{-Z(r+r')\}\left[1+\frac{R(2Zr)}{Z}\right]^{\frac{1}{2}}\left[1+\frac{R(2Zr')}{Z}\right]^{\frac{1}{2}} \quad (2.6)$$

The equations (2.4) and (2.6) are very similar if $R/Z \ll 1$, and in fact direct calculation shows that in regions of appreciable electron density for $Z \sim 10$ this is true. Thus, over a large region of space, $\rho_B(\underset{\sim}{r}\,\underset{\sim}{r}')$ is a very useful approximation to $\gamma(\underset{\sim}{r}\,\underset{\sim}{r}')$. In an Appendix, some results are also sketched for the potential $V_B(\underset{\sim}{r})$ in this system and in the H_2 molecule, and also for the momentum density.

II.2. Improved Form for First-Order Density Matrix

This encourages us to proceed to determine corrections, which we have argued above should be small except far from the diagonal, to convert $\rho_B(\underset{\sim}{r}\,\underset{\sim}{r}')$ into a good approximation to $\gamma(\underset{\sim}{r}\,\underset{\sim}{r}')$. To do this in a crystal, and in particular in a simple metal, we shall argue that we have built in the electronic band structure fully through $\rho_B(\underset{\sim}{r}\,\underset{\sim}{r}')$ and some account of correlations through the use of the exact density $\rho(\underset{\sim}{r})$ on its diagonal. We now argue that we can take over effects of electron interactions in a uniform gas to the inhomogeneous case by writing

$$\gamma(\underset{\sim}{r}\,\underset{\sim}{r}') = \rho_B(\underset{\sim}{r}\,\underset{\sim}{r}') + \left[\gamma_0(\underset{\sim}{r}\,\underset{\sim}{r}'\,k_f) - \rho_0(\underset{\sim}{r}\,\underset{\sim}{r}'\,k_f)\right]\Big|_{k_f=\left(3\pi^2\rho\left(\frac{\underset{\sim}{r}+\underset{\sim}{r}'}{2}\right)\right)^{\frac{1}{3}}} \quad (2.7)$$

Here

$$\rho_0(\underset{\sim}{r}\,\underset{\sim}{r}'\,k_f) = \frac{k_f^3}{\pi^2}\,\frac{j_1(k_f|\underset{\sim}{r}-\underset{\sim}{r}'|)}{k_f|\underset{\sim}{r}-\underset{\sim}{r}'|} \;:\; j_1(x)=\frac{\sin x - x\cos x}{x^2} \quad (2.8)$$

with

$$\gamma_0(\underset{\sim}{r}\,\underset{\sim}{r}'\,k_f) = \frac{k_f^3}{\pi^2}\int P_0(k,k_f)\,e^{i\underset{\sim}{k}\cdot(\underset{\sim}{r}-\underset{\sim}{r}')}\,d\underset{\sim}{k}\,. \tag{2.9}$$

Various workers have calculated the momentum distribution $P_o(k, k_f)$ in a uniform electron gas (see reference (1))so that we can regard γ_o as known to good accuracy.

We emphasize that, in a given crystal, we can check the accuracy of (2.7) in at least two ways:

(i) γ^2 must be less than γ, otherwise a small numerical adjustment should be make, to preserve intact the Pauli conditions on the first order density matrix.

(ii) The integral $\int\left\{\left[\gamma_0(\underset{\sim}{r}\,\underset{\sim}{r}'\,k_f) - \rho_0(\underset{\sim}{r}\,\underset{\sim}{r}'\,k_f)\right]\Big|_{k_f=(3\pi^2\rho(\frac{\underset{\sim}{r}+\underset{\sim}{r}'}{2}))^{\frac{1}{3}}}\right\}e^{i\underset{\sim}{p}\cdot\underset{\sim}{r}-\underset{\sim}{r}'}\,d\underset{\sim}{r}\,d\underset{\sim}{r}'$ can be computed, and compared with $P(\underset{\sim}{p}) - P_B(\underset{\sim}{p})$, with $P(\underset{\sim}{p})$ taken from Compton line measurements.

Having discussed an approximation to the exact first-order density matrix which relates $\rho(\underset{\sim}{r})$ and $P(\underset{\sim}{p})$, and which determines the correlation kinetic energy in the many-body problem, we turn to consider the exchange and correlation hole around an electron in an inhomogeneous electron gas.

III EXCHANGE AND CORRELATION HOLE

We define first the second-order density matrix $\Gamma(\underset{\sim}{r}_1\,\underset{\sim}{r}_2\,\underset{\sim}{r}'_1\,\underset{\sim}{r}'_2)$ in terms of the many-body wave function Ψ by,

$$\Gamma(\underset{\sim}{r}_1\,\underset{\sim}{r}_2\,\underset{\sim}{r}_1'\,\underset{\sim}{r}_2') = \int \Psi^*(\underset{\sim}{r}_1\,\underset{\sim}{r}_2\,\underset{\sim}{r}_3\cdots\underset{\sim}{r}_N)\,\Psi(\underset{\sim}{r}_1'\,\underset{\sim}{r}_2'\,\underset{\sim}{r}_3\cdots\underset{\sim}{r}_N)\,d\underset{\sim}{r}_3\cdots d\underset{\sim}{r}_N\,. \tag{3.1}$$

We shall focus attention on the pair function $\Gamma(\underset{\sim}{r}_1\,\underset{\sim}{r}_2\,\underset{\sim}{r}_1\,\underset{\sim}{r}_2) \equiv \Gamma(\underset{\sim}{r}_1\,\underset{\sim}{r}_2)$ using an approximation which parallels that for $\gamma(\underset{\sim}{r}\,\underset{\sim}{r}')$. Thus, we use $\rho_B(\underset{\sim}{r}_1\,\underset{\sim}{r}_2)$ to calculate $\Gamma_B(\underset{\sim}{r}_1\,\underset{\sim}{r}_2)$, which is given as usual in a single-particle theory by

$$\frac{\Gamma_B(\underset{\sim}{r}_1\,\underset{\sim}{r}_2)}{\rho(\underset{\sim}{r}_1)} = \rho(\underset{\sim}{r}_2) - \frac{1}{2}\frac{\{\rho_B(\underset{\sim}{r}_1\,\underset{\sim}{r}_2)\}^2}{\rho(\underset{\sim}{r}_1)}\,. \tag{3.2}$$

The first term is the exact electron density at the position $\underset{\sim}{r}_2$ of the electron while the second term represents the hole surrounding it. It is clear from (3.2) and the idempotency of ρ_B that this localized charge density normalizes precisely to unity. However, (3.2) does not account correctly for the Coulomb repulsions between antiparallel spin electrons.

We therefore attempt to correct $\Gamma_B(\underset{\sim}{r}_1\,\underset{\sim}{r}_2)/\rho(\underset{\sim}{r}_1)$ using a similar philosophy to that adopted for $\gamma(\underset{\sim}{r}\,\underset{\sim}{r}')$. In particular we write

$$\frac{\Gamma(\underset{\sim}{r_1}\underset{\sim}{r_2})}{\rho(\underset{\sim}{r_1})} - \frac{\Gamma_B(\underset{\sim}{r_1}\underset{\sim}{r_2})}{\rho(\underset{\sim}{r_1})} = \left\{\frac{k_f^3}{3\pi^2}\left[g_c^{\,c}(|\underset{\sim}{r_1}-\underset{\sim}{r_2}|,k_f) - g_c(|\underset{\sim}{r_1}-\underset{\sim}{r_2}|,k_f)\right]\right\}_{k_f=\left\{3\pi^2\rho\left(\frac{\underset{\sim}{r_1}+\underset{\sim}{r_2}}{2}\right)\right\}^{\frac{1}{3}}} \qquad (3.3)$$

where g_o is the usual Fermi hole

$$g_c(r) = 1 - \frac{9}{2}\left(\frac{j_1(k_f r)}{k_f r}\right)^2 \qquad (3.4)$$

while $g_c^{\,c}$ includes correlations between antiparallel spins in a uniform electron gas and is taken from the work of Singwi et al (4).

Again we stress that a check of the normalization of the exchange and correlation hole, namely

$$\int \frac{\Gamma(\underset{\sim}{r_1}\underset{\sim}{r_2}) - \Gamma_B(\underset{\sim}{r_1}\underset{\sim}{r_2})}{\rho(\underset{\sim}{r_1})}\, d\underset{\sim}{r_2} = 0 \qquad (3.5)$$

should be made, and again small adjustments made to (3.3) to satisfy (3.5).

In this way, for a given density $\rho(\underset{\sim}{r})$, the correlation potential energy can be calculated by direct integration.

IV DENSITY MATRIX ρ_B AS FUNCTIONAL OF $\rho(\underset{\sim}{r})$

However, while the calculation of $\rho_B(\underset{\sim}{r}\,\underset{\sim}{r}')$ can be carried out by established band theory methods, an important development of the theory outlined here would be to get γ and the pair function as functionals of $\rho(\underset{\sim}{r})$. The only way we can do this analytically at present is via a gradient series. We will sketch below the main results for ρ_B, and some consequences of these.

The work of March and Murray (5) and Stoddart and March (6), (see also references (7) and (8)) showed that in single-particle theory the Dirac matrix ρ_B is a unique functional of the diagonal component $\rho(\underset{\sim}{r})$. The most useful result obtained so far is the density gradient expansion for ρ_B, and the first few terms were given by Stoddart, Beattie and March (7) as

$$\begin{aligned}\rho_B(\underset{\sim}{r}+\underset{\sim}{x},\underset{\sim}{r},E_f) &= \rho_o(\underset{\sim}{x},B)\\ &+\left[\frac{1}{72}\int_0^1 da\,\rho^{-2}(\underset{\sim}{r}+a\underset{\sim}{x})\left(\nabla_{\underset{\sim}{r}}\rho(\underset{\sim}{r}+a\underset{\sim}{x})\right)^2 - \frac{1}{36}\int_0^1 da\,\rho^{-1}(\underset{\sim}{r}+a\underset{\sim}{x})\,\nabla_{\underset{\sim}{r}}^2\rho(\underset{\sim}{r}+a\underset{\sim}{x})\right]\rho_o'(\underset{\sim}{x},B)\\ &+F_1(\underset{\sim}{r},\underset{\sim}{x})\,\rho_o''(\underset{\sim}{x}\,B) + F_2(\underset{\sim}{r},\underset{\sim}{x})\,\rho_o'''(\underset{\sim}{x},B) + \cdots\end{aligned} \qquad (4.1)$$

with

$$F_1(\underset{\sim}{r}, \underset{\sim}{x}) = -\int_0^1 da\, \frac{a(1-a)}{2} \frac{(3\pi^2)^{\frac{2}{3}}}{2} \left[\frac{2}{9} \rho^{-\frac{4}{3}}(\underset{\sim}{r}+a\underset{\sim}{x})(\nabla_{\underset{\sim}{r}} \rho(\underset{\sim}{r}+a\underset{\sim}{x}))^2 - \frac{2}{3}\rho^{-\frac{1}{3}}(\underset{\sim}{r}+a\underset{\sim}{x}) \nabla^2_{\underset{\sim}{r}} \rho(\underset{\sim}{r}+a\underset{\sim}{x})\right] \quad (4.2)$$

$$F_2(\underset{\sim}{r}, \underset{\sim}{x}) = \frac{(3\pi^2)^{\frac{4}{3}}}{8} \int_0^1 da\, a^2 \int_0^1 db_1 \int_0^1 db_2\, b_1 b_2\, \frac{4}{9}\, \rho^{-\frac{1}{3}}(\underset{\sim}{r}+ab_1\underset{\sim}{x})\, \rho^{-\frac{1}{3}}(\underset{\sim}{r}+ab_2\underset{\sim}{x}) \times (\nabla_{\underset{\sim}{r}} \rho(\underset{\sim}{r}+ab_1\underset{\sim}{x})).(\nabla_{\underset{\sim}{r}} \rho(\underset{\sim}{r}+ab_2\underset{\sim}{x})) . \quad (4.3)$$

Here

$$\mathcal{B} = \frac{(3\pi^2)^{\frac{2}{3}}}{2} \int_0^1 da\, \rho^{\frac{2}{3}}(\underset{\sim}{r}+a\underset{\sim}{x}) \quad (4.4)$$

and

$$\rho_0(x, E) = \frac{(2E)^{\frac{3}{2}}}{\pi^2} \left[\frac{j_1((2E)^{\frac{1}{2}} x)}{(2E)^{\frac{1}{2}} x} \right]. \quad (4.5)$$

Thus we have directly from (2.7) and (4.1) the first terms in a gradient development in $\rho(\underset{\sim}{r})$ for the first-order density matrix.

In a one-body theory we write the exchange energy in terms of ρ_B as

$$E_x = -\frac{1}{4} \iint d\underset{\sim}{r}\, d\underset{\sim}{x}\, \frac{[\rho_B(\underset{\sim}{r}+\underset{\sim}{x}, \underset{\sim}{r}, E_f)]^2}{x} \quad (4.6)$$

and substitution of (4.1) into (4.6) gives a density gradient series for the exchange energy. This was discussed by Stoddart, Beattie and March (7) where it was shown that the first two terms in a perturbation expansion for E_x could be used to sum an important subset of terms in this gradient series. The result has the form

$$E_x = -\frac{3}{4}\left(\frac{3}{\pi}\right)^{\frac{1}{3}} \int d\underset{\sim}{r}\, \rho^{\frac{4}{3}}(\underset{\sim}{r}) - \frac{1}{2} \iint d\underset{\sim}{r}\, d\underset{\sim}{r}'\, B_x(\underset{\sim}{r}', \rho(\underset{\sim}{r})) \left[\rho(\underset{\sim}{r}+\tfrac{1}{2}\underset{\sim}{r}') - \rho(\underset{\sim}{r}-\tfrac{1}{2}\underset{\sim}{r}')\right]^2 \quad (4.7)$$

where the kernel B_x is given explicitly in reference (7).

We believe that a form like (4.7) will be useful both for the analysis of the exchange energy, and as we discuss below, the correlation energy. As yet unfortunately, no numerical application of (4.7) has been carried through.

As ρ_B is a unique functional of $\rho(\underset{\sim}{r})$ we see from

$$P_B(q) = \int d\underset{\sim}{r}\, d\underset{\sim}{x}\, e^{i\underset{\sim}{q}\cdot\underset{\sim}{r}} \rho_B(\underset{\sim}{r}+\underset{\sim}{x}, \underset{\sim}{r}) \tag{4.8}$$

that $P_B(q)$ is also a unique functional of the exact charge density.

We may use (4.1) to obtain $P_B(q)$ directly as a density gradient series. If we neglect all density gradients, using only the first term in (4.1), together with $B \sim 0.5\,(3\pi^2\rho)^{\frac{2}{3}}$ then we obtain the usual semi-classical relation between $P_B(q)$ and a spherical $\rho(r)$, namely

$$P_B(q) = 4\pi\int_0^\infty dr\, r^2\, \Theta\left(\frac{(3\pi^2)^{\frac{2}{3}}}{2}\rho^{\frac{2}{3}}(r) - \frac{q^2}{2}\right). \tag{4.9}$$

Obviously correction terms to (4.9) can be constructed from (4.1). However these obviously become rapidly more complicated to handle. This can be illustrated by a calculation of the correction terms to the semi-classical result (4.9). We assume that the coefficients of ρ_0', ρ_0'' and ρ_0''' in (4.1) do not depend strongly on $\underset{\sim}{x}$, and take their values at $\underset{\sim}{x} = 0$. We then obtain to $O((\nabla\rho)^2)$,

$$\begin{aligned} P_B(q) &= 4\pi\int_0^\infty dr\, r^2\, \Theta\left(B(r) - \frac{q^2}{2}\right) \\ &- \frac{4\pi}{q}\frac{\partial}{\partial q}\int_0^\infty dr\, r^2\left[\frac{1}{72}\rho^{-2}(\underset{\sim}{r})(\nabla_{\underset{\sim}{r}}\rho)^2 - \frac{1}{36}\rho^{-1}(\underset{\sim}{r})\nabla_{\underset{\sim}{r}}^2\rho\right]\Theta\left(B(r) - \frac{q^2}{2}\right) \\ &- \frac{4\pi}{q^2}\frac{\partial^2}{\partial q^2}\int_0^\infty dr\, r^2\,\frac{(3\pi^2)^{\frac{2}{3}}}{24}\left[\frac{2}{9}\rho^{-\frac{4}{3}}(\underset{\sim}{r})(\nabla_{\underset{\sim}{r}}\rho)^2 - \frac{2}{3}\rho^{-\frac{1}{3}}\nabla_{\underset{\sim}{r}}^2\rho\right]\Theta\left(B(r) - \frac{q^2}{2}\right) \\ &- \frac{12\pi}{q^5}\frac{\partial}{\partial q}\int_0^\infty dr\, r^2\,\frac{(3\pi^2)^{\frac{4}{3}}}{216}\rho^{-\frac{2}{3}}(r)(\nabla_{\underset{\sim}{r}}\rho)^2\,\Theta\left(B(r) - \frac{q^2}{2}\right) \\ &+ \frac{12\pi}{q^4}\frac{\partial^2}{\partial q^2}\int_0^\infty dr\, r^2\,\frac{(3\pi^2)^{\frac{4}{3}}}{216}\rho^{-\frac{2}{3}}(r)(\nabla_{\underset{\sim}{r}}\rho)^2\,\Theta\left(B(r) - \frac{q^2}{2}\right) \\ &- \frac{4\pi}{q^2}\frac{\partial^3}{\partial q^3}\int_0^\infty dr\, r^2\,\frac{(3\pi^2)^{\frac{4}{3}}}{216}\rho^{-\frac{2}{3}}(r)(\nabla_{\underset{\sim}{r}}\rho)^2\,\Theta\left(B(r) - \frac{q^2}{2}\right). \end{aligned} \tag{4.10}$$

The form of the gradient development (4.1) suggests that a more practical way of improving (4.9) might be effected by putting

$$\rho(\underset{\sim}{r}+\underset{\sim}{x}, \underset{\sim}{r}, E_f) = \rho_0(\underset{\sim}{x}, E_f - \Gamma(\underset{\sim}{r}, \underset{\sim}{x}, E_f)). \tag{4.11}$$

In the semi-classical theory we have

$$\Gamma_{sc}(\underset{\sim}{r}\,\underset{\sim}{x}\,E_f) = E_f - \frac{(3\pi^2\rho(\underset{\sim}{r}))^{\frac{2}{3}}}{2} \tag{4.12}$$

but this theory gets Γ correct only at $\underset{\sim}{x} = 0$, where of course (4.11) is an identity. A refined theory, which at the very least sums a subset of gradient terms is obtained from the first term on the right-hand side of (4.1). This gives

$$\Gamma(\underset{\sim}{r}\,\underset{\sim}{x}\,E_f) = E_f - \frac{(3\pi^2)^{\frac{2}{3}}}{2}\int_0^1 d\alpha\, \rho^{\frac{2}{3}}(\underset{\sim}{r}+\alpha\underset{\sim}{x}). \tag{4.13}$$

We may thus through (4.5) and (4.13) relate $P(\underset{\sim}{q})$ and $\rho(\underset{\sim}{r})$.

V EXCHANGE AND CORRELATION ENERGY DENSITY

Finally we note that we can obtain a generalization of the one-body result (4.7), to give the exchange and correlation energy E_{xc} as a functional of $\rho(\underset{\sim}{r})$.

In the inhomogeneous electron system we suppose that E_{xc} can be written as a series in the displaced charge, $\Delta(\underset{\sim}{r}) = \rho(\underset{\sim}{r}) - \rho_0$ where ρ_0 is the density in the homogeneous gas. To second order we write

$$E_{xc} = E_{xc}^{c}(\rho_0) + \iint d\underset{\sim}{r}\, d\underset{\sim}{s}\, B_{xc}(\underset{\sim}{r}-\underset{\sim}{s}, k_f)\,\Delta(\underset{\sim}{r})\Delta(\underset{\sim}{s}) + O(\Delta^3) \tag{5.1}$$

where $E_{xc}^{c}(\rho_0)$ is the exchange and correlation energy in a homogeneous gas of density ρ_0.

It may be shown (see reference (7), appendix 2), that the kernel B_{xc} is related to the static dielectric function $\epsilon(q)$ by

$$B_{xc}(\underset{\sim}{r}, k_f) = \frac{1}{(2\pi)^3}\int d\underset{\sim}{q}\, e^{i\underset{\sim}{q}\cdot\underset{\sim}{r}}\, \tilde{B}_{xc}(\underset{\sim}{q}, k_f) \tag{5.2}$$

with

$$\tilde{B}_{xc}(\underset{\sim}{q}, k_f) = \frac{1}{4}K(\underset{\sim}{q}, k_f) - \frac{2\pi}{q^2(1-\epsilon(q))} \tag{5.3}$$

and

$$K(x) \equiv K\left(\frac{q}{2k_F}\right) = \left[\frac{1}{2} + \frac{1}{4}\,\frac{(1-x^2)}{x}\,\ln\left|\frac{1+x}{1-x}\right|\right]. \tag{5.4}$$

Equations (5.1) - (5.4) may be used to sum an infinite subset of terms in the gradient expansion for E_{xc}, giving an expression which is correct in both the slowly varying density limit, and to $O(\Delta^3)$ in a perturbation series. The method has been discussed in reference (7) and takes the same form as (4.7), namely

$$E_{xc}[\rho] = E_{xc}^{c}[\rho] - \frac{1}{2}\iint d\underset{\sim}{r}\,d\underset{\sim}{r}'\,B_{xc}(\underset{\sim}{r}', \rho(\underset{\sim}{r}))\left[\rho(\underset{\sim}{r}+\tfrac{1}{2}\underset{\sim}{r}') - \rho(\underset{\sim}{r}-\tfrac{1}{2}\underset{\sim}{r}')\right]^2 \tag{5.5}$$

where

$$B_{xc}(\underset{\sim}{r}', \rho(\underset{\sim}{r})) \equiv B_{xc}(\underset{\sim}{r}', k_f)\Big|_{k_f = (3\pi^2\rho(\underset{\sim}{r}))^{\frac{1}{3}}} \tag{5.6}$$

Thus, provided we know $\epsilon(q,k_f)$ to insert in (5.3), we may sum the dominant terms in the gradient series for the energy E_{xc} and the corresponding one-body potential defined by Kohn and Sham (10). Of course, the exact $\epsilon(q, k_f)$ is still not known and approximate forms must be used (for example, Singwi et al, (4)). A particularly simple model has been developed by Overhauser (9) in an investigation of correlation effects in metals.

Defining $Q(x)$ by

$$\left[\frac{1}{2} + \frac{1}{4}\,\frac{(1-x^2)}{x}\,\ln\left|\frac{1+x}{1-x}\right|\right] = \frac{\pi q^2}{4k_f}\,Q(x) \tag{5.7}$$

his form for the dielectric function is

$$\epsilon(q) = 1 + \frac{Q(q/2k_f)}{1 - G(q/2k_f)\,Q(q/2k_f)} \tag{5.8}$$

where $G(x)$ contains the dependence of $\epsilon(q)$ on exchange and correlation effects. In Overhauser's simplified model $G(x)$ takes the form

$$G(x) = \frac{1\cdot 1\, x^2}{(1 + 10x^2 + 1\cdot 5\, x^4)^{\frac{1}{2}}} \tag{5.9}$$

which is correct at both large and small x and interpolates reasonably well between these limits. Thus, through eqns. (5.2) to (5.9) we have a method for constructing a simplified model for the exchange and correlation energy functionals and the corresponding one-body potential. It would seem to be of considerable interest to carry out a full self-consistent calculation on this basis for some simple metal, though major computational facilities will, of course, be required.

APPENDIX

TWO-ELECTRON SYSTEMS WITH ELECTRON-ELECTRON INTERACTIONS

Although we are primarily concerned with crystals, the following two-electron systems are of interest, as we are able to illustrate directly in these the one-body potential V_B discussed in the text.

(A.1) Two-electron ion with large atomic number

The electron density for this system has been given by Schwartz (2), who obtained

$$\rho(r) = \frac{2Z^3}{\pi} \exp(-2Zr)\, G(r) \qquad (A.1)$$

with,

$$G(r) = 1 + \frac{R(2Zr)}{Z} \qquad (A.2)$$

and with $R(2Zr)$ given by (2.5).

As both electrons can be in the same orbital $\psi(r)$ then we may write

$$\psi(r) = \left\{\frac{\rho(r)}{2}\right\}^{\frac{1}{2}} \qquad (A.3)$$

and we can obtain immediately from the Schroedinger equation the one-body potential $V_B(r)$ which will reproduce the exact charge density. (For the general theory of one-body potential theory including electron interactions, see references (7) and (10)). Thus we have

$$V_B(r) = E + \frac{1}{2\psi(r)} \cdot \nabla_r^2 \psi(r) \qquad (A.4)$$

and with $E \sim -Z^2/2$ determined by the condition $V_B(r) \to 0$ as $r \to \infty$.

We obtain using (A.1) to (A.4)

$$\Delta V_B(r) = V_B(r) + \frac{Z}{r} = \frac{1}{2}\left(\frac{1}{r} - Z\right)\frac{G'(r)}{G(r)} + \frac{G''}{4G} - \frac{1}{8}\left(\frac{G'}{G}\right)^2 \qquad \text{(A.5)}$$

where $G' = \partial G/\partial r$ etc.

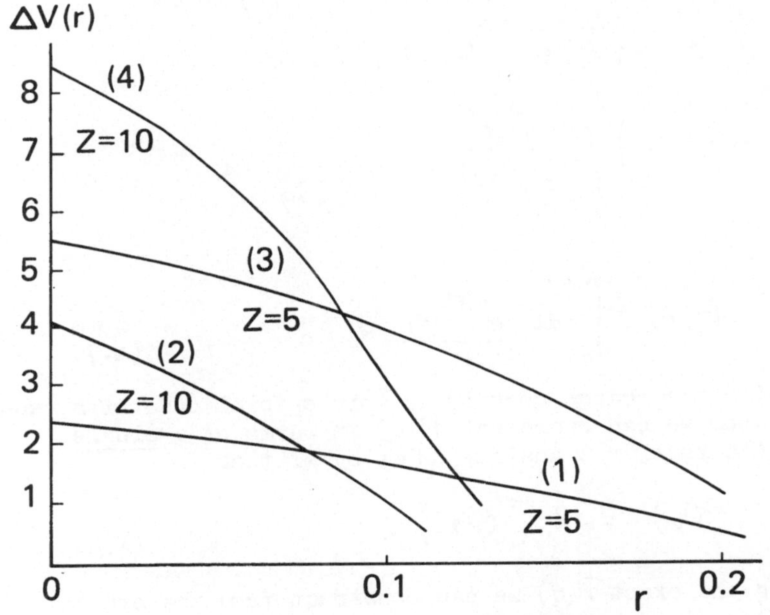

The corrections V to the ionic potential $-Z/r$. Cruves (1) and (2) are ΔV_B defined by (A.5) for $Z = 5$ and 10 respectively. Curves (3) and (4) are ΔV_c defined by (A.9) for $Z = 5$ and 10 respectively.

In the Figure we have plotted $V_B(r)$ from (A.5) for two values of Z. For the range of r in which the electron density is appreciable, ΔV_B, representing the effect of electron-electron interactions on the one-body potential, is small, being always $< 2\%$ of Z/r.

For the two-electron atom, the first-order density matrix has been obtained by Hall, Jones and Rees (3), and is given by (2.4). From (2.4) and (2.3) we obtain the momentum density $\mathcal{P}(q)$ as

$$P(q) = \frac{8}{\pi}\frac{(4\pi)^2 Z^5}{(Z^2+q^2)^4} + \frac{8(4\pi)^2 Z^4}{\pi(Z^2+q^2)^2}\left[\frac{5Z^2}{(Z^2+q^2)^3} - \frac{3}{2(Z^2+q^2)^2} - \frac{3}{2(9Z^2+q^2)^2} - \frac{3}{16Z^2(9Z^2+q^2)} + \frac{3}{16Z^2(Z^2+q^2)} + \frac{3}{8Zq}\int_0^\infty dr\, r \sin qr\, e^{-Zr} E(r)\right] - \left(\frac{19}{4} - 3\ln 2\right)\frac{(4\pi)^2 Z^4}{\pi(Z^2+q^2)^4} \tag{A.6}$$

where

$$E(r) = \int_0^{Zr} dt\, \frac{e^{-2t}-1}{t}. \tag{A.7}$$

As with the charge density in this special case of a two-electron atom we can represent $P(q)$ in terms of a single orbital with Fourier transform $\phi(q)$ by writing

$$P(q) = 2\,(\phi(q))^2. \tag{A.8}$$

Using the exact $P(q)$ we can construct from the orbital and the Schrodinger equation a 'Compton potential' $V_c(r)$ as in (A.4) which will reproduce $P(q)$ through a one-body equation. For $2Zr < 1$ we obtain for the leading terms

$$\Delta V_c(r) = V_c(r) + \frac{Z}{r} = \frac{9Z}{8}\left(1+\frac{0{\cdot}67}{Z}\right)^{\frac{1}{2}} - \frac{3Z^2 r}{4}\left(1+\frac{0{\cdot}67}{Z}\right)^{-\frac{1}{2}}. \tag{A.9}$$

In the same limit, from (A.5) we obtain

$$\Delta V_B(r) = V_B(r) + \frac{Z}{r} = \frac{3Z}{8}\left(1+\frac{0{\cdot}67}{Z}\right) - \frac{Z^2 r}{8}\left(1+\frac{0{\cdot}67}{Z}\right) + \cdots \tag{A.10}$$

The correction terms ΔV_c and ΔV_B are not the same, and the difference between them is evidently directly reflecting the deviations from independent-particle behaviour. In the Figure we illustrate this point by plotting ΔV_B and ΔV_c for two values of Z.

The two-electron atom however is a very special case in which we can represent the charge and momentum densities using just one orbital in each case. When we deal with more than one orbital, the situation is more complex, and we have not been able to prove that there is a 'Compton potential' V_c which will reproduce the exact momentum density in the general case.

(A.2) Hydrogen Molecule in Heitler-London Limit

As the simplest example involving binding of atoms, we shall finally consider the hydrogen molecule in the strong-coupling limit in which the separation R between nuclei A and B becomes very large. We are then in the Heitler-London regime and a good approximation to the spatial part of the ground-state wave function is given by

$$\Psi(\underset{\sim}{r_1}\,\underset{\sim}{r_2}) = N\left[\psi_A(1)\psi_B(2) + \psi_A(2)\psi_B(1)\right]. \qquad \text{(A.11)}$$

The normalized electron density is then given by

$$\rho(\underset{\sim}{r_1}) = \int |\Psi(\underset{\sim}{r_1}\,\underset{\sim}{r_2})|^2 d\underset{\sim}{r_2}$$
$$= \frac{1}{1+S^2}\left[\psi_A^2(1) + 2S\psi_A(1)\psi_B(1) + \psi_B^2(1)\right] \qquad \text{(A.12)}$$

where S is the overlap integral

$$S = \int d\underset{\sim}{r_2}\, \psi_A(2)\psi_B(2). \qquad \text{(A.13)}$$

At sufficiently large R we may choose

$$\psi_A(1) = \frac{1}{\pi^{\frac{1}{2}} a_0^{\frac{3}{2}}} e^{-r_{1A}/a_0} \qquad \text{(A.14)}$$

with a similar choice for

Now we follow exactly the procedure of (A.1). We construct a one-body (Bragg) orbital $\psi_B(\underset{\sim}{r})$ such that

$$\psi_B(\underset{\sim}{r}) = \left\{\frac{\rho(\underset{\sim}{r})}{2}\right\}^{\frac{1}{2}} \qquad \text{(A.15)}$$

with $\rho(\underset{\sim}{r})$ given by eqn. (A.12). The corresponding one-body potential is again obtained from the Schrodinger equation, as in equation (A.4). This one-body potential can thus be calculated for large R. To illustrate briefly the results, we have considered its form along the internuclear axis for r << R with r the distance from nucleus A assuming exp (-R) is small. We find for the leading term

$$V(r) = \frac{1}{a_o r} + \frac{S}{a_o}\left(\frac{1}{r} - \frac{1}{R}\right)\exp\left(-\frac{R}{a_o} + \frac{2r}{a_o}\right) \qquad (A.16)$$

where the overlap integral S is

$$S = e^{-R/a_o}\left(1 + \frac{R}{a_o} + \frac{1}{3}\frac{R^2}{a_o^2}\right) \qquad (A.17)$$

Again for large R, the corrections to the ionic potential are small and repulsive.

REFERENCES

(1) March N. H., Young W. H. and Sampanthar S. 1967, The Many-Body Problem in Quantum Mechanics, Cambridge University Press.

(2) Schwartz C., 1959, Ann. Phys. 2, 156.

(3) Hall G. G., Jones L. L. and Rees D., 1965, Proc. Roy. Soc., A283, 194.

(4) Singwi K. S., Sjölander A., Tosi M.P. and Land R. H., 1970, Phys. Rev. B1, 1044.

(5) March N. H. and Murray A. M., 1961, Proc. Roy. Soc. A261, 119.

(6) Stoddart J. C. and March N. H., 1967, Proc. Roy. Soc. A299, 279.

(7) Stoddart J. C., Beattie A. M. and March H. H., 1971, Inter. J. Quantum Chem. 4, 35.

(8) Beattie A. M., Stoddart J. C. and March N. H., 1972, Proc. Roy. Soc. A (in the press).

(9) Overhauser A. W., 1971, Phys. Rev. B3, 1888.

(10) Kohn W. and Sham L. J., 1965, Phys. Rev. 140 A1133.

LOCAL EXCHANGE-CORRELATION POTENTIALS

B.I. Lundqvist and S. Lundqvist

Institute of Theoretical Physics
Chalmers University of Technology.
S-402 20 Göteborg 5, Sweden

1. INTRODUCTION

This review will summarize some recent discussions about the problem how to choose a local potential to approximate the effects of exchange and correlation in electron structure problems. We shall limit ourselves to simple aspects of the problem and emphasize explicit results which may be directly useful for electron structure calculations.

In principle, we have just to study the electron self-energy Σ to obtain the exchange and correlation effects on the ground state energy and charge density as well as on the excitation spectrum. However, the theory for ground state properties developed by Hohenberg, Kohn and Sham /1, 2/ has considerable advantages, particularly for the calculation of charge densities. We shall accordingly discuss two types of local potentials, one for the ground state properties based on the Hohenberg-Kohn-Sham theory and one for excitations based on the electron self-energy Σ.

2. A LOCAL POTENTIAL FOR GROUND STATE PROPERTIES

On the basis of the general results and theorems by Hohenberg and Kohn /1/, Kohn and Sham /2/ derived a one-body equation

$$\left[-\hbar^2\nabla^2/(2m)+ V(\underline{r}) + v_{xc}(\underline{r})\right] \psi_i(\underline{r}) = \varepsilon_i \psi_i(\underline{r}). \qquad (1)$$

$V(\underline{r})$ is the ordinary Hartree potential and v_{xc} gives the exchange and correlation contributions. The electron density ρ is obtained by the usual independent particle expression, i.e. by summing over the set of occupied states,

$$\rho(\underline{r}) = \sum_{i,\text{occ.}} |\psi_i(\underline{r})|^2. \tag{2}$$

Equations (1) and (2) should be solved self-consistently using the proper functional dependence of v_{xc} on ρ. The exact form of this functional is not known. However, in the limit of a weakly varying density, v_{xc} is simply a function of ρ,

$$v_{xc}(\underline{r}) = \mu_{xc}\,(\rho(\underline{r})), \tag{3}$$

where $\mu_{xc}(\rho)$ is the exchange-correlation part of the chemical potential for a uniform electron gas at density ρ. Attempts to improve on this approximation by including terms depending on the density gradient have been done (see e.g. Refs. /3/ and /4/), but we shall not discuss such corrections here.

In most applications using this scheme one has considered only exchange and replaced μ_{xc} by μ_x,

$$\mu_x = -e^2k_F/\pi = -e^2/\pi\ (3\pi^2\rho)^{1/3}, \tag{4}$$

sometimes called the Kohn-Sham potential. This approximation is unnecessarily crude and one can make use of results for the correlation energy of a uniform electron gas to obtain a considerably more accurate result. Although a completely satisfactory theory of an electron gas in the intermediate density regime does not exist, the various results and interpolation schemes agree closely enough with each other and with general criteria to determine the quantity μ_{xc} within 0.005-0.010 Ry (/5/ and /6/), which is a small uncertainty compared to the difference between μ_{xc} and μ_x. It is convenient to introduce a parameter β and use the well-known electron gas parameter $r_s \sim \rho^{1/3}$, i.e. $\mu_{xc} = \beta(r_s)\mu_x$.

In References /5/ and /6/ the parameter β has been determined from the recent results by Singwi et al. /7/ (SSTL) for the correlation energy per electron ε_c, using the relation

$$\mu_c = \varepsilon_c - r_s/3\ (d\varepsilon_c/dr_s). \tag{5}$$

Recently Vashishta and Singwi /8/ have presented an improved calculation, which almost exactly satisfies the compressibility sum rule and at the same time gives a reasonably good pair correlation function. Their results for the correlation energy are quite close to those obtained earlier by Lundqvist and Samathiyakanit /9/ from the spectral weight function using the Galitskii-Migdal formula /10/ for the total energy. A good fit to the results by Vashishta and Singwi corresponds to the formula

$$\beta(r_s) = 1 + 0.0316\ r_s \ln \frac{r_s + 24.3}{r_s}. \tag{6}$$

The results for the parameter β are illustrated in Fig. 1, using both the previous results by Singwi et al. /7/ (upper curve) and the most recent ones by Vashishta and Singwi /8/ (lower curve). At high densities the effect of correlation is small, while at metallic densities (r_s = 2-5) there is an appreciable effect and the values of the potential are intermediate between the Kohn-Sham potential (μ_x) and the Slater potential ($^3/2\ \mu_x$).

To our knowledge there are no explicit applications of this scheme using $\beta(r_s)$ according to Eq. (6) available yet. However, calculations with effectively only slightly different r_s-dependent β of such widely varied properties as atomic densities /11/, metallic lattice parameters /11, 12/ and metallic work functions /13/ have been relatively successful. There is also a whole school of calculations using r_s-independent β-factors and then adjusting the value of β to some external criterion Mostly these are performed for atoms, indicating good results for ground state properties with β slightly above one at atomic densities /14/. A calculation of cohesive properties of metallic aluminium /15/ gives the correct 0°K equilibrium density for β = 1.07, a value somewhat smaller than according to Eq. (6) for r_s = 2.08.

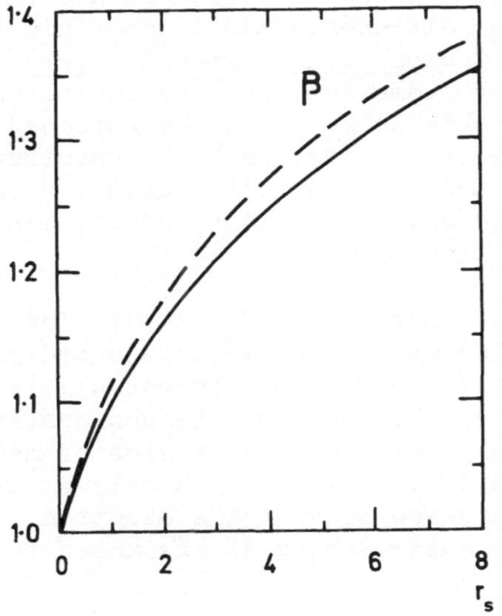

Fig. 1. The parameter β describes correlation effects in the local potential for ground state properties as $\mu_{xc} = \beta\ \mu_x$. The upper curve is deduced from Ref. /7/ and the lower from Ref. /8/.

3. THE SELF-ENERGY AS A NONLOCAL ENERGY-DEPENDENT POTENTIAL

The Hohenberg-Kohn-Sham scheme is strictly limited to ground state properties of the system and the energy eigenvalues ε_i of Eq. (1) should not be identified with one-electron eigenvalues. In principle, the energy value ε_i has a physical meaning only when it equals the Fermi energy. For physical, excited states of electrons and holes, as needed for e.g. in photoemission and optical properties, we may start from the equation

$$\left[-\hbar^2\nabla^2/(2m) + V(\underline{r})\right] \phi_k(\underline{r}) + \int d^3r' \, \Sigma(\underline{r}, \underline{r}'; E_k) \, \phi_k(\underline{r}') = E_k \, \phi_k(\underline{r}), \tag{7}$$

where E_k are energies and ϕ_k amplitudes for one-electron or one-hole excitations, and where Σ is the electron self-energy. Σ is nonlocal, complex and energy-dependent.

We shall later discuss how to approximate Σ with a local potential, using the results for a homogeneous electron gas. In this section we shall briefly comment on the nonlocal properties of the self-energy Σ_h for a homogeneous electron gas.

We will make use of the investigations by Hedin and Lundqvist, which are reviewed in Ref. /16/, where all references to the original papers can be found. These calculations are based on the RPA-approximation for the self-energy and is conveniently separated into two contributions, a Coulomb hole term and a screened exchange contribution. It should be remarked that the Coulomb hole contribution is generally considerably larger than the screened exchange. We stress the importance of considering both contributions in applications of the theory, and that the often used procedure to consider only screened exchange gives a serious underestimate of the effect of exchange and correlation.

The results for $\Sigma_h(\underline{r}-\underline{r}',\omega)$ by Hedin /17/ for different values of r_s and for particles on the Fermi surface are reproduced in Fig. 2 and compared with the Hartree-Fock potential. It is striking how the interactions reduce the range of the nonlocality compared to the Hartree-Fock potential over the whole region of metallic densities. The results show an appreciable strength only at distances smaller than the average separation between the electron, whereas the Hartree-Fock potential is quite strong at distances far beyond that separation.

In order to show some light on the non-locality effect for excited states we wish to report on some preliminary calculations by A.K. Das of the real part of the self-energy for energies away from the Fermi energy. In these calculations use has been made of the property of the dielectric function $\varepsilon(q,\omega)$ of being well given by the

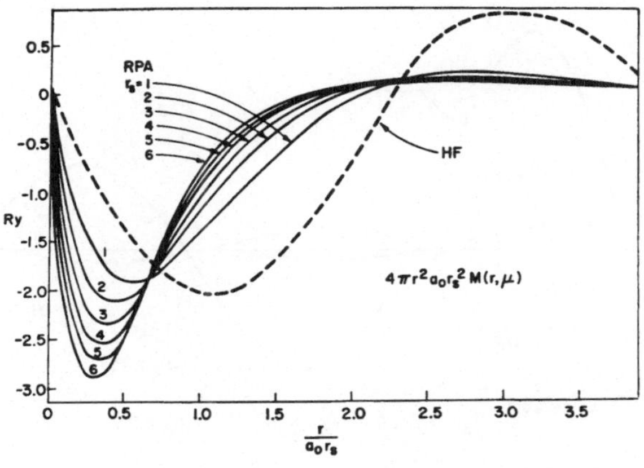

Fig. 2. Self-energy operator M (=Σ) as a non-local potential for particles on the Fermi surface, from L. Hedin, Phys.Rev. 139, A796 (1965).

simple plasmon-pole approximation, discussed in Refs. /16/ and /18/:

$$1/\varepsilon(q,\omega) = 1 + \omega_p^2 \,/\, (\omega^2 - \omega^2(q)\,), \tag{8}$$

where $\omega(q)$ approaches the plasma frequency ω_p for long wave lengths and approaches the free particle energy at short wave lengths. The calculations by Hedin were based on the full Lindhard dielectric function. The simple approximation in Eq. (8) should be good enough to show the main trends for $\Sigma_h(r,\omega)$ as a function of the energy ω, but seems to exaggerate the attractive portion of the self-energy for small values of r.

The results of this approximate calculation show for moderate energies ω the same qualitative behaviour as in Fig. 2, i.e. the potential has a strong attractive part at small values of r and tends rapidly and monotonously to zero, when r increases. As illustrated in Fig. 3, there is more structure for higher ω-values. The nonlocality range increases and some long range oscillatory contributions start to build up. For ω around the plasma frequency ω_p these oscillations have large amplitudes, the system literally opens up and the self-energy has a truly long range character. The long range and oscillatory nature of the self-energy will probably be characteristic features in the whole regime $\omega \gtrsim \omega_p$, although we have at the moment no numerical results to substantiate this conjecture.

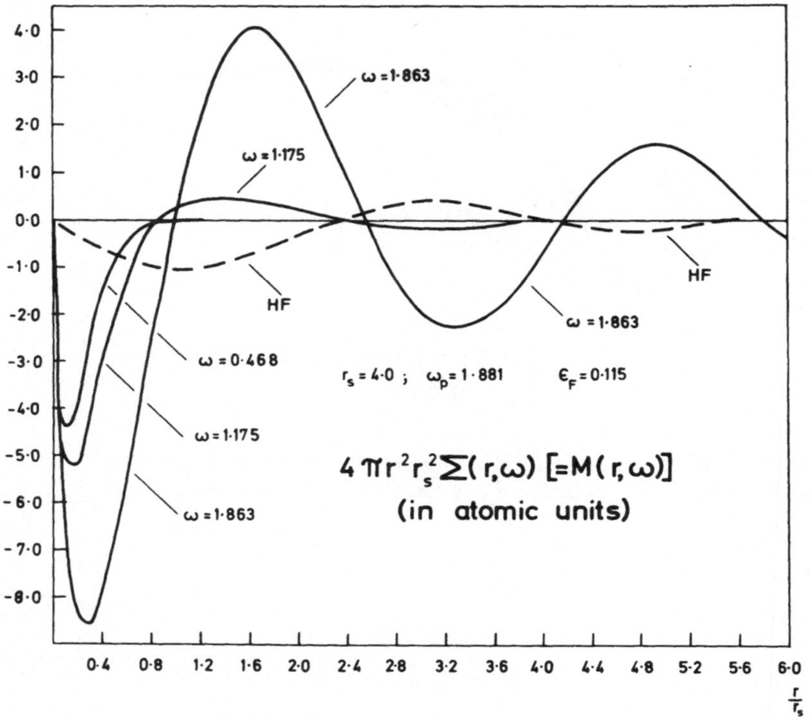

Fig. 3. Characteristic behaviour of the real part of $\Sigma(r,\omega)$ in the plasmon-pole approximation for three values of ω above the Fermi energy ε_F. The zero value for ω in this figure corresponds to the bottom of the Fermi sea.

At a first sight, the behaviour of $\Sigma(r,\omega)$ in the regime $\omega \gtrsim \omega_p$, in particular the long range of $\Sigma(r,\omega)$, seems to exclude for these energies the construction of a suitable local density approximation. However, this conclusion is not necessarily true for the construction of a local density approximation for the quasiparticle excitations, which are the excitations of prime interest. We interprete the structure of $\Sigma(r,\omega)$ around and above the plasmon threshold to be at least partly connected with the complex electron-plasmon sattellite structures also present in the spectrum /16/ and not just connected with the quasiparticle branch. We have not succeeded to identify and eliminate the quasiparticle part of $\Sigma(r,\omega)$ in r-space. In the next section we will describe a scheme, in which such an identification has been made by going over to the Fourier space and studying the behaviour of the spectral weight function $A(\underline{k},\omega)$ as a function of $\underline{k}$ and ω.

However, for moderate energies, i.e. $\omega \lesssim \omega_p$, this yet incomplete study suggests the alternative possibility of using an approximate exchange-correlation potential in r-space chosen as the self-energy of a uniform electron gas of the same density as the local density $\rho(\underline{r})$. The energy argument has to be taken in the way described in section 5. For moderately excited states the preliminary results indicate that the potential has a narrow range and a weak energy dependence. The reason why we do not recommend such a potential for quasiparticles with $\omega \gtrsim \omega_p$ is the above-mentioned coupling to the collective electron-plasmon satellite structure. We believe that such a nonlocal potential may provide a useful alternative to other schemes. Work is now in progress to provide numerical data over an appropriate range of densities and energies.

4. THE MOMENTUM DEPENDENT POTENTIAL FOR ELECTRONS IN AN ELECTRON GAS

We shall briefly review and slightly extend the above-mentioned results by Hedin and Lundqvist. These calculations were based on the RPA approximation for the self-energy $\Sigma(\underline{k},\omega)$, using the full Lindhard dielectric function rather than the simple plasmon-pole formula given in Eq. (8).

In order to obtain the quasi-particle energy one has to solve the Dyson equation involving the self-energy. After solving this equation one can write the resulting quasiparticle energy in the form

$$E(\underline{k}) = \varepsilon(\underline{k}) + V_{xc}(\underline{k}), \tag{9}$$

where $V_{xc}(\underline{k})$ can be interpreted as an effective exchange and correlation potential. Figure 4 shows the results over a range of metallic densities, taken from the peak of the spectral function $A(k,\omega)$/6/.

It is remarkable how the interactions wipe out essentially all k-dependence and leave a potential that is almost constant for all states below the Fermi momentum k_F and in a considerable range above k_F. This means essentially that the effective interaction in ordinary space is almost local. This is of course in agreement with our discussion in the preceding section where we found that the range of the non-local potential is small for moderate energies. The relative constancy with respect to k provides some posteriori justification for simplified procedures in energy band calculations, where effects of exchange and correlation have been simulated using a local potential, such as e.g. the Slater potential. For applications, where it is accurate enough to have such a k-independent potential, the potential (3) with β according to Eq. (6) should be the most appropriate.

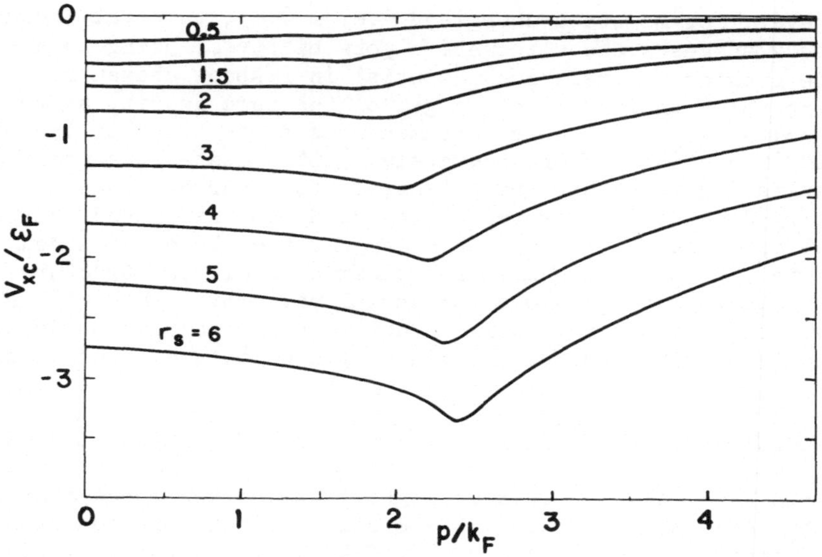

Fig. 4. The effective exchange-correlation potential $V_{xc}(k)$ in units of the Fermi energy $\varepsilon_F = \hbar^2 k_F^2/(2m)$ as a function of the wave-vector k/k_F. The numbers on the curves are the r_s-values.

In the limit of high densities ($r_s \to 0$) we know that the effective potential $V_{xc}(k)$ must approach the Hartree-Fock potential $V_x(k)$. In Fig. 5 we have extended the calculation and give results for some small values of r_s. The most startling feature of Fig. 5 is how important the correlation effects are in reducing the k-dependence of the potential even at very high densities and particularly for electrons deep in the Fermi sea. The Hartree-Fock theory seems indeed to be an extremely high-density theory, the quantitative validity of which seems to be confined to the region $r_s << 0.1$.

The results shown in Figs. 4 and 5 form the basis for the construction of a local potential to be described in the following section. Before finishing this section we would like to comment briefly on a recent paper by Overhauser /19/ on a simplified model for the contribution from correlation to one-electron spectra. The simplifying idea in this paper, to replace the full density fluctuation spectrum by a single mode, i.e. applying Eq. (8), has already been applied in Ref. /18/ (a similar form of $\varepsilon(q,\omega)$ can be found e.g. already in the pioneering paper by Lindhard /20/). His results and conclusions agree in all essentials with those previously published in Refs. /17/ and /18/, reviewed in Ref. /16/.

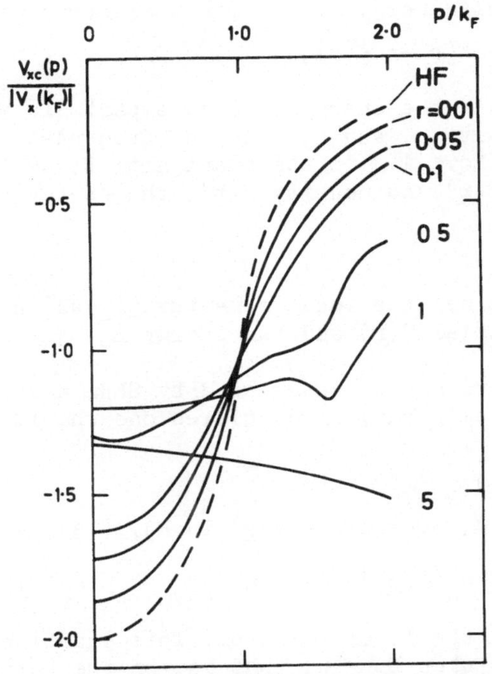

Fig. 5. The effective exchange-correlation potential for small r_s-values in units of $V_x(k_F)$ as a function of the wave-vector k/k_F. The numbers on the curves are the r_s-values.

5. A LOCAL POTENTIAL FOR EXCITATION SPECTRA

The results obtained for the self-energy of an electron gas can be utilized to construct a local energy-dependent potential $V_{xc}(\underline{r})$ following a procedure outlined by Sham and Kohn /21/. Because the self-energy is a ground state property, it is a unique functional of the density ρ; according to the theorem by Hohenberg and Kohn /1/ Sham and Kohn suggested the local density approximation

$$\Sigma(\underline{r},\underline{r}',E) \simeq \Sigma_h(\underline{r}-\underline{r}',E-V(\underline{r}_o);\ \rho(\underline{r}_o)), \tag{10}$$

where Σ_h is the self-energy of a homogeneous electron gas at the electron density ρ, and where $\underline{r}_o = (\underline{r}+\underline{r}')/2$. The energy argument in Σ is chosen to make Eq. (10) exact in the limit of almost constant density. In order to approximate the non-local self-energy by a local potential they used a WKB argument, introducing a local momentum $p(\underline{r})$, thus obtaining

$$\int\Sigma_h(\underline{r}-\underline{r}',E-V(\underline{r}_o);\rho(\underline{r}_o))\ \phi_k(\underline{r}')d\underline{r}'\simeq\Sigma_h(p(\underline{r}),\ E-V(\underline{r});\rho(\underline{r}))\phi_k(\underline{r}). \tag{11}$$

The quasiparticle self-energy in a uniform electron gas is given by

$$\Sigma_h(p,\ E(p);\rho) = E(p) - \varepsilon(p), \tag{12}$$

where $E(p) = E(p;\rho)$ is the energy of a quasiparticle with momentum p in a gas with density ρ, and $\varepsilon(p)$ is the free particle energy $\hbar^2p^2/(2m)$. As Σ_h in Eq. (11) is the quasiparticle self-energy at the local momentum $p(\underline{r})$ and density $\rho(\underline{r})$, the energy argument should be given by

$$E(p;\rho) = E_k - V(\underline{r}). \tag{13}$$

This equation determines the local momentum $p(\underline{r})$which thus depends on the Hartree potential $V(\underline{r})$ and the energy E_k.

A possible refinement was suggested by Sham and Kohn /21/, who proposed that for weakly varying densities one should use the approximate relation

$$\mu = V(\underline{r}) + \mu_h(\rho), \tag{14}$$

where $\mu_h(\rho) = E(k_F,\rho)$, to replace $V(\underline{r})$ in Eq. (13), giving

$$E(p(\underline{r});\ \rho(\underline{r})) - \mu_h(\rho(\underline{r})) = E_k - \mu \tag{15}$$

for determination of the local momentum. This equation guarantees that $p(\underline{r}) = k_F(\rho(\underline{r}))$, when $E_k = \mu$, that is, on the Fermi surface. It also guarantees that V_{xc} and the potential μ_{xc} discussed in section 2 will give the same Fermi surface. In Ref. /6/ there are given arguments in favour of Eq. (15) for the determination of the local momentum, however, without the restriction in Eq. (14). These arguments are based on a calculation of the electron gas vertex function /22/ and are concerned with how the result in the almost constant density limit is approached.

With the procedure just outlined we arrive at the equation

$$\{-\hbar^2\nabla^2/(2m) + V(\underline{r}) + V_{xc}(\underline{r})\}\ \phi_k(\underline{r}) = E_k\phi_k(\underline{r}) \tag{16}$$

where the effective exchange-correlation potential is given by

$$V_{xc}(\underline{r}) = \Sigma_h\{\ p(\underline{r}),\ E(p(\underline{r}));\ \rho(\underline{r})\} \tag{17}$$

taking $p(\underline{r})$ from Eq. (15). To obtain $V_{xc}(\underline{r})$ for a particular $\underline{r}$ one has to solve Eq. (15) for $p(\underline{r})$ and to get $\Sigma_h\ (p(\underline{r}),\ r_s(\underline{r}))$ by interpolation. The necessary data for applying this procedure have recently been published and can be found in Refs. /5/ and /6/.

It should be observed that the charge density represents an input parameter to the calculation and must be obtained by other means, e.g. a self-consistent calculation using the ground state scheme discussed in section 2.

From the discussion of Fig. 4 in section 4 it is clear that the $p(\underline{r})$-dependence of the potential is very weak for moderate excitation energies at metallic densities. For electron energies above the plasmon threshold, however, the momentum dependence is not negligible. This has bearings on the energy dependence of the "inner potential" found in LEED-data, as shown by several calculations /23/.

The strong momentum dependence at high densities shown in Fig. 5 is consistent with the empirical experience that atomic core excitation energies are better described by the Slater potential than by the Kohn-Sham potential. As atomic densities correspond to small r_s-values, Eq. (15) will give momentum values in the bottom of the local Fermi sea. Figure 5 shows that the potential then can be up to two times the Kohn-Sham value $\mu_x = V_x(k_F)$.

6. SPIN-DEPENDENT POTENTIALS

The ground state theory of Hohenberg and Kohn can directly be extended to include the effects of spin interaction /2/. The local-density approximation can be extended to a local-spin density approximation, in practice by putting spin indices on υ_{xc}, ψ_i, ε_i and ρ in Eqs. (1) and (2):

$$\left[-\hbar^2\nabla^2/(2m)+V(\underline{r})+\upsilon_{xc,s}(\underline{r})\right]\psi_{i,s}(\underline{r}) = \varepsilon_{i,s}\,\psi_{i,s}(\underline{r}) \tag{18}$$

giving the spin density as

$$\rho_s(\underline{r}) = \sum_{i,occ.}|\psi_{i,s}(\underline{r})|^2. \tag{19}$$

In the local-spin density approximation the problem is just to find the exchange and correlation energy of a spin-polarized homogeneous electron gas. The literature on this model problem is rather limited. The result in the Hartree-Fock approximation /24/ is of course well-known, the homogeneous electron gas becoming ferromagnetic at r_s=5.45. To indicate the general behaviour of the spin-dependent potentials, we show in Fig. 6 the Hartree-Fock results /25/ for spin-polarizations ζ varying from the paramagnetic (ζ=0) to the saturated ferromagnetic (ζ=1) limits. The tendency to favour parallell alignment of the spins is obvious.

Correlation effects are expected to reduce this tendency, although the explicit calculations of this effect are few. Rajagopal /26/ employs the staticly screened exchange approximation, which has been commented on in section 3. In the subsequent paper Hedin will present results in the random phase approximation. We are presently computing the correlation energy of the spin-polarized electron gas, extending the method of Ref. /9/. This calculation in-

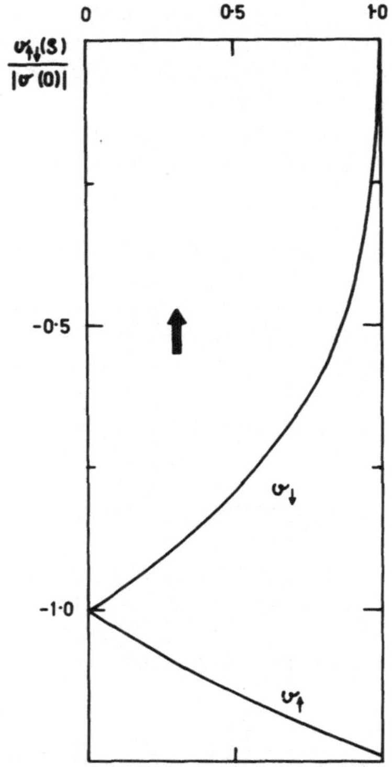

Fig. 6. Spin-dependent potentials applying Hartree-Fock data /25/ for the spin-polarized electron gas. The filled arrow indicates the direction of the majority spins.

volves a determination of the self-energy in the approximation described in section 2 and an integration over the one-electron spectra /10/ for the different spins.

The approach just described will not only give the ground state properties but the self-energy can also be used to construct spin-dependent potentials for excitation energies.

7. CONCLUDING REMARKS AND SUMMARY

This paper has summarized some aspects of the recent discussion about exchange correlation potentials to be used in energy band calculations. We have adopted the Hohenberg-Kohn-Sham theory to express the exchange and correlation contributions in a local density approximation, and made use of the extensive investigations of a uniform electron gas to provide numerical results directly applicable for energy band calculations.

For the calculation of ground state properties, like the total energy and the charge density, we suggest the use of the quantity μ_{xc}, discussed in section 2. For the calculation of excitation spectra as needed in, for example, photoemission and optical properties we suggest the use of the potential $V_{xc}(\underline{r})$ discussed in section 5. The necessary data for using this potential in computations are published in Refs. /5/ and /6/.

In section 3 we indicate another possible choice of a potential, starting from the self-energy of an elelctron gas in real space. Preliminary calculations indicate that the non-locality range is small for moderate energies and that the energy dependence is weak. Work is in progress to provide numerical data over an appropriate range of densities and energies.

The extension of the Hohenberg-Kohn-Sham approach to spin polarized system is of considerable interest. Work is in progress to calculate the spin dependent potential using the Galitskii-Migdal formula. We believe that this work will be a useful complement to the results presented by Hedin at this conference.

We are not aware of any complete calculation making use of the potentials discussed in this paper and it is therefore impossible to say anything about the real usefulness of these ideas. Instead of the many calculations where one tries to fit empirical data with a local exchange potential with flexible strength parameter, it seems worth trying to apply this approach to electron structure calculations.

ACKNOWLEDGEMENT

We wish to thank Dr. A.K. Das and Mr. O. Gunnarsson for their helpful assistance in providing and preparing material for this paper.

REFERENCES

1. P. Hohenberg and W. Kohn, Phys. Rev. 136, B864 (1964).
2. W. Kohn and L.J. Sham, Phys. Rev. 140, A1133 (1965).
3. F. Herman, J.P. Van Dyke, and I.B. Ortenburger, Phys. Rev. Letters 22, 807 (1969).
4. I.B. Crtenburger and F. Herman in "Computational Methods in Band Theory" (Eds. P.M. Marcus, J.F. Janak and A.R. Williams, Plenum Press, New York 1971).

5. L. Hedin, B.I. Lundqvist and S. Lundqvist, Solid State Comm. 9, 537 (1971).

6. L. Hedin and B.I. Lundqvist, J.Phys. C (in press).

7. K.S. Singwi, A. Sjölander, M.P. Tosi and R.H. Land, Phys. Rev B1, 1044 (1970).

8. K. Singwi and P. Vashishta (to be published).

9. B.I. Lundqvist and V. Samathiyakanit, Phys. Kondens. Materie 9, 231 (1969).

10. V.M. Galitskii and A.B. Migdal, Sovjet Phys. JETP (English translation) 7, 96 (1958).

11. B.Y. Tong in "Computational Methods in Band Theory" (Eds. P.M. Marcus, J.F. Janak and A.R. Williams, Plenum Press, New York 1971), and references therein.

12. D.A. Liberman, Phys. Rev. B 3, 2081 (1971).

13. N.D. Lang, Solid State Comm. 7, 1047 (1969); N.D. Lang and W. Kohn, Phys. Rev. B1, 4555 (1970); 3 1215 (1971).

14. See e.g. E. Kmetko, Phys. Rev. A1, 37 (1970).

15. M. Ross and K.W. Johnson, Phys. Rev. B2,4709 (1970).

16. L. Hedin and S. Lundqvist, "Solid State Physics" (Eds. H. Ehrenreich, F. Seitz, and D. Turnbull, Academic Press, New York 1970), 24, 1.

17. L. Hedin, Phys. Rev. 139, A796 (1965).

18. B.I. Lundqvist, Phys. Kondens. Materie 6, 206 (1967).

19. A.W. Overhauser, Phys. Rev. B3, 1888 (1971).

20. J. Lindhard, Dan. Math. Phys. Medd. 28, No. 8 (1954).

21. L.J. Sham and W. Kohn, Phys. Rev. 145, 561 (1966).

22. M. Watabe and H. Yasuhara, Phys. Letters 28A, 329 (1968).

23. R.O. Jones and J.A. Strozier, Jr., Phys. Rev. Letters 22, 1186 (1969); 25, 516 (1970); Phys. Rev. B3, 3228 (1971); S.Y. Tong and T.N. Rhodin, Phys. Rev. Letters 26,711 (1971).

24. F. Bloch, Z. Physik 57, 545 (1929).

25. C. Herring, in "Magnetism" (Eds. G.T. Rado and H. Suhl, Academic Press, New York 1966), Vol. 4.

26. A.K. Rajagopal, Nuovo Cimento Suppl., 5, 807 (1967).

EXCHANGE-CORRELATION POTENTIALS

Lars Hedin

Dept. of Theoretical Physics

University of Lund, Lund, Sweden

I. INTRODUCTION

In the preceding paper Stig Lundqvist discussed properties of the exchange-correlation potentials in the paramagnetic case. I will build on that discussion and derive results in linear response theory. I will also report some recent results for the spin-polarized case, obtained together with Ulf von Barth. A more detailed account of these results will be given in forthcoming papers (1,2).

The purpose of treating linear response theory is not to obtain accurate response functions but to test the local density potentials in a situation where more sophisticated calculations have been made. These tests indicate that the local density theory should have a surprisingly large range of applicability. In the spin polarized case we find that the spin-dependent potential can be well represented by a shifted and rescaled Hartree-Fock potential. Finally I will report a few simple calculations which serve further to illustrate the range of applicability of local density schemes.

The conclusion of the discussion is that it seems well motivated to do large scale bandcalculations using the proposed exchange-correlation potentials, to learn how far ab initio calculations now can take us.

II. THE SPIN PROBLEM

To discuss the paramagnetic case Hohenberg and Kohn (3)

consider a Hamiltonian

$$H = T + U + V \qquad (1)$$

where T describes the kinetic energy of the electrons, U the electron-electron interactions ($1/r_{12}$ terms) and V the interactions with a one-body potential $v(\bar{r})$

$$V = \sum_{i=1}^{N} v(\bar{r}_i) = \int \psi^{+}(\bar{r}) v(\bar{r}) \psi(\bar{r}) d^3\bar{r}. \qquad (2)$$

In the last expression, $\Psi(\bar{r})$ is the field operator of the second quantized formulation. Hohenberg and Kohn find that for a non-degenerate ground state there is a unique relationship between the one-body potential $v(\bar{r})$ and the charge density $\rho(\bar{r})$, and that the ground state energy is a functional of $\rho(\bar{r})$ which is stationary with respect to variations of ρ which conserve the total number of electrons.

The local density problem in the spin-polarized case has so far received less attention than in the paramagnetic case. Kohn and Sham (4) discussed the spin susceptibility of a homogeneous electron gas, and Stoddart and March (5) discussed Hubbard-like models in a Hartree-Fock approximation.

We consider a Hamiltonian which besides the terms in Eq (1) contains the interaction V_M between the electron spins $\bar{S} = (h/2)\sigma$ and an arbitrary magnetic field $\bar{B}(\bar{r})$

$$V_M = \mu_B \sum_{i=1}^{N} \bar{B}(\bar{r}_i)\bar{\sigma}_i = \mu_B \sum_{\alpha\beta} \int \psi_\alpha^{+}(\bar{r}) \bar{B}(\bar{r}) \sigma_{\alpha\beta} \psi_\beta d^3\bar{r}. \qquad (3)$$

Here μ_B is the Bohr magneton, $\bar{\sigma}_{\alpha\beta}$ the vector of Pauli spin matrices and $\Psi_\alpha(\bar{r})$ the two component spinor field operator. For non-degenerate ground states there is for precisely the same reasons as put forward in the Hohenberg and Kohn paper, a unique relationship between the four quantities $(v,\bar{B})$ and the 2 x 2 spin density matrix

$$\rho_{\alpha\beta}(\bar{r}) = \langle \psi_G | \psi_\beta^{+}(\bar{r}) \psi_\alpha(\bar{r}) | \psi_G \rangle. \qquad (4)$$

The ground state energy is now a functional of $\rho_{\alpha\beta}$, which is stationary with respect to variations that conserve the total number of electrons

$$N = \int (\rho_{\uparrow\uparrow}(\bar{r}) + \rho_{\downarrow\downarrow}(\bar{r})) d^3\bar{r}. \qquad (5)$$

An unsolved question is what class of functions $\rho_{\alpha\beta}(\bar{r})$ can be written of the form given in Eq. (4), i.e., that actually correspond to a many-particle antisymmetric wave function Ψ_G satisfying the Schrödinger equation. This "N-representability problem" which we also have for $\rho(\bar{r})$, should be further investigated. However, it seems much less severe than the much-discussed N-representability problem for the density matrix $\rho(\bar{r},\bar{r}') = \langle \Psi_G|\Psi^+(\bar{r}')\ \Psi(\bar{r})|\Psi_G\rangle$, which is a function of two space variables rather than one.

We may consider adding to Eq (1) not only Eq (3) but also a term that couples $\bar{B}$ to the orbital angular momentum. This would indeed be necessary for a full discussion of magnetic susceptibilities. Here, however, we are primarily interested in ferromagnetic and other ordered magnetic states and $\bar{B}$ is then only an artifice that splits degeneracies.

To discuss ground state properties Kohn and Sham (4) wrote the energy functionas as (schematically)

$$E[\rho] = T_s[\rho] + \frac{1}{2}\int \rho g \rho + \int v \rho + E_{xc}[\rho], \tag{6}$$

where g is the Coulomb interaction $(1/r_{12})$, $T_s[\rho]$ the functional for the kinetic energy of non-interacting electrons, and $Exc[\rho]$ the exchange-correlation energy. Variation of E with respect to ρ gives

$$\frac{\delta T_s[\rho]}{\delta \rho(\bar{r})} + V_H(\bar{r}) + v(\bar{r}) + v_{xc}(\bar{r}) = \lambda, \tag{7}$$

where λ is the Lagrange parameter coming from the subsidiary condition of particle conservation, V_H is the Hartree (or Coulomb) potential and $v_{xc}(\bar{r}) = \delta E_{xc}[\rho]/\delta\rho(\bar{r})$ an exchange-correlation potential. Kohn and Sham then note that Eq (7) also follows if we study non-interacting electrons in a given potential $V_H + v + v_{xc}$. Thus we obtain the exact charge density $\rho(\bar{r})$ by solving the one-electron problem

$$\left[-\frac{\hbar^2}{2m}\nabla^2 + V_H(\bar{r}) + v(\bar{r}) + v_{xc}(\bar{r})\right]\varphi_i(\bar{r}) = \varepsilon_i\,\varphi_i(\bar{r}), \tag{8}$$

and adding the one-electron charge densities

$$\rho(\bar{r}) = \sum_{i=1}^{N} |\varphi_i(\bar{r})|^2. \tag{9}$$

In the spin polarized case Eq (7) is replaced by

$$\frac{\delta T_s[\rho_{\alpha\beta}]}{\delta \rho_{\alpha\beta}(\bar{r})} + \left(V_H(\bar{r}) + v(\bar{r})\right)\delta_{\alpha\beta} + \mu_B \bar{B}(\bar{r})\bar{\sigma}_{\alpha\beta} + V_{xc}^{\alpha\beta}(\bar{r}) = \lambda\, \delta_{\alpha\beta}, \tag{10}$$

where $V_{xc}^{\alpha\beta}(\bar{r}) = \delta E_{xc}[\rho_{\alpha\beta}]/\rho_{\alpha\beta}(\bar{r})$. Again this is the same equation as for non-interacting electrons, but now in a spin-dependent potential. We thus obtain the spin density matrix $\rho_{\alpha\beta}$ by solving the two coupled equations

$$\sum_\beta \left[-\frac{\hbar^2}{2m}\nabla^2 \delta_{\alpha\beta} + V_{eff}^{\alpha\beta}(\bar{r})\right] \varphi_i^\beta(\bar{r}) = \varepsilon_i \varphi_i^\alpha(\bar{r}), \quad \alpha = \uparrow, \downarrow, \tag{11}$$

for the spinor $\phi_i^\alpha(\bar{r})$ and then forming

$$\rho_{\alpha\beta}(\bar{r}) = \sum_{i=1}^{N} \varphi_i^\alpha(\bar{r})\, \varphi_i^{\beta *}(\bar{r}). \tag{12}$$

Here $V_{eff}^{\alpha\beta}$ is the effective potential $(V_H + v)\delta_{\alpha\beta} + \mu_B \bar{B}\, \bar{\sigma}_{\beta\alpha} + V_{xc}^{\alpha\beta}$.

For a slowly varying density we may write

$$E_{xc}[\rho_{\alpha\beta}] = \int \rho(\bar{r})\, \varepsilon_{xc}(\rho_{\alpha\beta}(\bar{r}))\, d^3r, \tag{13}$$

where ε_{xc} is the exchange-correlation energy of a uniform electron gas with spin-density matrix $\rho_{\alpha\beta}$ and ρ is the total density. At each point $\bar{r}$, the Hermitian matrix $\rho_{\alpha\beta}$ can be diagonalized; we assume that ε_{xc} depends only on the eigenvalues which we denote $\rho_\uparrow$ and $\rho_\downarrow$.

In the following we will only consider the special case when $\rho_{\alpha\beta}$ is diagonal. First we look at the Hartree-Fock approximation. We then have (6)

$$\varepsilon_x(\rho_\uparrow, \rho_\downarrow) = c\,(\rho_\uparrow^{\frac{4}{3}} + \rho_\downarrow^{\frac{4}{3}})/\rho, \tag{14}$$

where $c = -(3e^2/2)(3/4\pi)^{1/3}$ and $\rho = \rho_\uparrow + \rho_\downarrow$. The spin dependent potential becomes

$$V_x^\uparrow(\bar{r}) = \frac{\delta E_x(\rho_\uparrow, \rho_\downarrow)}{\delta \rho_\uparrow(\bar{r})} = c\left(\tfrac{4}{3}\right)\rho_\uparrow^{\frac{1}{3}}. \tag{15}$$

This is the standard $\rho^{1/3}$ potential which in the paramagnetic case reduces to the "Kohn-Sham" potential.

An exact calculation of the exchange-correlation energy ε_{xc} can be based on the dielectric function $\varepsilon(k\omega)$ (7)

$$\varepsilon_{xc} = \frac{1}{2\rho}\int_0^{e^2}\frac{d\lambda}{\lambda}\int\frac{d^3\bar{k}}{(2\pi)^3}\left[\int \mathrm{Im}\left(\frac{-1}{\varepsilon(\bar{k},\omega)}\right)\frac{d\omega}{2\pi} - \rho\,\frac{4\pi\lambda}{k^2}\right], \tag{16}$$

where in $\varepsilon(k\omega)$, e^2 is replaced by λ, and we have an integration over interaction strengths. We have evaluated Eq (16) with a straight forward generalization of the Lindhard dielectric function (the "bubble" approximation), introducing different spin up and spin down contributions. The numerical results were to an accuracy of better than 1% represented by the expression

$$\varepsilon_{xc}(r_s,x) = \varepsilon^{P}_{xc}(r_s) + A(r_s)\left[x^{\frac{4}{3}} + (1-x)^{\frac{4}{3}} - \left(\tfrac{1}{2}\right)^{\frac{1}{3}}\right], \tag{17}$$

where x describes the spin polarization, $x = \rho_\uparrow/\rho$, and r_S is the usual measure of the electron density, $4\pi r_S^3/3 = 1/\rho$. We note that for $x = \frac{1}{2}$, ε_{xc} reduces to the para-magnetic result ε^P_{xc}. We also note that the x-dependence in Eq (17) is the same as for the Hartree-Fock result in Eq (14).

Knowing ε_{xc} we can calculate $V^\uparrow$

$$\begin{aligned} V^\uparrow(r_s,x) &= \frac{\partial}{\partial\rho_\uparrow}\left[\rho\,\varepsilon_{xc}(\rho_\uparrow,\rho_\downarrow)\right] = \left[1 - \frac{r_s}{3}\frac{\partial}{\partial r_s} + (1-x)\frac{\partial}{\partial x}\right]\varepsilon_{xc} \\ &= \mu^{P}_{xc}(r_s) + \tfrac{4}{3}A\left[x^{\frac{1}{3}} - \left(\tfrac{1}{2}\right)^{\frac{1}{3}}\right] - \tfrac{1}{3}\left[A + r_s\frac{dA}{dr_s}\right]\left[x^{\frac{4}{3}} + (1-x)^{\frac{4}{3}} - \left(\tfrac{1}{2}\right)^{\frac{1}{3}}\right]. \end{aligned} \tag{18}$$

The last term contributes less than 1% to $V^\uparrow$ and can be neglected. Thus also $V^\uparrow$ has the same dependence on x as in the Hartree-Fock case Eq (15), however, with strongly changed coefficients. Numerical results are given in table 1. First the results for the paramagnetic potential obtained with the LIndhard (RPA) and the more refined theory by Singwi et al (8) are compared; the differences are not very large. Next we illustrate the variation of $V^\uparrow$ by giving the maximum and minimum values. In the last column we give the variation of $V^\uparrow$ in percent of the Hartree-Fock result. We note that the $x = 0$ value of $V^\uparrow$ is appreciably different from zero and that at metallic densities the variation of $V^\uparrow$ is only about half that in the Hartree-Fock theory.

Table 1

r_s	μ^P_{xc} (Ry)		$V^{\uparrow}$ (Ry)		Variation
	RPA	STLS	x = 0	x = 1	of $V^{\uparrow}$ (%)
1	-1.40	-1.36	-0.40	-1.59	78 %
2	-0.75	-0.72	-0.34	-0.87	68 %
3	-0.53	-0.50	-0.26	-0.58	61 %
4	-0.41	-0.39	-0.23	-0.45	56 %
5	-0.34	-0.32	-0.20	-0.36	52 %
6	-0.29	-0.27	-0.18	-0.31	49 %

III. LINEAR RESPONSE RESULTS IN THE SPINLESS CASE

We will discuss three different dielectric functions. First we have the dielectric function $\varepsilon(q)$ that gives the induced charge density ρ_e in terms of an "external" charge density ρ_{ext}

$$\rho_e(q) = \left[\frac{1}{\varepsilon(q)} - 1\right] \rho_{ext}(q) . \tag{19}$$

This dielectric function also gives the Hartree potential $V_H(q) = g(q)\ (\rho_e(q) + \rho_{ext}(q))$ in terms of the perturbing potential $V_{ext}(q) = g(q)\ \rho_{ext}(q)$, thus

$$V_H(q) = \frac{V_{ext}(q)}{\varepsilon(q)} , \tag{20}$$

where $g(q)$ is the Coulomb interaction $4\pi e^2/q^2$. Next we have the dielectric function $\tilde{\varepsilon}(q)$ describing changes in the gound state potential

$$V_H(q) + \mu_{xc}(q) = \frac{V_{ext}(q)}{\tilde{\varepsilon}(q)} , \tag{21}$$

and finally the dielectric function $\tilde{\tilde{\varepsilon}}(q,E)$ for the excitation spectrum potential

$$V_H(q) + V_{xc}(q,E) = \frac{V_{ext}(q)}{\tilde{\tilde{\varepsilon}}(q,E)} . \tag{22}$$

As a reference we also introduce the Hartree (or Bardeen, or Lindhard or RPA) approximation

$$\varepsilon_0(q) = 1 - P_0(q)\, g(q) . \tag{23}$$

Approximating $E_{xc}[\rho]$ as $\int \rho \varepsilon_{xc}(\rho)\, d\bar{r}$ we obtain (1)

$$\varepsilon(q) = 1 - \frac{P_0(q)\, g(q)}{1 + G(q) P_0(q)\, g(q)}\,, \qquad (24)$$

$$\tilde{\varepsilon}(q) = 1 - (1 - G(q)) P_0(q)\, g(q),$$

where

$$G(q) = \frac{1}{4}\gamma q^2 \,. \qquad (25)$$

The coefficient γ is related to the compressibility ratio, $\kappa_F/\kappa = 1 - \gamma(\alpha r_s/\pi)$. The third function $\tilde{\varepsilon}(q,E)$ coincides with $\tilde{\varepsilon}(q)$ at the Fermi level $(E = \mu)$ and approaches $\varepsilon(q)$ at large energies. In the $q \to 0$ limit the two functions $\varepsilon(q)$ and $\tilde{\varepsilon}(q)$ are quite different,

$$\varepsilon(q) \to 1 + \frac{\kappa}{\kappa_F}\frac{k_{TF}^2}{q^2}\,,$$

$$\tilde{\varepsilon}(q) \to 1 + \frac{k_{TF}^2}{q^2}\,, \qquad (26)$$

since the compressibility ratio κ/κ_F is quite different from unity in the metallic density range.

The important and critical approximation usually made in the Hohenberg-Kohn-Sham theory is to write the functional $E_{xc}[\rho]$ just as a function of ρ, $\int \rho\, \varepsilon_{xc}(\rho) d\bar{r}$, and neglecting all gradient terms. To test this approximation we compare in fig. 1 the results for G from different theories.

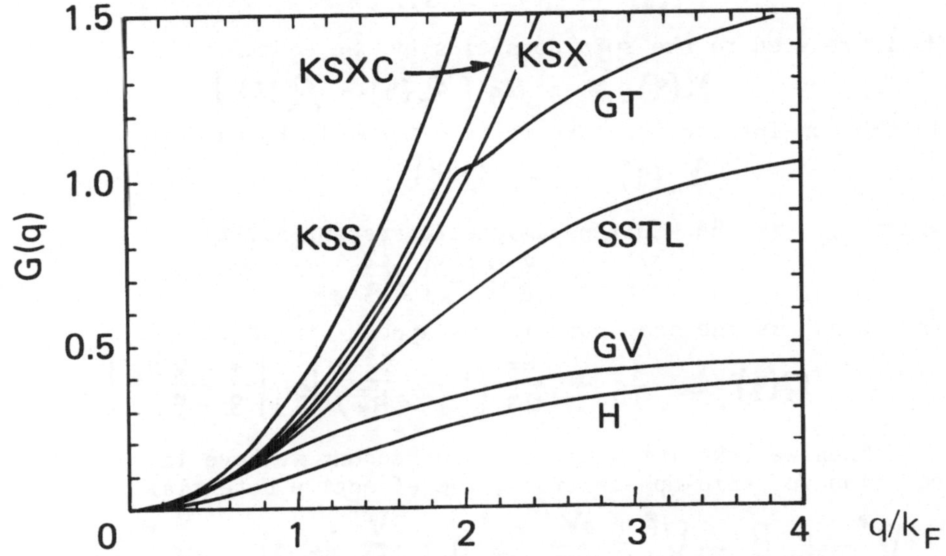

Fig. 1. Results for the G-function at $r_s = 4$ in different approximations, taken from reference (1).

The KSXC curve in fig. 1 gives the parabola from Eq (25) with a γ-value taken from the calculation of the compressibility by Singwi et al (8). In the KSX parabola the Hartree-Fock value for γ is used ($\gamma = 1$), and the KSS parabola gives the result from the Slater potential ($\gamma = 3/2$). The H and GV curves give early estimates made by Hubbard and by Geldart and Vasko while the GT and SSTL results are from recent, more refined calculations by Geldart and Taylor (9) and by Singwi et al (8). Singwi et al have suggested that the correct G-function should be somewhat larger than their result.

The correct G-curve starts like a parabola and the recent theoretical results indicate that it remains closely parabolic out to about $k = 2k_F$, then it flattens out. For larger q-values the G-factor plays only a minor role since then $g(q)\, P_o(q)$ drops quickly to zero, outphasing the effect of G. Thus the simple parabolic form of G predicted by a local density theory without gradient terms seems to give a quite remarkably good representation of the dielectric function.

IV. LINEAR RESPONSE THEORY IN THE SPIN POLARIZED CASE

We may also in this case discuss three different response functions, however, we will here limit ourselves to the usual susceptibility χ, which gives the relation between the magnetic field B and the induced magnetic moment M

$$M(q) = \chi(q)\, B(q). \tag{27}$$

M is related to the spin densities by the equation

$$M(q) = -\mu_B \left(\rho_\uparrow(q) - \rho_\downarrow(q) \right). \tag{28}$$

In the non-interacting case we have the well-known result (6)

$$\chi_o(q) = \chi_o\, U(q), \tag{29}$$

where χ_o is the Pauli paramagnetic susceptibility

$$\chi_o = \frac{3}{2} \left(\frac{\rho}{\varepsilon_F} \right) \mu_B^2, \tag{30}$$

and U(q) is the same function as appears in $P_o(q)$

$$U(q) = \frac{1}{2} + \frac{k_F}{2q} \left(1 - \frac{q^2}{4k_F^2} \right) \log \left| \frac{q + 2k_F}{q - 2k_F} \right|. \tag{31}$$

When we take interactions into account we have in e.g. the equation for spin-up electrons, the effective potential

$$V^\uparrow + \mu_B B = V^P + \frac{\partial V^\uparrow}{\partial \rho^\uparrow} \delta\rho^\uparrow(q) + \frac{\partial V^\uparrow}{\partial \rho^\downarrow} \delta\rho_\downarrow(q) + \mu_B B(q). \tag{32}$$

The first term V^P is constant and can be neglected. From general principles it follows that $\delta\rho(q) = \delta\rho^{\uparrow}(q) + \delta\rho_{\downarrow}(q) = 0$, and we can thus write

$$V^{\uparrow} + \mu_B B = \mu_B\left[B + \frac{1}{2\mu_B}\left(\frac{\partial V^{\uparrow}}{\partial \rho^{\uparrow}} - \frac{\partial V^{\uparrow}}{\partial \rho_{\downarrow}}\right)(\delta\rho^{\uparrow} - \delta\rho_{\downarrow})\right]. \tag{33}$$

We have one-electron equations to solve with an effective B given by the square bracket in Eq (33), and we can hence write the solution as

$$M(q) = \chi_0(q)\left[B(q) + \frac{1}{2\mu_B}\left(\frac{\partial V^{\uparrow}}{\partial \rho^{\uparrow}} - \frac{\partial V^{\uparrow}}{\partial \rho_{\downarrow}}\right)(\delta\rho_{\uparrow} - \delta\rho_{\downarrow})\right]. \tag{34}$$

Substituting from Eq (28) and comparing with Eq (27) we have for the susceptibility

$$\chi(q) = \frac{\chi_0(q)}{1 + \frac{1}{2}(\chi_0(q)/\mu_B^2)\left(\frac{\partial V^{\uparrow}}{\partial \rho^{\uparrow}} - \frac{\partial V^{\uparrow}}{\partial \rho_{\downarrow}}\right)} \; . \tag{35}$$

From Eq (18) we obtain

$$\frac{\partial V^{\uparrow}}{\partial \rho^{\uparrow}} - \frac{\partial V^{\uparrow}}{\partial \rho_{\downarrow}} = \frac{1}{\rho}\frac{\partial V^{\uparrow}}{\partial x} = \left(\frac{4^{4/3}}{9}\right)\left(\frac{A}{\rho}\right), \tag{36}$$

and we can hence write χ in the compact form

$$\chi(q) = \frac{\chi_0 U(q)}{1 + \eta\, U(q)}, \tag{37}$$

$$\eta = \left(\frac{4^{4/3}}{9}\right)\left(\frac{A}{\rho}\right) = 0.144\, r_s^2 A,$$

where A is given in units of Ry.

In table 2 we give values for the paramagnetic susceptibility ratio $\chi(o)/\chi_o = 1/(1 + \eta)$ in different approximations. We first give the Hartree-Fock result

Table 2. Values of the susceptibility enhancement

r_s	HF	HL	DG	RPA
1	1.20	1.15		1.15
2	1.50	1.28	1.31	1.29
3	2.00	1.40	1.47	1.44
4	2.96	1.49	1.65	1.60
5	5.9	1.56	1.85	1.77

($\eta = -0.166\, r_s$), then the estimate by Hedin and Lundqvist (10), which is very close to the results obtained by Rice, next the results by Dupree and Geldart (11) which are very close to those obtained by the Singwi method (12), and in the last column the results obtained here. These RPA values are somewhat larger than the HL values which seems reasonable since the HL values essentially are RPA values corrected for exchange effects. The DG values are larger than the RPA values; if they give the right rrend as the experimental results seem to indicate, than the exchange corrections calculated by Hedin and Lundqvist and by Rice go in the wrong direction. The differences between the estimates of correlation effects are, however, small as compared to the Hartree or the Hartree-Fock results.

The dimensionless quantity η is proportional to the "internal field correction" I(q) discussed by Singwi et al (8). They find that I(q) has a very gentle q-dependence. The local density theory which takes I(q) to be a constant might thus be a reasonable approximation in many cases.

V. RESULTS OF A FEW SIMPLE CALCULATIONS

To test the exchange-correlation potential for excitation energies a few simple calculations were made on solium and potassium atoms with boundary conditions for free atoms and for atoms in a box. In the latter case the valence electron wavefunction was taken to have zero derivative on a sphere corresponding to the Wigner-Seitz cell in the metal, i.e., a Wigner-Seitz calculation was made for the bottom of the conduction band.

The results for the core levels were poor in the sense that the large difference between the Koopmans' theory result and the correct result, could not be accounted for. In retrospect this is not surprising, the experimental core level energies have large contributions from relaxation effects of the valence electrons; these effects, however, only enter the local density theory through the very small contribution to the charge density in the core region, that comes from the valence electrons.

The results for the valence electron energies in a free atom were poor. As an example we give in table 3 the energy of the 4s electron in K calculated with different potentials. The indices c and v stand for core and valence contributions.

Table 3. Energy of the 4s electron in a K atom

$V_H^c + V_H^v + V_{xc}(\rho^c + \rho^v)$	-0.186 Ry
$V_H^c + V_{xc}(\rho^c)$	-0.457 Ry
V_{HF} (non-local)	-0.295 Ry

In this case the Hartree-Fock result is quite close to the experimental value, while the two conceivable local density approaches fail badly.

The results for the solid state problem look quite good. As an example we give in table 4 the results for the bottom of the conduction band in Na.

Table 4. Energy of the bottom of the conduction band in Na

V_H^c	-0.507 Ry	
$V_H^c + V_{xc}$ (Seitz)	-0.610	0.103
$V_H^c + V_H^v + V_{xc}(\rho^v)$	-0.314	
$V_H^c + V_H^v + V_{xc}(\rho^c + \rho^v)$	-0.417	0.103

The difference between the two first values gives the exchange-correlation contribution to the conduction electron from the ion-core. We see that the local density approximation can account for this effect very accurately (the precise agreement is of course spurious). To obtain the Fermi level we have to add the Fermi energy $\varepsilon_F = 0.238$ Ry, giving

$$\mu = -0.417 + 0.238 = -0.179 \text{ Ry}$$

This agrees quite well with the experimental result for the work function, $\phi = -0.168$ Ry (the dipole layer contribution is believed to be small in sodium).

From these calculations we are led to conclude that the local density theory should have considerably better possibilities to work for solids (particularly for metals) than for atoms. However, addition of gradient terms could possibly also make the theory useful for atoms and for solids with an open structure.

Acknowledgements

The results for the spin polarized problem were obtained in cooperation with Ulf von Barth. I am grateful to him and to Bengt Kjöllerström for discussions of the spin problem. The simple calculations reported in the last section were done at IBM, San Jose in September 1970; I am indebted to Frank Herman for making this visit possible and for stimulating discussions. I am also grateful to Stig and Bengt Lundqvist for many stimulating discussions of these problems through a number of years.

References

1. L. Hedin and B.I. Lundqvist, J. Phys. C: Solid St. Phys. in press

2. U. von Barth and L. Hedin, to be published

3. P. Hohenberg and W. Kohn, Phys. Rev. 136, B 864 (1964)

4. W. Kohn and L.J. Sham, Phys. Rev. 140, A 1133 (1965)

5. J.C. Stoddart and N.H. March, Annals of Physics 64, 174 (1971)

6. C. Herring in Magnetism (ed. Rado and Suhl), Academic Press 1966.

7. J. Hubbard, Proc. Roy. Soc. A243, 336, (1957)

8. Singwi, Sjölander, Tosi and Land, Phys. Rev. B1, 1044 (1970)

9. D.J.W. Geldart and R. Taylor, Can. J. Phys. 48, 167 (1970)

10. L. Hedin and S. Lundqvist in Solid State Physics (eds Ehrenreich, Seitz and Turnbull) vol. 23, 1969, Academic Press

11. R. Dupree and D.J.W. Geldart, Sol. State Comm. 9, 145 (1971)

12. Pizzimenti, Tosi and Villari, Lett. al Nuova Cim. 2, 81 (1971)

AN IMPROVED STATISTICAL EXCHANGE APPROXIMATION

Frank Herman and Karlheinz Schwarz[†]

IBM Research Laboratory, San Jose, California 95114

Various forms of the statistical exchange approximation[1-3] (abbreviated XA) have been used in atomic structure[4-6] and energy band[7,8] calculations over the past 20 years. In view of the continuing importance of the XA in atomic and solid state investigations and its increasing prominence in molecular studies,[9-11] it would be desirable to have a version of the XA which (a) is simple to deal with in a wide variety of applications, (b) is at least as satisfactory from a physical point of view as existing versions, and (c) is universal in character, in the sense that its key parameters are all independent of atomic number. The object of this short note is to propose such a form of the XA. For additional details and supporting information, the reader is referred to a recent paper.[12]

The optimized Xα version[3] of the XA is indeed extremely simple in form. Unfortunately, the optimum value of α is Z dependent, varying from nearly 1 for very small Z to nearly 2/3 for very large Z.[13-15] This Z dependence can become quite troublesome in studies of polyatomic molecules[9-11] or crystals containing more than one atomic species in the unit cell.

For example, suppose we want to construct optimized Xα exchange potentials for crystalline PbS and HgO. Using the results of Kmetko,[13] we would choose the following optimized values of α: for PbS: $\alpha(\mathrm{Pb}) = 0.70$, $\alpha(\mathrm{S}) = 0.765$; for HgO: $\alpha(\mathrm{Hg}) = 0.70$, $\alpha(\mathrm{O}) = 0.84$. If we adopt a muffin-tin crystal potential model, we could use these values of α within the inscribed spheres, but then what would

---

† Permanent address: Institute for Physical Chemistry, University of Vienna, Vienna, Austria.

we use for α in the interstitial regions? Whatever we do, we would be forced to introduce unphysical discontinuities in the magnitude of the exchange potential at the inscribed sphere boundaries.

Moreover, if we attempt to carry out first-principles self-consistent calculations, we would have to introduce arbitrary assumptions regarding the spatial variation of α throughout the unit cell, a variation implied by the Z dependence of α already mentioned. In practical terms, the introduction of these assumptions would vitiate the first-principles character of the study and transform it into a semi-empirical study, since the natural tendency would be to choose assumptions that would lead to the best agreement between theory and experiment.

Having thus made a case for the undesirability of an optimized Z-dependent XA, we now ask: Can we find a universal (Z-independent) XA that is "better" than the optimized Xα version[3] for all (or most) individual atoms, and, by implication, for (most) molecules and crystals as well?

Before proposing such an improved approximation, it is desirable to distinguish three different points of view concerning the physical significance of the XA: (a) Empirical viewpoint. The success of an XA should be judged by the degree to which theoretical predictions based on it agree with experiment; the closer and more far-ranging the agreement between theoretical predictions and experiment, the better. (b) Approximate Hartree-Fock viewpoint. The success of an XA should be judged by the degree to which it can be used as a replacement for the rigorous Hartree-Fock (HF) exchange approximation in numerical calculations, particularly where the object is to obtain "Hartree-Fock" results with considerably less effort. The merits of an XA should be judged, for example, by the smallness of the difference between the actual Hartree-Fock total energy (E_{HF}) and the value obtained by substituting the XA atomic orbitals into the Hartree-Fock expression for the total energy E_{HF}^{XA}). Other tests of "success" include comparisons of HF and XA electronic charge densities, oscillator strengths, and one-electron binding energies. (c) Heuristic viewpoint. Although a given XA starts out as an attempt to approximate the HF exchange operator, it turns out that the XA also takes account of electron correlation effects. In fact, various authors have shown that approximations which attempt to take both exchange and correlation effects into account lead to theories that closely resemble one or another version of the XA.[16-20] Accordingly, let us see what progress we can make by adopting the XA for practical computations, allowing for the possibility that the XA may represent correlation as well as exchange effects.

In the present paper, we will proceed in the spirit of viewpoint (b), although this does not rule out the use of our results (by ourselves and others) in terms of (a), (b), or (c).

It will be recalled that in earlier work[21-23] we introduced the second-order exchange inhomogeneity correction $\beta G(\rho)V_{XS}$ into the statistical exchange approximation,

$$V_{X\alpha\beta} = [\alpha + \beta G(\rho)]V_{XS} \quad , \tag{1}$$

where V_{XS} is the Slater exchange approximation:[1] $V_{XS} = -6\left(\frac{3}{8\pi}\rho\right)^{1/3}$,

$$G(\rho) = \frac{1}{\rho^{2/3}}\left[\frac{4}{3}\left(\frac{\nabla\rho}{\rho}\right)^2 - 2\,\frac{\nabla^2\rho}{\rho}\right] \quad , \tag{2}$$

and α and β are as-yet undetermined parameters. Because $G(\rho)$ diverges at very small and very large values of r, certain convergence factors were introduced to eliminate these divergences. Adopting the minimum $E_{HF}^{\alpha\beta}$ criterion, we varied α and β for each of several representative atomic systems, and found that the minimum values of $E_{HF}^{\alpha\beta}$ were obtained for $\alpha = 2/3$ (the Kohn-Sham value), and $\beta = 0.005 \pm 0.001$. The minimum values of $E_{HF}^{\alpha\beta}$ obtained in this manner were consistently closer to E_{HF} than the corresponding values of E_{HF}^{α} based on optimized choices of α.

We were greatly encouraged by the improvement in $E_{HF}^{\alpha\beta}$ brought through the inclusion of the exchange inhomogeneity term, as well as by the fact that the optimum value of turned out to be very close to 2/3, the value expected on theoretical grounds for a homogeneous free-electron gas.[2] We interpreted[21] the Z dependence of α in the optimized Xα approximation as being due to the single αV_{XS} term having to take into account not only the homogeneous contribution to exchange, but also inhomogeneity contributions which were different for different atoms. By treating the homogeneous and inhomogeneous parts separately, through αV_{XS} and $\beta G(\rho)V_{XS}$, respectively, we were able to demonstrate the universality of the homogeneous term ($\alpha = 2/3$ for all atoms studied), as well as the relatively weak Z dependence of β.

Subsequent work by Boring[24] showed that one obtained nearly the same value for β for several atoms located throughout the periodic table, provided one used the virial theorem criterion[25] rather than the minimum $E_{HF}^{\alpha\beta}$ criterion in choosing β (with α set equal to 2/3). According to the virial theorem criterion, one chooses β such that $\eta = -\langle V\rangle/\langle T\rangle = 2$, where $\langle V\rangle$ and $\langle T\rangle$ are the expectation values of potential and kinetic energies, obtained by inserting the X$\alpha\beta$ atomic orbitals into the Hartree-Fock expressions for these energies. In practice, the value for $E_{HF}^{\alpha\beta}$ obtained on the basis of the virial theorem criterion lies slightly above the $E_{HF}^{\alpha\beta}$ obtained from the minimum $E_{HF}^{\alpha\beta}$ criterion, and considerably below the optimized value of E_{HF}^{α}. (See Figures 1 and 2.)

In recent months we returned to the X$\alpha\beta$ approximation (cf. Eqs. (1) and (2)) and modified the convergence factor[21,22] so that the

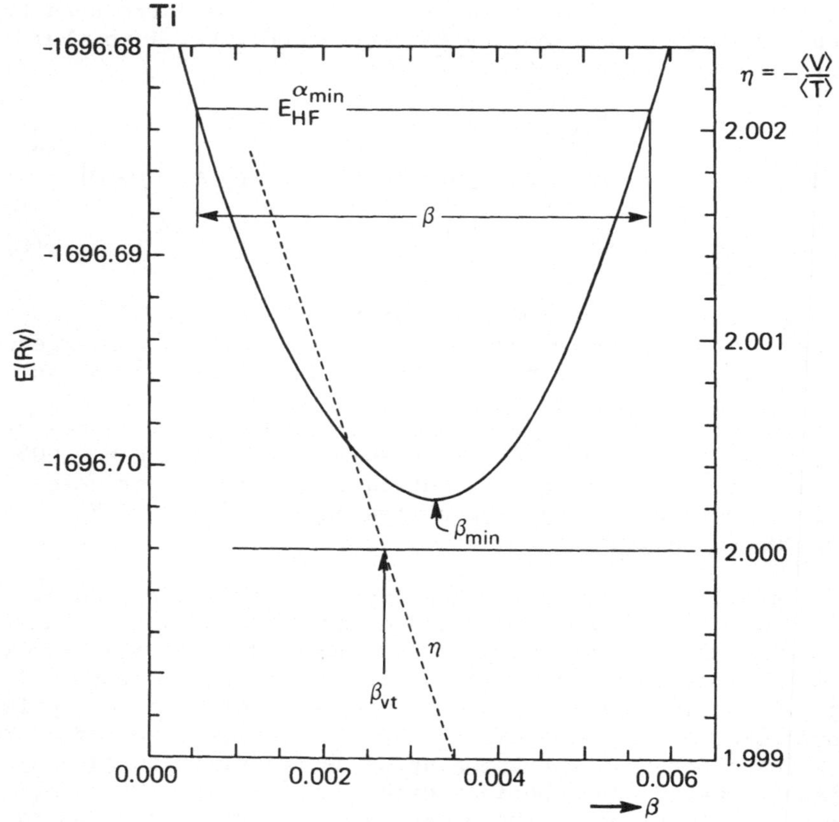

Figure 1. Determination of the optimum value of β for atomic Ti on the basis of new Xαβ exchange approximation.[12] Here α has been set equal to 2/3, and the total energy (Hartree-Fock expression) is displayed as a function of β. The total energy given by the optimized Xα approximation lies near the top of the figure, where it is denoted by $E_{HF}^{\alpha min}$. In the Xαβ approximation, with α = 2/3, any value of β in the range 0.0006 to 0.0057 yields a Hartree-Fock total energy lower than $E_{HF}^{\alpha min}$. The minimum value of $E_{HF}^{\alpha\beta}$, which occurs at α = 2/3, β_{min} = 0.0033, is only 0.038 Ryd above the rigorous Hartree-Fock total energy of -1696.74 Ryd,[26] a fractional deviation of 2.4 x 10^{-5}. The dependence of the virial coefficient, η, on β is shown by the dashed line. The virial theorem criterion leads to a value of β_{min} = 0.0027, and a corresponding $E_{HF}^{\alpha\beta}$ which is only 0.002 Ryd above the absolute minimum of $E_{HF}^{\alpha\beta}$ (given by β_{min} = 0.0033). The presently recommended universal value, β = 0.003, represents a compromise between the minimum $E_{HF}^{\alpha\beta}$ and the virial theorem criteria, with greater weight given to the latter. It is important to bear in mind that the choice β = 0.003 is keyed to Eq. (3), i.e., to the hyperbolic tangent convergence factor, which is considered superior to that used earlier.[21-24]

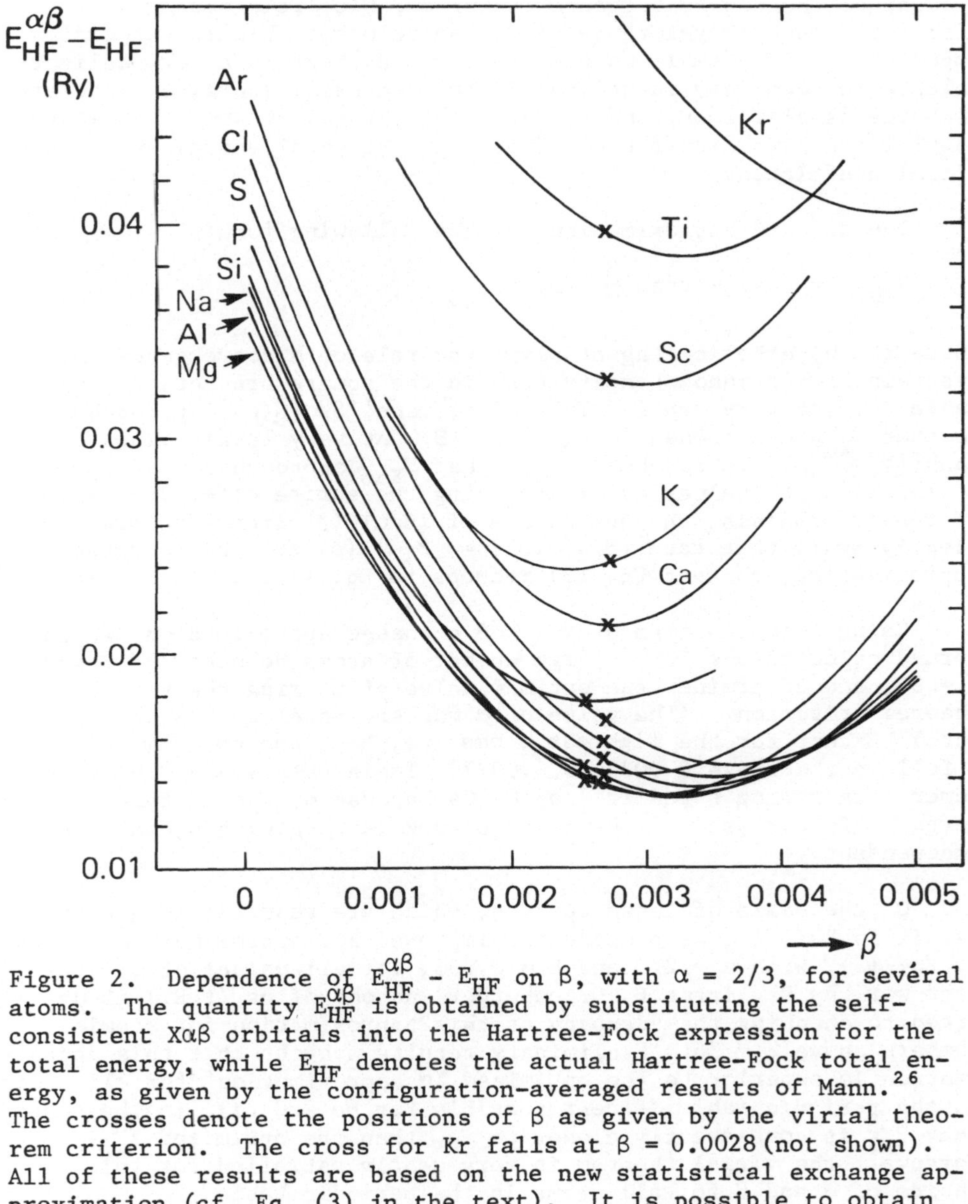

Figure 2. Dependence of $E_{HF}^{\alpha\beta} - E_{HF}$ on β, with α = 2/3, for several atoms. The quantity $E_{HF}^{\alpha\beta}$ is obtained by substituting the self-consistent Xαβ orbitals into the Hartree-Fock expression for the total energy, while E_{HF} denotes the actual Hartree-Fock total energy, as given by the configuration-averaged results of Mann.[26] The crosses denote the positions of β as given by the virial theorem criterion. The cross for Kr falls at β = 0.0028 (not shown). All of these results are based on the new statistical exchange approximation (cf. Eq. (3) in the text). It is possible to obtain another set of optimized β parameters by finding the values of β for which $E_{STAT}^{\alpha\beta} = E_{HF}$, where $E_{STAT}^{\alpha\beta}$ denotes the total energy in the statistical exchange approximation. As already found in earlier work by one of us,[15] this set of optimized β parameters is nearly identical to that given by the virial theorem criterion. For all the atoms shown, the minimum value of the fraction $(E_{HF}^{\alpha\beta} - E_{HF})/E_{HF}$ is of order 10^{-5}. This minimum fraction ranges from about 3×10^{-5} for the lightest atoms shown to 0.7×10^{-5} for Kr.

product of the new convergence factor and $G(\rho)$ remained equal to $G(\rho)$ for a much larger range of r than before. This is especially important at very small values of r, where there is a very delicate balance between very large kinetic and potential energies associated with the 1s electrons, and where slight changes in the exchange potential can have significant effects on the total energy and on the virial coefficient.

The new Xαβ approximation has the following form:

$$V_{X\alpha\beta} = \alpha[1 + \tanh\{\frac{\beta}{\alpha} G(\rho)\}] V_{XS} , \tag{3}$$

where the hyperbolic tangent plays the role of a convergence factor, restraining the inhomogeneity term in the square brackets to the range ±1. At very small values of r, $\tanh\{[\beta/\alpha]G(\rho)\}$ approaches +1, so that $V_{X\alpha\beta}$ approaches $2\alpha V_{XS}$ as $r \to 0$. At very large values of r, $\tanh\{[\beta/\alpha]G(\rho)\}$ approaches -1, so that $V_{X\alpha\beta}$ approaches 0 as $r \to \infty$. At intermediate values of r, including the entire effective range of the 1s orbitals, the deviations of $[\beta/\alpha]G(\rho)$ from zero are sufficiently small that $\tanh\{[\beta/\alpha]G(\rho)\} = [\beta/\alpha]G(\rho)$ to a high degree of approximation, so that Eq. (3) reduces to Eq. (1), as it should.

Using this new form of the Xαβ exchange approximation, we performed calculations for a large number of atoms between Z = 2 and Z = 36, and determined the optimum value of β using the virial theorem criterion. (The value of α was set equal to 2/3 throughout.) Except for the lightest atoms (He, Be), the optimum value of β fell in the range 0.0023 to 0.0027. These values are consistently lower than Boring's result[24] of 0.004 because of the different convergence factor used in the present work (as Boring has confirmed independently).

On the basis of these results, which are reported in greater detail in Ref. 12, we propose the improved Xαβ approximation, given by Eq. (3), with α = 2/3 and β = 0.003, for all values of Z. Current work by Dr. Irene B. Ortenburger and one of us (F.H.) is devoted to checking the adequacy of this approximation for atomic numbers above Z = 36. Preliminary results suggest that this approximation is superior to the optimized Xα approximation[3] for all atoms in the periodic table (except possibly for He and Li), in the sense that $E_{HF}^{\alpha\beta}$ is consistently closer to E_{HF} than the optimized E_{HF}^{α}. Moreover, the virial theorem is very nearly satisfied for this choice of α and β for all atoms in the periodic table.

Adoption of this improved Xαβ approximation will lead to total energies closer to the Hartree-Fock values, at the same time avoiding the difficulties with spatially-dependent (or discontinuous) exchange potentials inherent in the optimized Xα approximation.[3] Inclusion of the exchange inhomogeneity term is relatively simple from the computational point of view. It is hoped that this improved Xαβ approximation will find increasing use in atomic, molecular, and solid state calculations in the future.

The authors wish to thank Dr. Michael E. Boring and Dr. Irene B. Ortenburger for instructive discussions and useful exchanges of information.

REFERENCES

1. J. C. Slater, Phys. Rev. 81, 385 (1951). (XS method)
2. W. Kohn and L. J. Sham, Phys. Rev. 140, A1133 (1965); and R. Gaspar, Acta Phys. Acad. Soc. Hung. 3, 263 (1954). (XKSG method)
3. J. C. Slater, T. M. Wilson and J. H. Wood, Phys. Rev. 179, 28 (1969); J. C. Slater and J. H. Wood, Intern. J. Quantum Chem. 4S, 3 (1971); and J. C. Slater, Advances in Quantum Chemistry, Vol. 6. P.-O. Löwdin, editor (Academic Press, New York), to be published. (Xα method)
4. F. Herman and S. Skillman, Atomic Structure Calculations. (Prentice Hall, Englewood Cliffs, NJ, 1963).
5. D. Liberman, J. T. Waber and D. Cromer, Phys. Rev. 137, A27 (1965); and D. Liberman, D. Cromer and J. T. Waber, Computer Phys. Commun. 2, 107 (1971). Also J. B. Mann, unpublished.
6. J. P. Desclaux, Computer Phys. Commun. 1, 216 (1970); and C. C. Lu, T.A. Carlson, F. B. Malik, T. C. Tucker and C. W. Nestor, Jr., Atomic Data 3, 1 (1971).
7. B. Alder, S. Fernbach and M. Rotenberg, editors, Methods in Computational Physics. (Academic Press, New York, 1968), Vol. 8.
8. J. O. Dimmock, Solid State Phys. 26, 104 (1971).
9. K. H. Johnson and F. C. Smith, Jr., Computational Methods in Band Theory, edited by P. M. Marcus, J. F. Janak, and A. R. Williams (Plenum Press, New York, 1971), p. 377.
10. K. H. Johnson, J. Physique, 1972, in press (Menton Conference Proceedings).
11. K. H. Johnson, Advances in Quantum Chemistry, Vol. 7, edited by P.-O. Löwdin (Academic Press, New York) to be published.
12. K.-H. Schwarz and F. Herman, J. Physique, 1972, in press (Menton Conference Proceedings).
13. E. Kmetko, Phys. Rev. A1, 37 (1970).
14. J. H. Wood, Intern. J. Quantum Chem. 3S, 747 (1970).
15. K.-H. Schwarz, Phys. Rev., in press.
16. L. Hedin and S. Lundqvist, Solid State Phys. 24, 1 (1970).
17. L. Hedin and S. Lundqvist, J. Physique, 1972, in press (Menton Conference Proceedings).
18. B. I. Lundqvist and S. Lundqvist, this volume.
19. L. Hedin, this volume.
20. A. W. Overhauser, Phys. Rev. B2, 874 (1970).

21. F. Herman, J. P. van Dyke and I. B. Ortenburger, Phys. Rev. Letters 22, 807 (1969).
22. F. Herman, I. B. Ortenburger and J. P. van Dyke, Intern. J. Quantum Chem. 3S, 827 (1970). See also N. O. Folland, Phys. Rev. A3, 1535 (1971); and J. C. Stoddart, A. M. Beattie, and N. H. March, Intern. J. Quantum Chem. 4S, 35 (1971).
23. I. B. Ortenburger and F. Herman, Computational Methods in Band Theory, edited by P. M. Marcus, J. F. Janak and A. R. Williams, (Plenum Press, New York, 1971) p. 469.
24. M. E. Boring, Phys. Rev. B2, 1506 (1970). Boring used the same convergence factors as we did in Ref. 21.
25. M. Berrondo and O. Goscinski, Phys. Rev. 184, 10 (1969); and D. J. McNaughton and V. H. Smith, Jr., Intern. J. Quantum Chem. 3S, 775 (1970).
26. J. B. Mann, "Atomic Structure Calculations. I. Hartree-Fock Energy Results for the Elements Hydrogen to Lawrencium," Los Alamos Scientific Laboratory Report LA-3690 (1967), unpublished.

COHESIVE ENERGY OF THE LITHIUM METAL BY THE AMO METHOD

Gunnar Sperber and Jean-Louis Calais

Quantum Chemistry Group, University of Uppsala

Uppsala, Sweden

The total energy per atom of the lithium metal has been calculated as a function of the internuclear distance with several versions of the alternant molecular orbital method (AMO). This method is applicable to alternant systems like the body-centered cubic lattice of the alkali metals. The purpose of the method is to allow electrons with different spins to occupy different orbitals in order to account at least partially for correlation.

The AMO's can be regarded either as linear combinations of orbitals which are doubly occupied and non-occupied in the ordinary MO scheme, or, as linear combinations of the MO's associated with the simple cubic sublattices. In the first case the AMO method is considered as a limited configuration interaction, whereas in the second case the property of the AMO's of being semi-localized shows up clearly. It is thanks to this property that the potential energy is lowered.

The calculations have been carried out within an LCAO version of the AMO method. The overlap between atomic orbitals centered at different lattice sites has been accounted for by means of a combination of successive and symmetric orthogonalization. Clementi's double-zeta functions for the free Li atom were used as basis functions.

The main computational difficulty--the many-center integrals--has been circumvented by a procedure developed by Sperber. In the two-electron integrals three of the six integrations have been carried out by expanding the basis functions in Gaussians. All other integrations have been carried out numerically over 1/48 of the Wigner-Seitz cell.

The results depend in a sensitive way on the shape of the Fermi surface. For densities near the experimental equilibrium distance a spherical Fermi surface gives the lowest energy. The most remarkable result is that with this shape of the Fermi surface and for these densities the AMO and MO potential curves coincide. For smaller densities the AMO curve tends to a limit intermediate between the high MO value and the energy of the free atoms.

With a cubic Fermi surface one obtains a repulsive potential curve which tends to the energy of free atoms for large distances. It joins the MO-curve corresponding to a cubic Fermi surface for a considerably higher density than in the case of a spherical surface.

ACKNOWLEDGEMENT

These computations have been made possible thanks to an IBM World Trade fellowship to one of us (G.S.), which is gratefully acknowledged. A more detailed report will be published elsewhere.

IV

Solid State Astrophysics

INTRODUCTORY REMARKS

Ludwig Biermann

Max-Planck-Institut für Physik und Astrophysik

München, Deutschland

In order to introduce the subject of this session, I have first to explain why we feel, that a discussion of solid state astrophysics has become worthwhile, though for many years astrophysics and solid state physics co-existed with each other without any real interaction to speak of. To this end I should make first some very general observations.

Most of the observable matter in the universe is found to be present in the form of stars; our sun is a good example of a medium-aged star of rather normal mass, of a type of which many millions exist in our galaxy. As is well known, all ordinary stars consist of ionized gas, that is of plasma, maintained at high temperatures by the nuclear processes which take place in the deep interior. Only stars with a mass less than about 1/15 $M_\odot$ ($M_\odot$ mass of the sun, $2 \cdot 10^{33}$ gr) will become luminous after sufficient contraction without nuclear processes setting in; such stars must therefore eventually cool down again with their original chemical composition and become what is called a black dwarf. These are bodies like our planets which are too faint for direct optical observation at stellar distances, but sometimes indirectly observable, by their gravitational action, as invisible components of double stars.

Within our own planetary system, the members of which have masses $\leq$ 1/1000 $M_\odot$, we have the two groups of the terrestrial and the major planets, which differ in their mass range, their chemical constitution and their solar distance. The problems of the internal constitution both of the terrestrial and the major planets lead of course to questions of phase transformation and density changes under high pressure - above 10^{12} dyn/cm^2 - by which even normal terrestrial matter may acquire characteristics of metals.[1] Such questions, however, have usually been left to the geophysicists or to interested theoretical chemists, since there was little relation to the central subjects of astrophysics, the constitution and the evolution of stars and stellar systems, and of the universe at large.

Solid matter is finally observed in cosmic spaces, outside the solar system, in the form of small dust particles, which partly absorb and to some extent polarize the light of all but the nearest stars. These dust particles now seem to some extent to be a by-product of the condensation of stars like our sun and (perhaps) of the formation of planetary systems. To their theory solid state physics has been of little help so far, the main tool still being classical electro-magnetic theory of light scattering.

As an exception I may mention the theory of the interaction of such particles with the intersteller gas and of the chemical processes on the surface of such dust grains, which is receiving increasing attention in connection with the observations of interstellar molecules in the radio frequencies and in the UV.

Solid state astrophysics actually came into being only after the very recent discovery (in 1968) of the pulsars or, to be precise, after it was realized that they must be rotating strong magnetized neutron stars (with periods of some 10^{-2} sec to several sec, radii $\approx$ 10 km, mass $\approx M_{\odot}$, $\approx 10^{12}$ Gauss at the surface). Such stars must necessarily possess an outer solid crust of ordinary material. The interpretation of the observations of young pulsars, particularly the small irregularities in the otherwise very smooth increase of the period, requires, as will be seen from Dr. Baym's and Dr. Börner's contributions, the knowledge of the crust's properties, for instance its electrical conductivity and its mechanical properties. Though such irregularities are measured in nanoseconds only, their interpretation, for instance in terms of star quakes, is by no means trivial. This appears to be the first real impact of solid state physics on a major astrophysical problem.

Before I call on Dr. Börner to present his companion paper, I should, however, explain perhaps in a little more detail why neutron stars are of such great interest. They represent a semi-final state of steller evolution[2], the existence of which had been suspected by Baade and Zwicky already in 1934; their theory based on that of the equation of state at sub-nuclear and nuclear densities, $\gtrsim 10^{11}$ and $\gtrsim 10^{14}$ gr/cm^3 respectively, has been developed over the last 30 years by many authors, of whom I mention Oppenheimer, Wheeler, and Cameron[3].

The state of stellar evolution has been called semi-final,

because it is final only with regard to the density distribution, the pressure having become almost independent of the temperature. The star may, however, still have sufficient energy, mainly (in the case of the pulsars) in the form of kinetic energy of rotation, to become, by producing energetic electrons, visible at least in the radio frequency range; the Crab pulsar, the youngest of all known pulsars, is visible also in the optical and still higher frequencies. The loss of rotational energy is measured by the observed slow increase of the period. The moment of inertia and the coupling between the crust and the fluid interior are, however, not necessarily strictly constant, as suggested by the irregularities in the steady, very slow increase of the period; it is at this point that the properties of the crust and solid state physics come in.

From the theoretical work mentioned before it appears likely, that the same processes which lead to the formation of a neutron star, may also lead to the formation of a so-called "black hole", a singularity in the language of the relativistic theory of gravitation. Such a "black Hole" would essentially be unobservable, though it would contribute to the gravitational field of the galaxy to which it belongs. The discovery of the pulsars makes it thus probable that some part of the mass of our galaxy is contained in black holes, though it is still quite impossible to say how large this may be.

The collapse of a star into either a neutron star or a black hole seems to manifest itself by a supernova event with a total energy release of the order of 10^{53} to 10^{54} erg, of which only a small fraction appears in the form of conventional kinetic or radiation energy. Taking into account the observed frequency of supernovae (one in 10^9 sec or 30 years), it is seen that these energies are comparable to the integral thermal radiation of all stars of our galaxy. Needless to say that this statement implies many more questions than answers.

References

1. cf. for instance A.H. Cook, 1971, Quart. J.R.A.S. 12, 154, and S.K. Runcorn (Ed.) "The Application of Modern Physics to the Earth and Planetary Interiors", Wiley-Interscience, 1969.

2. The somewhat better known semi-final state of the so-called "white dwarfs", which have densities only of the order of 10^6 gr/cm^3 and which may end up in a quasi-crystallized state, will, I understand, be included in Dr. Baym's talk.

3. See B. Harrison, K. Thorne, M. Wakano, J. Wheeler (Eds.),"Gravitation Theory and Gravitational Collapse", Univ. of Chicago Press 1965.

OBSERVATIONAL EVIDENCE FOR SOLID STATE PHENOMENA IN PULSARS

Gerhard Börner

Max-Planck-Institut für Physik und Astrophysik

München, Deutschland

What is the observational evidence supporting the view of pulsars as gigantic astrophysical solid state laboratories? Let us first of all review briefly the elimination process by which one arrives at the conclusion that pulsars have to be rotating neutron stars:

Initially there were essentially two alternatives for explaining the observed periods of pulsars: Pulsations of very dense stars, with mean densities in the range $10^8 \cdots 10^9$ g/cc (white dwarfs, $\sqrt{G\rho} \approx 10^{0.5} \cdots 10$ rad/sec), which was discussed by a number of authors. Secondly rotating neutron stars with $\Omega^2 \ll G\rho$ (T. Gold, E. Pacini). The further possibility of dense contact binary systems, which leads essentially to the same relation between the period and the density as pulsation in the fundamental mode, was soon ruled out by the very high stability of the periods. This stability showed that there cannot be an emission of large amounts of gravitational radiation. The white dwarf pulsation hypothesis was ruled out consequently with the discovery of the fine structure of the pulses ($\leq$ 10% of the period, 0.2 ms $\approx$ 60 km or 10^{-4} $R_\odot$ diameter of the emitting region), with that of the regular rapid change of the angle of polarization during each pulse, and particularly with the discovery of the two young pulsars in supernova remnants, with periods of less than 0.1 sec, which cannot be understood at all as stable pulsations of a white dwarf (Crab NP 0532 : 33 ms, Vela PSR 0833-45 : 89 ms). Finally the slow secular increase of the periods typical for pulsars is another point in favour of the rotating neutron star model, because one would expect the loss of rotational energy to lead to a slowing down of the rotation. For all these reasons the model of a rotating neutron star as an explanation for the pulsar phenomenon has been generally accepted.

The formation of a neutron star in the course of a supernova event would lead to a small object, with high density and a strong magnetic field: Because of the high electron number in such a star, the electrical conductivity will be very large, and the magnetic flux will be frozen in ($B \sim R^{-2}$)

$$(R : 10^{12} \rightarrow 10^{6};\ M = M_{\odot};\quad \rho : 10^{-3} \rightarrow 10^{15}\ \text{g/cc};\ B : 1 \rightarrow 10^{12}\ \text{gauss})$$

How does the rotating neutron star communicate with us? All that is observable is the electromagnetic radiation produced by charged particles accelerated in the strong magnetic fields around the pulsar, and reaching us as continuum radiation or in pulsed form. No convincing model of how the pulses are formed has been put forward, but some gross features have been explained quite well. Thus it was shown by Goldreich and Julian that despite the strong gravitational attraction from a neutron star, there cannot be a vacuum outside the star (they took the magnetic field aligned with the rotation axis; the oblique rotator was treated similarly by J. M. Cohen and E. T. Toton).

Assume an interior magnetic field, which will be frozen in, and which is consistent with an exterior dipole field. Because of the high conductivity of the neutron star interior, the condition that the electric field vanishes in the rest frame of the star is a good approximation. Consequently in the frame of an observer not rotating with the star, the electric field is given by

$$\vec{E} + \vec{V} \times \vec{B} = 0$$

Via div $E = \rho/\varepsilon_0$ the charge density associated with the electric field is given by

$$\rho = -\,2\,B_0\,\Omega\cos\Theta\,\varepsilon_0\ ; \qquad \frac{\vec{B}\cdot\vec{\Omega}}{B\cdot\Omega} = \cos\Theta$$

If it is now assumed that the neutron star is surrounded by a vacuum, the solution of Maxwell's equation in the vacuum outside has to be matched to the interior solution via continuity of the magnetic field component B_r normal to the surface, and of the tangential component of the electric field. It is found then that the quantity $\vec{E}\cdot\vec{B}$, which is 0 inside the star, does not vanish outside. On the contrary $\vec{E}\cdot\vec{B} \sim R\,\Omega\,B_0^{\,2}$.

Thus near the surface charge layer of the neutron star the electric force along the magnetic field exceeds the gravitational force by a large factor of the order of 10^{13} for electrons and 10^{10} for protons. These ratios were obtained by using parameters typical of the Crab pulsar ($B \simeq 10^{12}$ Gauss, $\nu \simeq 30$ Hz, $R \simeq 10$ km, $M \simeq M_{\odot}$). Thus if the surface region is ionized, the surface charge layer cannot be in dynamical equilibrium. A rotating magnetic neutron star must possess a magnetosphere, composed of charged particles traveling along the magnetic field lines.

The radiation from these charged particles, injected into the magnetosphere and accelerated, is what we observe. For 22 pulsars both frequency Ω, and change of frequency $\dot{\Omega}$ have been measured. Then by measuring their rate of loss of energy $\dot{E} = I\,\Omega\,\dot{\Omega}$, we could in principle determine the moment of inertia I of these neutron stars. This in turn would precisely fix mass and density profile of the star, according to the equation of state used. Lower limits on the momenta of inertia actually present in neutron stars might be deduced this way.

The observations are not exact enough to permit definite conclusions in this line of investigation. There are, however, 3 other observations which present us with direct evidence of the solid structure that should be present, as e.g., according to Gordon Baym, in neutron stars:

The first observation is the simple statement that the absolute constancy of the pulse shape in all pulsars indicates that the magnetic field has a very constant shape too, which it can only have if it is anchored in the crust composed of a Coulomb lattice of nuclei as discussed by Gordon Baym. If there was no crust, wiggles in the magnetic field would show up as irregularities in the pulse shapes. The second two observations referred to are the sudden speed ups observed in the two fastest pulsars.

On September 28, 1969, a sudden increase ("glitch") in the crab pulsar (NPO 532) frequency was observed, smaller but similar in nature to the speed up inferred for the Vela pulsar PSR 0833-45. The parameters for these two pulsars are as follows:

	Vela	Crab
$\Omega(s^{-1})$	70.5	190
T(years)	2.4×10^4	2.4×10^3
$\frac{\Delta\Omega}{\Omega}$	2.34×10^{-6}	6.9×10^{-9}
$\frac{\Delta\dot{\Omega}}{\dot{\Omega}}$	6.8×10^{-3}	8.5×10^{-4}

where $T = \frac{\Omega}{\dot{\Omega}} \frac{1}{(2\pi)^2}$

Further analysis showed that the post-speed-up behaviour looked very much like some sort of relaxation phonomenon, where the frequency seems to fall back exponentially to the steady state with a characteristic time of 1 year for Vela and 8 days for the crab. Thus the pulsars settled down to a long-term frequency increase of

$$\frac{\Delta\Omega}{\Omega} = 0.28 \times 10^{-9} \quad \text{for the Crab}$$

and of

$$\frac{\Delta\Omega}{\Omega} = 1.96 \times 10^{-6} \quad \text{for Vela.}$$

Baym, Pines, Pethick, Ruderman explained these observations in the following way ("starquake theory"):

The initially oblate crust, formed when the star was spinning comparatively fast, and stressed as the centrifugal force on it decreases, cracks when the internal stress exceeds the yield point. This results in a fractional decrease in its moment of inertia, and by conservation of its angular momentum in a speed up of the crust.

The behaviour following this starquake can be understood in terms of a simple two-component model. One component is the combined crust-charged particle system of moment of inertia I_c, rotating uniformly with angular velocity $\Omega(t)$. The second component is the neutron superfluid, with moment of inertia I_n rotating uniformly with angular velocity $\Omega_n(t)$. The neutron superfluid responds in a characteristic time τ_c to changes in the crust's angular velocity, via the scattering of the electrons by the "normal fluid" cores of the vortex lines in the rotating superfluid. One would expect the time τ_c, characterizing the coupling between the neutron liquid and the crust-charged particle system, to be microscopic because it does not depend on global processes. That it is of the order of years (Vela) or days (Crab) is taken as evidence that the interior of these pulsars is at least in part superfluid (Baym, Pines, Pethick, Rudermann). When both protons and neutrons are superfluid the coupling occurs via the magnetic moment interaction of the electrons with the normal cores of the rotational vortices in the neutron superfluid. This leads to (Baym, Pethick, Pines).

$$\frac{1}{\tau_c} \sim \frac{8\pi}{g} \left(\frac{e^2}{\hbar c}\right)^2 \left(\frac{T}{T_e}\right)^2 \chi^2 \; c \; K_f \; \frac{\rho_e}{\rho_{ep}} \frac{\Omega}{\Omega_{c_2}}$$

T_e : electron fermi temperature, K_f el. fermi wave number

Ω_{c_2} : critical angular velocity $\sim 10^{20}\ \text{sec}^{-1}$

χ : neutron magnetic moment; ρ_e electron, ρ_{cp} electron-proton mass density

Putting in reasonable numbers one finds

$$\tau_c \sim \left(\frac{1}{T_8}\right)^2 \quad \text{(years)}$$

T_8 : temperature in units of 10^8 °K.

τ is, as stated above, 8 days for the Crab, 1 year for Vela, which ties in perfectly with the fact that Vela is much older than the Crab (and colder). Recent observations after these big speedups show in both pulsars small erratic changes in period, and a possible explanation for these is that microquakes are occurring very frequently.

The explanation of starquakes as common happenings (every few years) meets with difficulties, because it is very hard to introduce big enough stresses in the crust in a few years. So both these pulsars have to have a very large crust indeed, they have to be neutron stars on the lower end of the stability region. If one takes the 1 parameter starquake theory by Baym and Pines, and requires speedups in the Crab to occur once a year, then $m_{crab} = \frac{1}{8} m_o$, quakes for Vela every 10 years would require this star to be lighter still. The situation is illustrated by the following table:

m/m_o	I_{44}	ω_q^2
.1	.61	
.15	.60	2.7×10^8
.20	.70	4.7×10^9
.23	.80	
.3	1.00	7×10^{10}
.46	1.50	5×10^{11}

I_{44} is the moment of inertia in units of 10^{44} g cm^2. ω_q^2 tells the time to the next quake of comparable magnitude t_q (in years) if one multiplies it by 6.7×10^{-10} for the Crab, and 10^{-5} for Vela. There are difficulties here with the theory because the stars have to be very light (almost contradicting the observations of the lower limit of the necessary energy supply to the Crab nebula). There is also one difficulty with the observations itself, namely that there are two big glitches observed right in the first 6 months of observations, and after that no more. Thus, if there is no speedup observed in the course of the next 5 years or so, doubts will grow on these events.

As usual, the theoretical situation is much better than the experimental one, because only two speedups have been observed but there are at least 5 different theories to explain them. The Vela glitch is always difficult to account for, in fact Pacini's plasma explanation does not work there at all.

The explanation that speedups are caused by planets (e.g., Rees, Trimble, Cohen), is ruled out by the observed post-glitch long-term frequency increase, and by the microquakes. Cameron and G. Greenstein want to explain the speedup by fluid instabilities occurring in the superfluid core, but their theory is still too vague to permit an evaluation of its merits and demerits. Börner and Cohen finally accept the two-component model for the post-glitch behaviour, but attribute the initial speedup to the infall of mass. In this accretion model a massive body would fall onto the pulsar, transferring its angular momentum to the crust and producing the observed initial spin-up. After that some time will be necessary for the mass-distribution of the pulsar to readjust itself as well as for the change in angular momentum to be communicated to the stellar interior, and then the pulsar will settle down to the observed long-term frequency increase. They find that by choosing a specific model of a neutron star, all the unknown quantities are determined, even the infalling mass Δm can be found. For the Crab glitch they obtain, assuming as a model a rotating neutron star of 1.44 $m_\odot$, that

$$\Delta m = \frac{2.8 \times 10^{-10}}{0.88} m_\odot$$

about 10% the moon's mass. The angular momentum transferred is 1.1×10^{38} g cm^2 sec^{-1}. An energy is set free equal to the difference between the initial and final configuration in gravitational mass (binding energy) of 7×10^{43} ergs. This just corresponds to the energy estimated by Harlan and Scargle for the activity observed in the wisps of the Crab nebula, accompanying the glitch.

For the Vela pulsar, a 1.44 $m_\odot$ star requires the infall of $\Delta m = 2 \times 10^{-6} m_\odot$ about 2/3 the earth's mass.

Börner's and Cohen's theory gives a lower limit for the mass of the Crab pulsar of 1 $m_\odot$.

NEUTRON STARS AND WHITE DWARFS*

Gordon Baym

Department of Physics, University of Illinois

Urbana, Illinois, 61801, U.S.A.

This talk reviews the structure of neutron stars and white dwarfs, and the role of solid state physics in determining their properties. The nature of the matter in neutron stars (matter essentially in its absolute ground state), and the determination of its equation of state are first discussed; this is followed by a description of the resulting stellar models. Superconductivity and superfluidity in the interior, and other solid state aspects of these stars are then reviewed.

I. INTRODUCTION

Astrophysics presents many interesting and exciting new problems in solid state physics. In neutron stars and white dwarfs, which contain practically all the solid in the universe, one meets condensed matter existing in a whole new range of parameters - magnetic fields $\sim 10^{12}$ gauss, and densities at which the nuclei are almost touching, to give but two examples. The recent theoretical interest in neutron stars is due to their having been observed as pulsars - the stars that blink on and off with extraordinary precision[1]. In this talk I shall review the structure of neutron stars and white dwarfs and try to sketch out the role played by solid state physics in determining their properties.

White dwarfs and neutron stars are two possible endstates of stellar evolution; the third is black holes. Masses of white dwarfs[2] are typically a solar mass ($M_\odot$) or less, with radii several thousand kilometers or more; typical mass densities in the interior are $\sim 10^8 g/cm^3$. The mass of the star is supported against the tremendous force of gravity by the degeneracy pressure of the electrons in the matter, exactly as electron degeneracy pressure supports an ordinary metal. Neutron star masses are also on the order of a solar mass, but their radii are in the range of 10 km; thus the densities found in their interiors are $\sim 10^{14}$-10^{15} g/cm^3. By comparison, the density of matter inside a large nucleus is $\sim 3 \times 10^{14}$ g/cm^3. Neutron stars are supported against gravity primarily by the pressure of free neutrons in their interiors.

A cross section of a typical neutron star is shown in Fig. 1.

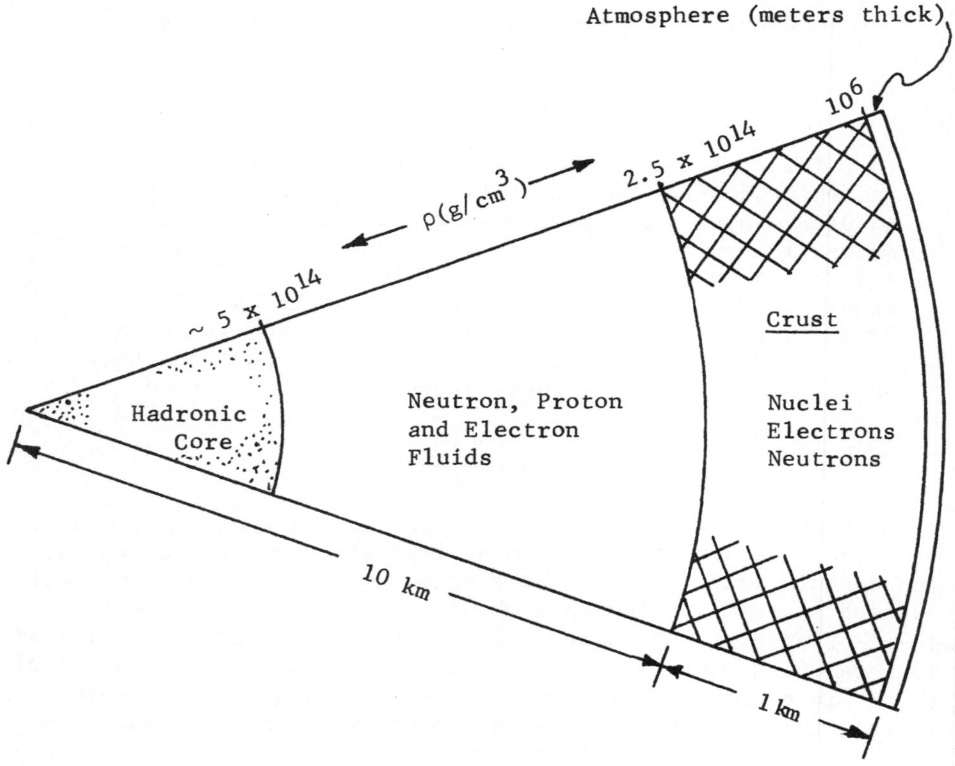

Fig. 1. Cross section of a typical neutron star.

The surface gravity g is about 10^{11} times that of the earth; any normal density object larger than about 10 cm would be torn apart by the tidal forces of the star.

Temperatures of young neutron stars are estimated from calculations[3] of cooling rates to be $\sim 10^8$ K in the interior and one to two orders of magnitude smaller at the surface. Interior temperatures are extremely uniform, as a consequence of the high thermal conductivity of degenerate matter. Measured on the scale of microscopic excitation energies, typically several MeV (1 MeV=10^{10} K), neutron stars are extremely cold, comparable to liquid He^3 in the millidegree range. The height of the atmosphere of a neutron star is thus no more than a few meters. (We shall see later that if the magnetic field is sufficiently large there may well be no atmosphsere at all.) The melting temperature T_m of the matter rises with density ρ as[4]

$$T_m \sim 3 \times 10^2 \rho^{1/3} Z^{5/3} \tag{1.1}$$

(in degrees K, with ρ in g/cm^3; Z is the nuclear charge). Not too far into the star the local T_m equals the ambient T, and at greater depths, $T_m > T$, that is, the matter is solid. The solid crust of a neutron star is on the order of a kilometer thick, although very light neutron stars can have entirely solid interiors.

The existence of the solid crust in neutron stars is responsible for pulsar timing being so regular. The pulsar clock mechanism is simply the rotation of the neutron star; the magnetic field couples the rotating star to the surrounding magnetosphere,where the electromagnetic emission takes place. Were the star not sheathed in solid crust, the field would not be firmly anchored in the star and forced to co-rotate rigidly with it. The stable pulsar clockwork can be taken as evidence for the existence of the solid crust.

Beyond $\rho \sim 10^4$ g/cm^3 the atoms are completely ionized; all the electrons are free to move throughout the lattice. At further depth, beyond $\rho = 4.3 \times 10^{11}$ g/cm^3, a gas of free neutrons, possibly superfluid, is also present in the lattice. The crust ends at a density $\sim 2.5 \times 10^{14}$ g/cm^3, at which point the nuclei are essentially touching. The matter in the interior is composed primarily of a degenerate neutron fluid, with a small admixture of degenerate proton and electron fluids; the neutrons and protons are likely to be superfluid here. Then deep in the interior there is a hadronic core in which various strange particles, such as Λ's and Σ's can exist stably.

White dwarfs have a much simpler structure. The maximum interior temperature reached in the evolution of a white dwarf depends on its total mass M as[5]

$$T_{max} \sim 10^9 \left(\frac{M}{M_\odot}\right)^{4/3} \left(\frac{A}{Z}\right)^{8/3} \tag{1.2}$$

where A and Z are the total number of nucleons and protons in a typical nucleus. As the star cools the temperature of its interior falls below the local melting temperature and the matter crystallizes. Thus older white dwarfs have a solid core, of uniform temperature, surrounded by a fluid surface, while younger white dwarfs are entirely fluid.

II. EQUATION OF STATE

In the evolution of a neutron star, nuclear processes take place sufficiently rapidly that the star, except possibly in the outermost layers, remains in complete nuclear equilibrium. Strong, electromagnetic, and weak interactions adjust the nuclear composition at each point to the thermodynamically most favorable one. Each time a proton is converted into a neutron (or vice versa) by weak interactions a neutrino is emitted, which leaves the star, carrying away energy and entropy. On a time scale of years from its birth in a supernova explosion, a neutron star cools to a relatively low temperature. To a first approximation we can regard the matter in a neutron star as being in its lowest possible energy state at each point, subject only to local charge neutrality and conservation of baryons.

What is the absolute ground state of matter like? Imagine that we specify that there be n_b baryons per unit volume in the matter and also overall charge neutrality, and allow for all possible strong, electromagnetic, and weak interactions. For matter at zero pressure the lowest energy state consists, first of all, of nucleons arranged into Fe^{56} nuclei, the most tightly bound nucleus; the nuclei are arranged in a lattice, to get the lowest solid state energy, and finally, the electrons in solid Fe have the lowest energy in a ferromagnetic state. The absolute ground state of matter at zero pressure is thus ferromagnetic, solid Fe^{56}.

We expect then that the outermost layers of a neutron star will be primarily Fe^{56} (though the matter will not be solid or ferromagnetic because here $T > T_m$). As one goes deeper into the star the matter becomes compressed by gravity and its character begins to change. When the density reaches a value ≈ 6.3 AZ g/cm^3 the atomic cores begin to touch. Beyond this point the electrons are no longer associated with a given atom, and the nuclei are completely ionized. This begins to occur for Fe^{56} at about 10^4 g/cm^3. By densities an order of magnitude beyond this the electron Fermi temperature is $\gg T$ and the electrons form a degenerate plasma, which is fully relativistic above a mass density $\sim 10^7$ g/cm^3.

For electron Fermi energies $\gtrsim 1$ MeV, or $\rho \gtrsim 8 \times 10^6$ g/cm^3, Fe^{56} is no longer the lowest energy state, since it is favorable for Fe^{56} nuclei to capture energetic electrons from the top of the Fermi sea and rearrange themselves into Ni^{62} nuclei. In the process of electron capture, protons (in nuclei) are converted into neutrons:

$$e^- + p \rightarrow n + \nu \quad ; \tag{2.1}$$

the neutrino escapes, lowering the energy of the system.

Ni^{62} persists as the most favorable nucleus until 2.7×10^8 g/cm^3, when Ni^{64} becomes preferred. Through inverse beta decay processes the matter becomes composed of more and more neutron rich nuclei as the density increases. The sequence of equilibrium nuclei up to 4.3×10^{11} g/cm^3 is as follows:[6]

$$\begin{aligned} &Fe^{56},\ Ni^{62},\ Ni^{64}, \\ &Se^{84},\ Ge^{82},\ Zn^{80},\ Ni^{78},\ Fe^{76}, \\ &Mo^{124},\ Zr^{122},\ Sr^{120},\ Kr^{118} \quad . \end{aligned} \tag{2.2}$$

The nuclei on the second line have 50 neutrons, while those on the third line have 82 neutrons; both of these numbers are closed shell configurations and have particularly low energy.

Across a phase transition from one nuclide to the next in (2.2) the pressure remains continuous. Since the pressure is due almost entirely to electrons (the lattice contributes negligibly to the pressure here) their density, and hence the number of protons per cm^3, must be continuous. Since Z/A decreases at each transition, the density of neutrons must abruptly increase at each transition. Thus each phase transition is accompanied by a density increase

$$\frac{\Delta\rho}{\rho} \approx -\frac{\Delta(Z/A)}{Z/A} \quad . \tag{2.3}$$

For example in the Fe^{56}- Ni^{62} transition, $\Delta\rho/\rho \approx 2.9\%$.

In the sequence (2.2), Kr^{118} is so neutron rich that the last neutron is barely bound. Beyond a density $\rho_d = 4.3 \times 10^{11}$ g/cm^3, the neutrons begin to leak out of the nuclei, and form a degenerate liquid. In this regime, called the "neutron drip regime," the nuclei become more and more neutron rich with increasing mass density, remaining always in a lattice. The neutron drip regime continues until a density $\sim 2.5 \times 10^{14}$ g/cm^3 when the nuclei essentially merge together into a continuous fluid of neutrons, protons and electrons.

The very neutron rich nuclei, as well as the neutrons, are stabilized against beta decay by the presence of the electron Fermi sea. As long as the electron in the decay process

$$(A,Z) \rightarrow (A,Z+1) + e^- + \bar{\nu}$$

has energy below the electron Fermi energy, the process is forbidden by the exclusion principle.

Let me say a few words about calculations of the equation of state of the matter. Before neutron drip the problem of determining the equilibrium nucleus present is simply that of minimizing the total energy per unit volume at fixed baryon density, and overall charge neutrality. The number of nucleons A in a given nucleus is determined primarily by a balance between the nuclear surface energy ($\sim A^{-1/3}$ per nucleon) which favors large nuclei, and the nuclear Coulomb energy ($\sim A^{2/3}(Z/A)^2$ per nucleon) which favors small nuclei. The lattice energy (the net Coulomb interaction energy of all the nuclei and electrons), given by

$$E_{latt} = -1.82\ Z^2e^2/a \tag{2.4}$$

per nucleus for a bcc lattice of lattice constant a, shifts this balance slightly in favor of larger nuclei. This is the first example of solid state physics affecting the types of nuclei present. If one neglects the lattice energy (2.4) in calculating the nucleus present, the sequence before neutron drip becomes:

$$\begin{aligned}&Fe^{56},\ Ni^{62},\ Fe^{58},\ Ni^{64},\ Ni^{66},\ Ni^{68}\\&Se^{84},\ Ge^{82},\ Zn^{80},\ Ni^{78},\ Fe^{76},\ Cr^{74}\\&Kr^{118}\ .\end{aligned} \tag{2.5}$$

Above neutron drip, where the lattice constant varies from about 1/10 to 1 times the nuclear radius, the lattice Coulomb energy becomes more and more significant in determining the type of nucleus present. In this regime the solid state energies are comparable to the nuclear energies. Note that were the nuclei to fill all of space, the total Coulomb energy, nuclear plus lattice, would vanish.

Before neutron drip, nuclear energies can be taken from tables of extrapolated nuclear masses. The presence of free neutrons, however, modifies the simple phenomenological nuclear models. The effect is two-fold: one, the neutrons exert a pressure on the nuclei; two, when the free neutron density becomes comparable to the density inside the nuclei, the nuclei are very neutron rich; thus, the matter inside and the matter outside the nuclei are very similar, and this causes a great reduction of nuclear surface energy. These problems and the detailed calculation of the equation of state here are discussed by Baym, Bethe and Pethick in Ref. 7.

The transition from the nuclear phase to the fluid phase at higher density takes place as follows: by 2.5×10^{14} g/cm^3 the nuclei have become very large, and begin to touch. As the density is raised the density inhomogeneity begins to smooth out, disappearing, as it must[7,8], in a first order transition. The crucial point is that the higher density liquid becomes unstable against finite wavelength density fluctuations as its density is lowered.

The region from the dissolving of the nuclei to about 5×10^{14} g/cm^3 can be handled by the standard techniques of nuclear matter theory[7,9]. The matter is primarily neutrons here with ~ 4% protons. There is no theoretical evidence that the nucleons form a solid in this regime. At still higher densities it becomes energetically favorable to create other elementary particles through reactions such as

$$e^- + n \rightarrow \Sigma^- + \nu \quad ;$$

these particles can be stable in the matter, again as a consequence of the exclusion principle. A description of the matter in this regime is very difficult for several reasons. First, even if one knew all the forces between the various elementary particles, the high density requires new calculational techniques. Some progress here has been made recently by Bethe and Johnson (private communication) and by Pandharipande[10]. The second problem is that one doesn't know the interactions between the various elementary particles. Several models of noninteracting resonances have been proposed[11] but these all suffer from the defect of ignoring repulsion between the particles, with the consequence that the equation of state is unduly soft. Lastly, it is not clear that at high densities where the meson clouds of the baryons are strongly overlapping it makes sense at all to look for a description in terms of familiar elementary particles and resonances interacting via two-body forces.[12] The high density regime remains an outstanding problem.

I would like to jump back now to very low densities and describe briefly some recent work[13] on the properties of matter in the enormous magnetic fields $\sim 10^{12}$ gauss one expects in a neutron star. The basic point is that if the radius r_L of the lowest Landau orbit of a free electron in the magnetic field is very small compared with the Bohr radius Z/a_0 of the most tightly bound atomic electron (in zero field) then the magnetic field will dominate the electron motion perpendicular to the field, causing the atom to become a compact cylinder with axis along the field. Such atoms under zero pressure will form a solid of density $\sim 10^4$ g/cm^3 for a field $B \sim 5 \times 10^{12}$ gauss. The binding of the solid is due not to the attractions between the atomic electric quadrupole moments, but rather to a quantum mechanical sharing of electrons between atoms in a linear chain parallel to the field; this produces chains of extraordinary tensile strength. Adjacent chains are bound together by Coulomb attraction with the nuclei forming a practically bcc lattice.

The existence of this new state of matter implies that the density at the surface of a neutron star with a strong magnetic field falls discontinuously from $\sim 10^4$ g/cm^3 to zero; there is no atmosphere dwindling continuously to zero density. At higher densities, once the nuclei are much closer than r_L, the structure of the matter is little affected by the field.

III. MODELS OF NEUTRON STARS AND WHITE DWARFS

I would like to turn now from the microscopic problem of neutron star structure, namely the determination of the equation of state, to the engineering aspects - the construction of stellar models. The basic procedure is to integrate (numerically) the general-relativistic equation of hydrostatic balance, the Tolman-Oppenheimer-Volkoff equation

$$\frac{\partial P(r)}{\partial r} = -G\frac{[\rho(r)+P(r)/c^2][m(r)+4\pi r^3 P(r)/c^2]}{r^2[1-2Gm(r)/rc^2]} \tag{3.1}$$

to determine the pressure $P(r)$ and mass density $\rho(r)$ at radius r in a spherical star. In Eq. (3.1), $G = 6.67 \times 10^{-8}$ erg cm/gm^2 is the gravitational constant and

$$m(r) = \int_0^r d^3r'\rho(r') \tag{3.2}$$

is the mass contained within radius r. If one lets $c^2 \to \infty$, Eq. (3.1) reduces to the familiar non-relativistic equation of hydrostatic balance. The pressure P and density ρ are related by the equation of state[14] and the central density $\rho_c = \rho(r=0)$, the integration constant for Eq. (3.1), uniquely specifies the star, at T = 0.

In order for a star to be stable against radial oscillations it is necessary, non-relativistically, that the adiabatic index,

$$\Gamma = \frac{\rho}{P}\frac{\partial P}{\partial \rho} \tag{3.3}$$

averaged (with respect to P) over the star, be greater than 4/3

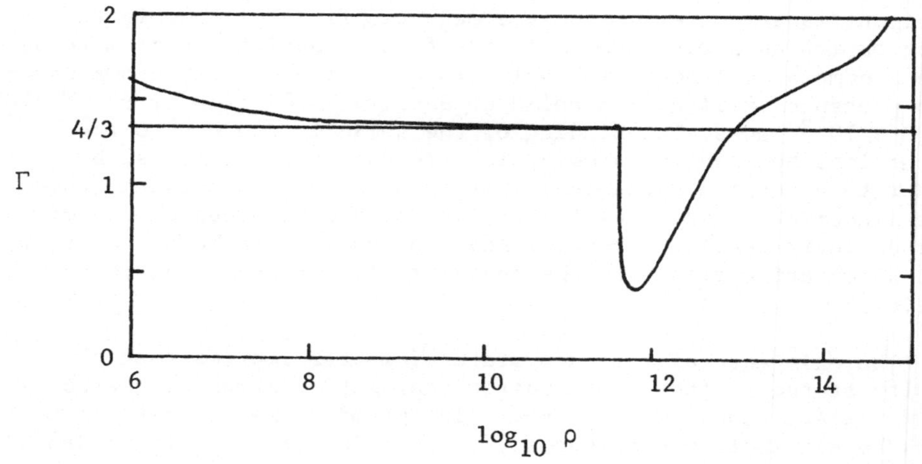

Fig. 2. Adiabatic index as a function of mass density.

A graph of Γ is shown in Fig. 2. For matter at zero pressure, $\Gamma = \infty$ since Γ is proportional to $1/P$; for an ideal monatomic gas it equals 5/3, and for an ideal relativistic gas such as photons or relativistic electrons it equals 4/3. Thus, before neutron drip, Γ is greater than 4/3, but it is very close to 4/3 by the point of neutron drip. Just above neutron drip, however, Γ falls well below 4/3; the reason that the matter is so soft here is that as the mass density increases, the main effect is an increase in the number of free neutrons. But free neutrons, when they first appear, have essentially zero momentum and do little to increase the pressure; thus just above drip $\Delta P/\Delta\rho$ is generally $< P/\rho$. The singularity at neutron drip is in fact of the form[7]

$$\Gamma - \Gamma_{drip} \sim -(\rho-\rho_{drip})^{\frac{1}{2}} \quad .$$

As the density of free neutrons begins to grow, the matter becomes stiffer and Γ begins to rise, becoming greater than 4/3 again when the neutron Fermi pressure becomes on the order of the electron pressure; this occurs at about 7×10^{12} g/cm^3.

We can see now the reason why there are two classes of dense, cold stars - white dwarfs and neutron stars - and not a continuum in between. Basically, most of the matter in a star is at densities close to the central density, and thus when $\Gamma(\rho_c)$ is $< 4/3$, the star will not be stable. There are stable stars with central densities in the region below neutron drip - these are white dwarfs - and stable neutron stars with high central densities, well above the point where Γ becomes greater than 4/3. The branch of stable neutron stars actually starts at $\rho_c = 1.55 \times 10^{14}$ g/cm^3. The microscopic phenomenon of the nuclei becoming so neutron rich that they begin to drip neutrons leads then to the existence of two branches of cold stars.

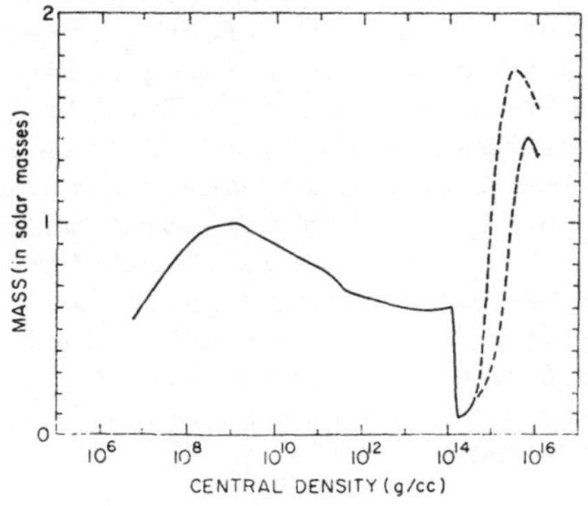

Fig. 3. Mass versus central density for stars composed of cold matter in complete nuclear equilibrium.

Figure 3 shows the masses of cold, equilibrium white dwarfs and neutron stars as functions of their central density. Stars to the left of the maximum at $\rho_c = 1.4 \times 10^9$ g/cm^3, the Chandrasekhar limit, are stable white dwarfs, while stars to the right of the minimum at $\rho_c = 1.55 \times 10^{14}$ g/cm^3 are neutron stars. For ρ_c just above this value the stars are entirely solid. The dashed portions of the curves are calculated for two different equations of state, both due to Pandharipande[10]. The equation of state for the lower curve assumed n,p,e^- and μ^- as well as the hyperons $\Lambda, \Sigma^\circ, \Sigma^\pm, \Delta^\circ$ and $\Delta^\pm$ to be present, while the upper curve was calculated with a pure neutron fluid equation of state. These two curves give one some idea of the range of uncertainty in our knowledge of the structure of massive neutron stars. Neutron stars beyond the maximum are again unstable. Stars with $M > M_{max}$ haven't sufficient pressure to resist gravitational collapse. The maxima here are at 1.41 $M_\odot$ ($M_\odot$ = solar mass) and 1.66 $M_\odot$. Generally, the stiffer the equation of state, the higher the mass of the most massive stable neutron star and the lower its central density.

The fact that the white dwarfs made of equilibrium T = 0 matter become unstable at the central density 1.4×10^9 g/cm^3 is an interesting reflection of nuclear shell structure. Just below this central density the stars are made primarily of Fe^{56}, Ni^{62} and Ni^{64}, with a small core of Se^{84}. As I mentioned earlier, as the nuclide present in the matter changes to one with a smaller Z/A, there must be a discontinuous change in the density in order for the pressure to remain continuous. In the Ni^{64} - Se^{84} transition there is a particularly large change, $\Delta\rho/\rho = 7.9\%$. That Se^{84} is the next nucleus in the sequence is a consequence of the fact that it has a closed shell structure with 50 neutrons, which is energetically quite favorable.

Now the presence in the matter of phase transitions with density discontinuities makes the matter effectively softer. To see this just imagine a closed vessel containing a liquid on the bottom in equilibrium with its vapor on top. Even if the liquid and gas are each separately incompressible, the system in the vessel would be compressible, since as one reduces the volume of the vessel, gas is converted to denser liquid. By the point where Se^{84} enters, $\Gamma - 4/3$ is ~ 0.01; once there is a large enough Ni^{64} - Se^{84} interface present the matter becomes soft enough for the effective Γ to drop below 4/3, and the stars become unstable. Incidentally, the limiting cold white dwarf mass is exactly 1.00 $M_\odot$. Actual white dwarfs never reach complete nuclear equilibrium and hence can have a higher maximum possible mass.

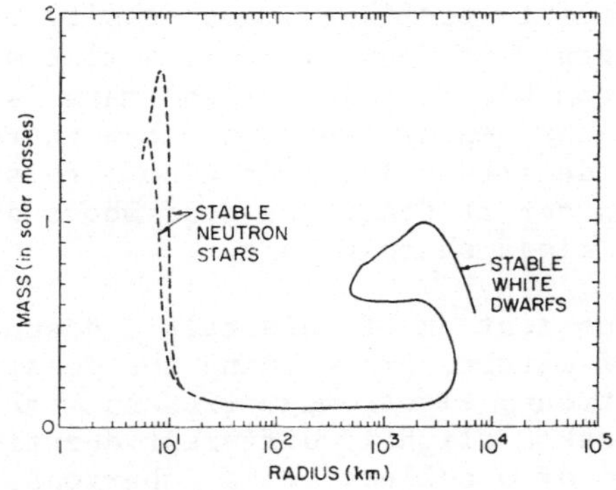

Fig. 4. Radius versus central density for stars composed of cold matter in complete nuclear equilibrium.

Fig. 4 shows the radius of the stars as a function of their central density. The dashed curves have the same meaning as in Fig. 3. The stable branches, which are indicated, terminate at the maxima and minimum. We see that white dwarfs have radii $\gtrsim 2 \times 10^3$ km, while neutron stars have radii primarily $\sim$ 10 km.

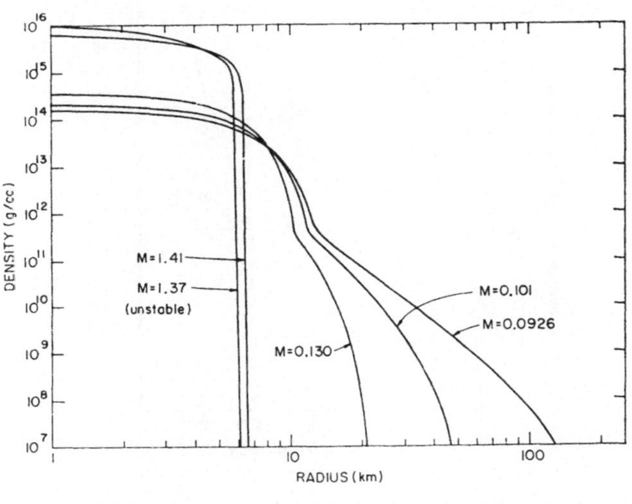

Fig. 5. Density profiles of five neutron stars. The two lightest stars are entirely solid.

Fig. 5 indicates density profiles (mass density versus radius) of several neutron stars. The density remains flat with increasing radius until a reasonable fraction of the mass is within the radius. In the heavy stars the density falls to zero extremely rapidly. Notice the knee at density 4.3×10^{11} g/cm^3; this is where neutron drip occurs. The matter at densities just above drip is very soft and so the density rises sharply here.

In the previous section of this talk I discussed the nature of the ground state of matter, prescribing the density of baryons and allowing for all strong, electromagnetic and weak interactions. I would like now to ask a slightly different question: what is the lowest energy state of a collection of B baryons, including gravitational interactions as well? The answer to this question is shown in Fig. 6 which graphs the total binding energy per baryon

$$b' \equiv -(E_{tot}/B - M_H c^2) \tag{3.4}$$

measured with respect to the hydrogen atom mass M_H.

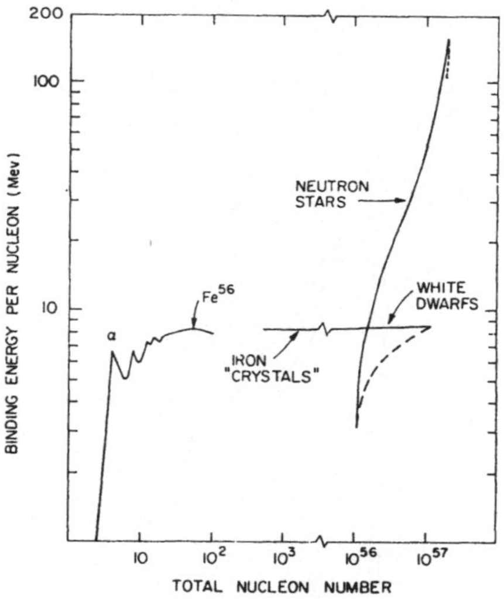

Fig. 6. Binding energy per baryon of a collection of B baryons in its ground state, including strong electromagnetic, weak, and gravitational interactions.

For $B \lesssim 10^2$ the lowest energy configuration is a single nucleus; the binding energy peaks at Fe^{56}. Beyond $\sim 10^2$ it becomes favorable for the nuclei to undergo fission, however, and by the time there are many thousand nucleons present the lowest energy state is a small iron crystal. Nothing dramatic happens to this state as B increases; the iron crystal grows and grows into a white dwarf star. Only by $B \sim 10^{56}$, one tenth the number of nucleons in the sun, does the gravitational contribution to the binding energy first become noticeable compared with the solid state contribution to the binding energy. The white dwarf branch continues until the Chandrasekhar limit, $B = 1.2 \times 10^{57}$. Above $B = 1.1 \times 10^{56}$ the baryons can also form a neutron star; however, it is not until $B \gtrsim 1.9 \times 10^{56}$, or $M \gtrsim 0.15\ M_\odot$, that neutron stars are more strongly bound than white dwarfs with the same B. (The decrease in the binding energy of lighter neutron stars is due to the matter being neutron rich; neutron-neutron interactions are less attractive than neutron-proton interactions, while the exclusion principle raises the mean kinetic energy inside the neutron-rich nuclei.) Because neutron stars have small radii, gravity very rapidly raises the binding energy with increasing B. The maximum energy one can extract from matter, ending up with a neutron star, is $\sim$ 100 MeV per baryon - about 10% of the rest mass. (By contrast, thermonuclear fusion is $\sim$ 1% efficient in converting mass into available energy.)

Actual white dwarfs are never sufficiently hot during their evolution ever to achieve nuclear equilibrium. All but the lightest stars become hot enough to convert H into He^4; the heavier the star the more the He^4 becomes converted to "α-particle nuclei" C^{12}, O^{16}, Ne^{20}, Mg^{24} and Si^{28}. Since these nuclei have one electron per two nucleons the equation of state in such stars is stiffer than in equilibrium stars which have fewer than $\frac{1}{2}$ an electron per nucleon. This effect raises the actual (Chandrasekhar) limiting white dwarf mass to greater than $1.00\ M_\odot$.

IV. SUPERFLUIDITY AND SUPERCONDUCTIVITY IN NEUTRON STARS

The neutrons in the crust and interior of neutron stars, as well as the protons in the interior are most likely superfluid. I would like to discuss now several consequences of their superfluidity, as well as recent calculations of pairing energies. Let us note first that it is very unlikely that the electrons will ever be superconducting. One can estimate, from the BCS weak coupling theory, the transition temperature $T_{c,e}$ for electron superconductivity by

$$T_{c,e} \sim T_{f,e}\, e^{-1/N(0)V} \; ; \qquad (4.1)$$

$T_{f,e}$ is the electron Fermi temperature (typically 1 to 100 MeV), expressed in terms of the electron Fermi momentum P_e by $kT_{f,e} = cP_e$ for fully relativistic electrons. The density of states at the electron Fermi surface is $N(0) = P_e^2/\pi^2 c\hbar^3$, since the electron effective mass is P_e/c. The mean net attraction V between electrons is a quantity of order $e^2(\hbar/P_e)^2$, so that $N(0)V \sim e^2/\hbar c$. Thus, $T_{c,e}$ is essentially zero; at any non-zero T the electrons are normal.

Superfluidity of neutrons, as well as protons, arises from pairing interactions, as in laboratory superconductors. Estimates of the energy gaps for ^1S_o pairing have been given by Yang and Clark[15] and for ^1S_o and $^3P_2$ pairing by Hoffberg *et al.*[16] whose results for the superfluid transition temperature are shown in Fig. 7.

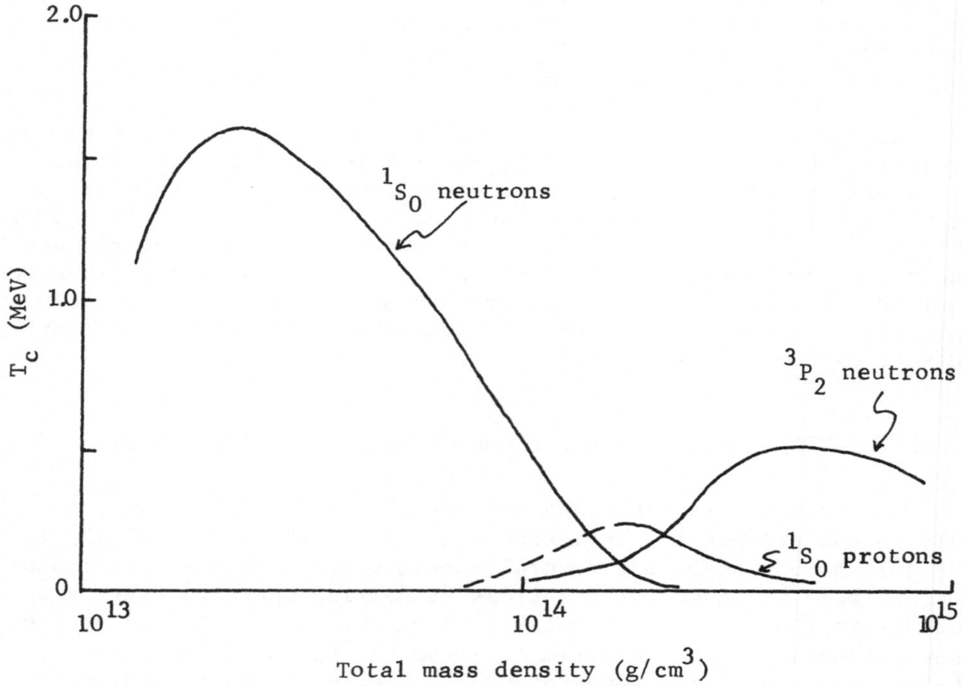

Fig. 7. Estimates of neutron and proton superfluid transition temperatures as a function of the total mass density of the matter. The dashed portion of the proton curve, below 2.5×10^{14} g/cm^3, the region of the crust, is that calculated for the uniform liquid state.

Also shown in Fig. 7 are estimates of the transition temperature for $^1S_0$ proton superconductivity, taken from the calculations of $T = 0$ proton energy gaps Δ_o by Chao, Clark and Yang[17]; we use here the BCS result $T_{c,p} \approx \Delta_p/1.76$. One should regard these calculations as only first estimates, since they leave out the effects of polarization of the neutron liquid (the sum of bubbles in diagrammatic language) on the effective interaction between pairs of protons, or pairs of neutrons. As emphasized by Pethick and Pines,[18] such polarization can greatly enhance the attraction between pairs, significantly strengthening the pairing. The transition temperatures generally are well above the expected ambient temperatures in all but very young neutron stars.

The calculations of Hoffberg et al. indicate that the neutrons in the crust should be paired in ^1S_o states, while in the liquid interior $^3P_2$ pairing should become preferable. Calculations of ^3P proton pairing have not, to my knowledge, been carried out.

One of the first consequences of neutron superfluidity is that because neutron stars rotate, the neutrons should form an array of quantized vortices, analogous to the vortices in rotating superfluid helium II. The quantum of circulation is $h/2m_n$, where $2m_n$ is the mass of a pair of neutrons. For a rotation period of 1/30 sec (that of the Crab pulsar) the vortex spacing is $\sim 10^{-3}$ cm, small compared with characteristic stellar dimensions. By contrast the vortices are widely spaced compared with the neutron coherence length

$$\xi_n = \frac{\hbar^2 k_n}{\pi \Delta_n m_n} \sim 10^{-12} \text{ cm} . \tag{4.2}$$

Here Δ_n is the neutron energy gap, and k_n the neutron Fermi wavenumber. The number of vortices present is always large enough that the moment of inertia of the neutron superfluid, in the case of ^1S_o pairing, is not reduced from the classical value. Vortices in a $^3P_2$ state superfluid carry a net spin angular momentum[19], which should imply an enhancement of the moment of inertia above the classical value.

The vortex lattice in a neutron superfluid has a mode of oscillation, first discussed by Tkachenko, in which the lines are displaced parallel to themselves. The propagation velocity of this mode is

$$C_T = (\hbar\Omega/8m_n)^{\frac{1}{2}} \sim 0.1 \text{ cm/sec} \quad , \tag{4.3}$$

where Ω is the rotation frequency. The period of the fundamental mode of a sphere of 10 km radius is ~ 4 months for $\Omega \sim 200$ sec^{-1} .

Tkachenko modes can play an interesting role in the rotational dynamics of pulsars. All the charged particles in a neutron star, the electrons, protons, nuclei, etc., must co-rotate since they are all tied to the enormous magnetic field. Any appreciable local differential rotation of positive and negative charges would lead to inordinately huge magnetic fields. The charged particles transfer their rotational energy via magnetic interactions to the magnetosphere surrounding the neutron star,where high energy particles and radiation are produced; the energy loss to the outside slows the rotation of the charged paricles. The neutrons, however, are the principal source of energy and angular momentum; the charged particles are simply intermediaries. The question is, how do the charged particles couple to the superfluid neutrons? The answer is through collisions of protons and electrons with the neutron vortex cores, in the process of which angular momentum and energy are extracted from the neutrons. The response of the vortices to the charged particles can be described in terms of the Tkachenko modes of the vortex lattice. One might hope to see evidence of these modes in the observed pulse repetition frequencies of pulsars.[21]

Let us consider now the properties of proton superconductivity. The critical magnetic field Hc (or Hc_1 for type II superconductivity) is of order 10^{16} gauss[17,22]. (Note that the scale of critical fields is $\sim (hc/2e)/4\pi r_p^2$, where r_p is the mean proton separation; for $r_p \sim 10^{-13}$ cm, this is $\sim 10^{17}$ gauss.) Since $B \sim 10^{12}$ gauss in a neutron star one would normally expect a Meissner effect in which the magnetic flux is expelled from the regions of superconducting protons. Such an expulsion would have a serious effect on the electromagnetic properties of pulsars. However, because of the enormous electrical conductivity of the normal state ($\sigma \sim 10^{29}\text{sec}^{-1}$) the time

$$\tau \approx \frac{4\pi\sigma}{c^2} R^2 \left(\frac{B}{2H_c}\right) \tag{4.4}$$

required to expel flux from a region of size $R \sim 1$ km is $\gg 10^8$ years. The superconductivity, rather than waiting so long, simply nucleates with the field present. While this is technically a metastable situation, the actual lifetime of the star can be shown to be orders of magnitude longer than the time τ in Eq.(4.4).

If the protons are a type I superconductor they will learn to live with the field by forming an intermediate state configuration, with alternate layers of superconducting material, which is field free, and normal material containing the field. On the other hand, for type II superconductivity, which as we shall see in a moment is more likely, the magnetic flux will be contained in an array of quantized vortices, each of a single flux quantum $\varphi_0 = hc/2e$. This is the same state as occurs in laboratory type II superconductors for $Hc_1 < B < Hc_2$, only now $B \ll Hc_1$.

The criterion for type II superconductivity is that the penetration depth λ should be greater than ξ_p, the proton coherence length, divided by $\sqrt{2}$. If we estimate λ from the London result

$$\lambda = \left(\frac{m_n c^2}{4\pi n_p e^2}\right)^{\frac{1}{2}} \quad , \tag{4.5}$$

where n_p is the proton number density, and ξ_p from

$$\xi_p = \frac{\hbar^2 k_p}{\pi \Delta_p m_n} \quad , \tag{4.6}$$

where k_p is the proton Fermi wavenumber, we find

$$\frac{\xi_p}{\sqrt{2}\lambda} = \left(\frac{8}{3\pi^2}\frac{e^2}{\hbar c}\frac{\hbar k_p}{m_n c}\right)^{\frac{1}{2}} \frac{\varepsilon_p}{\Delta_p} \quad ; \tag{4.7}$$

here $\varepsilon_p = \hbar^2 k_p{}^2/2m_n$. From the calculations of Ref.17, ε_p/Δ_p varies from about 10 at the outer surface of the liquid interior, to about 100 at 5×10^{14} g/cm^3. In this range $\xi_p > \sqrt{2}\lambda$, indicating that the protons should be a type II superconductor.

V. CONCLUSION

I would like to conclude this survey of neutron stars and white dwarfs with a brief mention of several topics in which the "solid state" structure of these stars plays an important role. The first is the influence of superfluidity on the cooling rates of neutron stars. The predominant cooling mechanism in the early stages of the star when the temperature is above 10^8-10^9 K is through neutrino emission. In each weak interaction such as $n \rightarrow p + e^- + \bar{\nu}$ or $p + e^- \rightarrow n + \nu$ a neutrino is lost from the star. The existence of energy gaps in the neutron excitation spectrum sharply reduces the available states for these processes when the temperature becomes much smaller than the gap; this reduction of the rate of neutrino emission tends to lengthen the cooling times. However, superfluidity is also accompanied by a reduction in the heat capacity of the matter, which means that a given energy loss is more efficient in lowering the temperature. A prediction of how these two effects together influence the thermal history of a neutron star requires detailed calculations; one such, by Tsuruta *et al.*[23], indicates that neutron superfluidity together with the effects of the strong magnetic field causes a significant increase in cooling rates.

A similar reduction in heat capacity occurs in white dwarf stars when, after they solidify, the temperature falls well below the Debye temperature

$$T_D \approx 4 \times 10^4 \rho^{1/3}$$

of the lattice. Then the heat capacity is of order $(T/T_D)^3$ times the classical value. The increased cooling rate in this region has been suggested as an explanation of why relatively few cooler white dwarfs are observed.[2]

The last topic I want to discuss is the influence of the elasticity of the crust of neutron stars on the observed behavior of pulsars. All pulsars slow down continuously, with rates consistent with the slowdown being due to a loss of rotational kinetic energy. However, in February, 1969 the pulsar PSR 0833-45 in the Vela constellation was observed to have sped up, by two parts in a million. (The details of the speeding up were not actually observed, but it certainly happened in less than a week.) This speedup was equal in magnitude to two weeks worth of slowing down. And, again, in September, 1969 the Crab pulsar increased its rotational frequency suddenly by about one part in 10^8. The Crab and Vela are the two fastest and youngest pulsars known.

There is a very simple explanation of pulsar speedup, proposed by Ruderman,[24] based entirely on the fact that the outer layer of a neutron star is solid. When the star is first formed, it is very hot and completely molten; it hasn't cooled enough for the crust to solidify. It is also spinning fairly rapidly, so that it is relatively oblate, by perhaps a part in 10^3 or 10^4 (comparable to the earth now). The star cools rapidly, freezing the crust in an oblate shape. As the star slows down the centrifugal force decreases, and gravity tries to pull the star into a more spherical shape. Because the crust is solid, it becomes strained as it is made more spherical; the strain keeps building up as the star slows down until the crust reaches its breaking point, at which point it fractures, and suddenly moves inward slightly, becoming more spherical. Ruderman called this sudden cracking a "starquake." As the crust moves in it decreases the moment of inertia, I, of the star slightly. Since the angular momentum $L = I\,\Omega$ is conserved in the starquake, as the moment of inertia decreases suddenly (by ΔI, say), the rotational frequency will suddenly increase by an amount

$$\frac{\Delta\Omega}{\Omega} = -\frac{\Delta I}{I} \quad . \tag{5.1}$$

This is just the ice skater effect - she pulls in her arms and spins faster. How far does the crust have to move inward to account for the observed increase in frequency? The required motion for Vela is one part in 10^6; since neutron stars are $\sim$ 10 km in radius, a motion of one cm would produce the observed speedup. For the Crab one needs only one two-hundreth of that, that is, only 50 microns. G. Börner, in the following talk, will have more to say about pulssr speedup.

The elastic properties of the crust determine both how frequently starquakes can occur, and the extent to which they can speed up the star. The crucial question is how much elastic strain energy can be stored in the crust compared with the change in gravitational energy of the star due to rotational deformation.[25] The crust must be fairly extensive in order that the two be comparable. Aside from contributions due to neutron-induced interactions between nuclei, the elastic constant C_{44}, the shear modulus, can be calculated from the standard Fuchs formula for a coulombic bcc lattice:

$$C_{44} = 0.3711 \frac{Z^2e^2}{a} n_N \quad , \tag{5.2}$$

where a is the lattice constant, and $n_N = 2/a^3$ is the density of nuclei. The Fuchs formula works better here than in normal metals, where electron screening leads to finite corrections. Typically, $C_{44} \sim 10^{30}$ ergs/cc over most of the crust. Breaking strains can be estimated from laboratory metals. We must also worry though about the extent to which the strain is relieved by plastic flow, or creep, in the crust. Even though the system is very cold, one can still have sufficient thermal activation of dislocations. A line of dislocations can move, forming various "jogs and kinks", and relieving strain at a rate proportional to an exponential of an activation energy over KT; this is the normal creep one has in very cold metals. The nature of the creep depends strongly on the particular on the particular lattice structure of the metal. One effect of possible relevance here is the increased plasticity of metals in the superconducting state. This has been explained on the basis of the motion of dislocations, unlike in the usual theories of creep, not being overdamped by interactions with the electrons.[26] Whether a similar increase in plasticity occurs here as a consequence of the very long electron mean free paths in neutron star crusts remains an interesting theoretical problem.

REFERENCES AND FOOTNOTES

* Much of the research described here has been supported by National Science Foundation Grant GP-16886.

1. A general review of pulsars and the arguments leading to their identification as neutron stars is given by A. Hewish, Ann. Rev. Astron. and Astrophys. 8, 265 (1970), while a theoretical review of neutron stars is found in A.G.W.Cameron, Ann.Rev.Astron. and Astrophys.8,179(1970).
2. The properties of white dwarfs are reviewed by V.Weidemann, Ann. Rev.Astron. and Astrophys. 6,351(1968) while recent theoretical developments are surveyed by J. Ostriker, Ann.Rev.Astron. and Astrophys. 9, 353(1971).
3. S. Tsuruta and A.G.W.Cameron, Can.J.Phys. 44,1895(1966).

4. L. Mestel and M. A. Ruderman, Mon.Not.R.Astr.Soc. 136, 27 (1967).
5. See Ya.B.Zeldovich and I.D.Novikov, Relativistic Astrophysics, v.1 (Univ. of Chicago Press, Chicago 1971), p.333.
6. This sequence is derived in G. Baym, C. Pethick, and P. Sutherland, Ap.J., Dec. 1971.
7. G. Baym, H. Bethe and C. Pethick, Nucl.Phys.A, (1971).
8. L. D. Landau, Phys.Z.Sowjet 11, 545 (1937) [JETP 7, 627 (1937)].
9. P. J. Siemens, Neutron Matter in Brueckner Theory (in press).
10. V. R. Pandharipande, Hyperonic Matter (in press).
11. For example, S. Frautschi, Phys.Rev.D 3, 2821 (1971).
12. W. Langer, A. Landé, F. Iachello (to be published) have recently constructed an interacting quark-like model to discuss the very high density regime.
13. R.Cohen, J. Lodenquai and M. Ruderman, Phys.Rev.Letters 25, 467 (1970); R. Mueller, A. R. Rau and L. Spruch, Phys.Rev.Letters 26, 1136 (1971); B. Kadomtsev and V. Kudryavtsev, Zh.Eksp.Teor.Fiz.Pis.Red 13, 15 (1971) [Sov.Phys.JETP Letters 13, 9 (1971)]; M. Ruderman, (to be published).
14. The equation of state used in the calculations described here, as well as the detailed calculations, are given in Ref. 6.
15. C.-H.Yang and J. W. Clark, Nucl.Phys.A (in press).
16. M. Hoffberg, A. E. Glassgold, R. W. Richardson and M. Ruderman, Phys. Rev. Letters 24, 775 (1970).
17. N.-C. Chao, J. W. Clark and C.-H. Yang, Nucl.Phys. A (in press).
18. D. Pines, Proceedings of the 12th International Conference on Low Temperature Physics (Keigaku, Kyoto 1971) p. 7.
19. R. N. Richardson, to be published.
20. K. Tkachenko, Zh.Exp.Teor.Fiz. 50, 1573 (1966); [Sov.Phys.JETP, 23, 1049 (1966)].
21. M. Ruderman, to be published; also Nature 225, 619 (1970).
22. G. Baym, C. Pethick and D. Pines, Nature 224, 673 (1969).
23. S. Tsuruta, V. Canuto, J. Lodenquai, and M. Ruderman (to be published).
24. M. Ruderman, Nature 223, 597 (1969).
25. G. Baym and D. Pines, Annals of Phys. (N.Y.) 66 (1971).
26. A. V. Granato, Phys.Rev.Letters 27, 660 (1971).

V

Lattice Dynamics

INTRODUCTORY REMARKS

T. R. Koehler

IBM Research Laboratory, San Jose, California 95114

The first paper in this section is a review by Balkanski of some of the highlights of the International Conference on Phonons, which was held at Rennes, France in late July 1971. Balkanski discusses a number of topics of current interest in lattice dynamics, including uncertainties in the determination of force constants from frequency measurements alone, calculation of phonon properties both from the microscopic and model theories, elastic interactions of defects in crystals, and anharmonic effects. Some of these topics are also discussed in other papers in this section, as well as in papers in the final section.

Following the review by Balkanski, which provides a useful overview of the subject, there are five original contributions selected for their relevance in computational solid state physics studies; they are devoted to aspects of the lattice dynamics of perfect crystals with the greatest emphasis being placed on phonon properties. Two topics constitute the dominant theme: the microscopic theory and anharmonicity. These are well-chosen themes for this conference because, for each, computational difficulties severely limit the application of well-developed theories. Thus, the proper comparison of theory and experiment is inhibited.

Accurate experimental measurements of phonon properties in perfect crystals by inelastic neutron scattering have provided a great impetus to the theoretical study of phonons. It is therefore highly appropriate that the paper by Cowley discusses inelastic neutron scattering.

He first treats the case of the purely harmonic crystal where the elementary excitations are rigorously a set of non-interacting phonons. Here, kinetic considerations alone enable one to obtain dispersion curves from the change in energy and wave vector of scattered neutrons, providing single scattering events can be distinguished. Such elementary considerations are adequate in many cases.

However, increased experimental precision and theoretical interest in more complicated processes have necessitated a more sophisticated analysis of the properties of phonons and of the results of neutron scattering experiments. Cowley considers the effect of anharmonicity on both aspects. Anharmonic interactions change the character of the phonons by giving them a finite lifetime and introduce interference effects in the experimental measurements. He also points out recent emphasis on the importance of multiple scattering effects.

The remaining papers in this section are concerned with the theoretical aspects of lattice dynamics, in particular with various techniques for the calculation of the properties of phonons.

The most fundamental approach is the microscopic theory as described by Pick. Here, after making the adiabatic approximation, one hopes to treat the many-electron problem exactly. This treatment is essential for the ultimate justification of the use of more computationally tractable models in calculations. It also clearly points out the macroscopic differences in the properties of metals and insulators and should ultimately provide the means for linking band structure and lattice dynamics calculations.

The stumbling block in microscopic theory calculations is the necessity of obtaining the inverse of an infinite matrix, the dielectric function $\varepsilon(q+k,q+k')$, for each q value, where k and k' run over all reciprocal lattice vectors. Pick discusses general approximation methods and critically reviews special ones appropriate to simple metals, semiconductors, transition metals, and insulators. It is clear that the microscopic theory is a challenging area both for pure theory and for computational efforts.

In order to expedite calculation related to systems which in the future will hopefully be treated on the basis of the microscopic theory, various models have been devised. These models anticipate the results of a true microscopic calculation and provide a physical picture of the crystal in terms of somewhat localized and overlap charge densities, with associated polarizability and deformability. The paper by Bilz discusses this approach and describes the mathematics and formulation of various models which have been applied to ionic crystals; for example, the rigid ion model, the breathing shell model, and the shell model. Bilz points out the importance of charge overlap effects and presents models which are appropriate to homopolar crystals and to highly polarizable crystals. He compares experimental and theoretical dispersion curves in some cases.

An area which has conventionally been treated in a spirit quite far from a rigorous microscopic approach is the lattice dynamics of simple Van der Waals crystals. Here one can hope to absorb all of the electronic effects into a phenomenological two-body potential, possibly with corrections from three-body or higher terms. Substances which should qualify for this treatment are the rare gase solids, solid helium and some simple molecular crystals. In these substances one is able to work with models which should be quite accurate. Hence, when discrepancies between the predictions of the quasi-harmonic theory and experiment are found, one is forced to concentrate on perturbation (anharmonic) corrections to the lowest order theory.

Two papers deal with these topics. The paper by Koehler treats the computational aspects of the quasi-harmonic theory and of a parallel but new theory, the self-consistent harmonic approximation, and of the evaluation of anharmonic corrections to each. The paper by Horner, while describing a new, general approach towards lattice dynamics, is primarily concerned with establishing a formalism which deals with the special problems encountered when perturbation theory breaks down. Thus, his work is especially applicable to the extreme anharmonicity induced when large amplitude atomic vibrations lead to hard-core overlap in solid helium and in solid rare gases at high temperatures.

An excellent supplement to the papers in this section is the book by W. Ludwig which is noted as Ref. 1C in the paper by Bilz. This book provides a more modern background to the subjects of this section than can be found in the classic books on lattice dynamics.

The theoretical papers in this section focus either on the proper application of the microscopic theory or on the adequate treatment of anharmonic effects. Both of these are formidable computational problems. Perhaps work in which both topics are treated simultaneously and in depth will be presented at a future conference on computational solid state physics.

NOTE ON THE RENNES CONFERENCE ON PHONONS

M. Balkanski

Laboratoire de Physique des Solides, Associé au C.N.R.S.,

Université Paris VI, Paris, FRANCE

The International Conference on Phonons held at Rennes, from July 26 to 29, 1971 followed a similar pattern to that of Copenhagen held in 1963. The scope of the Conference was to review recent progress in the field of lattice dynamics. Theory and experimental work were considered both for perfect and for imperfect crystals.

The Conference was opened by a general survey of the historical evolution of the idea of phonons given by H. Curien, General Director of the C.N.R.S.

It was natural to begin with the theoretical study of phonons in perfect crystals. From the different talks given in the attempt to demonstrate the merits of different models or microscopic theories, that of B. Szigetti distinguished itself by the fact that it sounded as a warning to those attempting to deduce force constants of a crystal from lattice frequencies alone. He showed that it is necessary to have detailed eigenvector measurements in order to prove that a set of force constants which reproduce the frequencies are the true force constants.

A solid has many more force constants than frequencies. Usually, one retains only the largest force constants and, using at least as many frequencies as the number of retained force constants, one calculates the latter. Szigetti showed that this procedure is incorrect because neglect of the small force constants limits the accuracy to which the frequency data can be used to calculate the retained constants. In fact, the number of data that can be extracted from the frequency measurements is always less than the number of force constants one wants to calculate. Therefore, if all the lattice frequencies were known exactly, they could still be satisfied with a very wide range of very different sets of force constants. A large proportion of these sets cannot be rejected on the basis of physical criteria alone.

It seems that the best hope for obtaining reliable force constants lies in the experimental determination of the eigenvectors (i.e., of the atomic displacements in the course of the various normal modes), combined with improvement in the technique of the frequency measurements.

The uncertainty in the determination of the force constants discussed by Szigetti goes far beyond that arising from alternative roots in the solution of the equations for the force constants which results in alternate, discrete sets of force constants as discussed by Herman, Brockhouse and others. Each such ambiguity can be cleared up by the approximate measurement of a single eigenvector. The uncertainty Szigetti discussed results in a wide and continuous range of possible sets of force constants and can only be resolved by obtaining at least $n_f - n_\omega$ reliable data from eigenvector measurements or from measurements of the changes which a substitution of isotopes produces in the lattice frequencies.

The rest of the theoretical papers essentially dealt with actual calculations of general phonon properties, comparing the results of model calculations with those obtained when starting with microscopic theory, i.e., from first principles. In the former calculations, models like the shell model, breathing shell model, rigid ion model and deformable bond model employing a certain number of adjustable parameters were used to calculate dispersion relations. In the latter case, Singha and Gupta showed that a self-consistent Born-Oppenheimer perturbation theoretical approach could be used to treat the lattice dynamics of a wide variety of crystals from first principles.

In such a theory, two aspects have to be considered, namely, the setting up of an effective pseudopotential for the evaluation of the electron-phonon matrix elements, and the inversion of the dielectric matrix for the solid in order to correctly account for local field effects.

The procedure used to calculate the electron-phonon matrix elements employs the replacement of the actual wave functions inside a certain volume V_0 surrounding the nuclei by pseudo wave functions. This has the effect of introducing a non-local pseudopotential acting between the pseudo wave functions as well as introducing into the electron response from the outset a "bound" portion which corresponds to the equilibrium electron density inside V_0 minus the pseudo electron density moving rigidly with the nucleus inside V_0. By suitably defining the pseudo wave functions, one can obtain a suitable pseudo potential for the electron-phonon matrix elements for free electron metals as well as tightly-bound solids such as ionic crystals.

To treat the dielectric crystals, namely the problem associated with the matrix inversion of the dielectric function, Singha and Gupta used a suitable factorization Ansatz for the dielectric function which yielded explicitly a general form for the dynamical matrix which can be identified in special cases with corresponding phenomenological models such as the shell model, breathing shell model, and so on.

Pick, to some extent, took the opposite way of thinking and tended to show that the microscopic theory of insulators may be cast into a form more general but essentially similar to that of the shall model and the breathing shell model. For that he used Wannier functions of the crystal in order to show the formal similarity between models and microscopic theory.

From the computational point of view, alkali metals yield satisfactory agreement with experiment without any adjustable parameters. For the polyvalent free electron metals and for semiconductors it appears also possible to define a local pseudopotential.

For silicon with only one adjustable parameter the agreement seemed to be reasonable, whereas for germanium the agreement was less satisfactory.

A high computational effort will be required to achieve a decent agreement between calculated dispersion curves and experimental results for polarizable homopolar crystals. W. Kress proposed to proceed in two steps: (a) find a model containing a few physical parameters obtained from macroscopic data, and (b) calculate the model parameters and give a physical interpretation in terms of electronic matrix elements.

Working on the first step only, a five-parameter shell model which allows for isotropic deformation of the shells as an additional degree of freedom, has been considered. For Si, Ge, and grey tin a good description was obtained for equal breathing and shell force constants. The isotropic deformation of the shells led to an additional term in the elastic constants which was compensated by the introduction of a stretch-stretch force constant to second nearest neighbors. The same good agreement was obtained in the case of alkali halides which gives the hope that such a model might be useful in the cases of III-V and II-VI compounds.

Reports on model calculations were numerous:
Shell Model: rutile (R. S. Katiyar); CdF_2 (J. Gobeau, M. Heuret, J. P. Mon); Metals (W. Hanke).
Breathing Shell Model: alkali halides (K. V. Namjoshi, S. S. Mitra,

J. F. Vetelino); cesium halides (G. Mahler, U. Schroder).
Rigid Ion Model: zinc-blend crystals (J. F. Vetelino, S. S. Mitra, K. V. Namjoshi); calcite crystals (M. Plihal).
Deformable Band Model: zinc-blend and diamond structure crystals (K. Kunc, M. Balkanski, M. Nusimovici); cinnabar (M. Nusimovici, G. Gorre).

Using the deformable bond model, Kunc has investigated the lattice dynamics of zinc-blend structure compounds. For thirteen such compounds for which the necessary experimental data are available, the static ionic charge was calculated and correlated with the ionicity of the bond. A linear dependence of the electric dipole moment per unit bond length and the ionicity of the bond was found. This followed closely the ionicity scale established by Pauling and no correlation was found when the ionicity scale of Phillips was used.

Experimental studies of phonons were also presented which used the techniques of x-ray scattering (Cd) and of inelastic scattering of neutrons: CaO, Be, $KCl(NH_4)$, $SrTiO_3$, solid hydrogen, and transition metal carbides.

As has become fashionable now, this Conference also considered the phonon spectra in amorphous solids and in glasses.

Experimental and theoretical studies on phonons in molecular crystals constituted a good part of the discussion. There the tentative approach was to extend Born-Von Karman and Huang theory to interpret the optical properties of lattice vibrations in molecular crystals, composed of electrically neutral molecules, bound by Van der Waals forces.

As it is generally admitted that lattice dynamics plays a dominant role in the understanding of phase transitions, it was not surprising to see a certain number of papers dealing with the problem of phase transition, but no particular emphasis on this aspect was given.

The dynamical properties of crystals containing defects was studied essentially for the cases of extended defects. Imperfections, such as screw dislocations, were considered and the localized vibration modes associated with such defects have been investigated.

A more detailed discussion on phonons in crystals containing imperfections will be given in my paper on "Localized Vibrational Modes in Crystals" at this Conference. I shall therefore briefly mention here only one aspect which I will not discuss in the above-mentioned paper--the elastic interactions of defects in crystals.

The interactions of defects in crystal lattices are considered to be generally of electrical origin. Besides, defects create elastic strain fields in the crystal which in turn lead to interaction between impurities. Introduction of impurities in the crystal gives rise to static displacements of atoms from their equilibrium positions. The elastic energy of the defect crystal can be expressed in terms of static displacements with respect to equilibrium positions of the perfect lattice.

The sign of interaction (repulsion or attraction) for anisotropic defects depends on the mutual orientation with respect to the symmetry axis of the crystal. It has been speculated that elastic impurity interaction can lead to an ordering of defects. In a crystal with several atoms per unit cell, a new component of static displacement appears, namely static displacement of sublattices.

Ipatova has shown that the distance-dependent contribution of optical branches is an impurity repulsion for parallel orientation of of centers. The repulsion itself can make an ordering of impurities; hence Ipatova concludes that this type of impurity interaction could be a possible reason for observed impurity ordering.

Other aspects of effects of imperfections on the phonon properties such as surface effects were discussed to some extent.

Anharmonic effects in perfect and imperfect crystals were also discussed extensively. It was well recognized that anharmonicity of the interatomic forces in a crystal gives rise to many effects which are of considerable theoretical and experimental interest. A. A. Maradudin discussed a certain number of such effects occurring in perfect and imperfect crystals.

The application of external forces such as strains and electric or magnetic fields which induce a lowering of the crystal symmetry leads to the so-called morphic effect. As an example of such an effect, the effect of static strain on the frequencies of the q=0 optical modes of a piezoelectric crystal of the zinc-blend structure was discussed together with the effect of strain on the transverse effective charge which governs the strength of the integrated infrared absorption in strained crystals.

These effects were interpreted by introducing cubic anharmonic terms in the crystal potential energy and the crystal's second order dipole moment.

In discussing the temperature dependence of the integrated absorption under localized and resonant modes peaks, anharmonic coupling of the impurity atom to the host lattice is also invoked.

The theory of two-phonon bound states was also developed in terms of cubic and quartic anharmonic interactions, and Maradudin produced calculations showing that the occurrence of true two-phonon bound states is relatively rare.

Other anharmonic effects were also mentioned and calculations were reported on the anharmonic vibrational properties of a certain number of crystals and impurity models. Anharmonicity at surfaces was also discussed.

This extremely brief and certainly incomplete survey of a large and well-furnished Conference cannot provide sufficient elements to evaluate the achievements at the Conference. It only aims to show a few general directions and to convey the impression of the Author that further, more elaborate calculations on different aspects of lattice dynamics will be undertaken. In the future, such calculations are more likely to be based on microscopic theories in spite of the fact that many difficulties still remain. Simple models have been useful and successful, but their essential interest in the future will be to serve as reference for comparison for the achievements and possibilities of microscopic theories in the study of general properties of phonons.

NEUTRON AND X-RAY INELASTIC SCATTERING BY LATTICE VIBRATIONS

R. A. Cowley

Department of Physics, University of Edinburgh,

SCOTLAND

ABSTRACT

Neutron and x-ray inelastic scattering measurements enable the frequencies and eigenvectors of the normal modes of vibration of crystals to be determined. The effects of the anharmonicity on these measurements are discussed. It is shown that they may give rise to both line shape and intensity anomalies, which may complicate the extraction of the parameters for the normal modes. Recent theoretical developments are described which suggest that the calculation of these effects may be extremely difficult even in weakly anharmonic crystals.

SCATTERING FROM HARMONIC CRYSTALS

The normal modes of vibration of a monatomic harmonic crystal may be described by their frequencies, $\omega(\underline{q}j)$, and eigenvectors, $\underline{e}(\underline{q}j)$ for each wavevector, $\underline{q}$, and branch index j. Inelastic coherent scattering occurs with conservation of energy and wavevector, so that if the frequency and wavevector transfers in the experiment are ω and $\underline{Q}$ and a single phonon is created, then it must have parameters given by

$$\omega = \omega(\underline{q}j) \quad , \qquad \underline{Q} = \underline{q} + \underline{\xi} \quad , \tag{1}$$

where $\underline{\xi}$ is a reciprocal lattice vector. The intensity of this one-phonon coherent inelastic scattering for a monatomic material is given by

$$|F(\underline{Q}j)|^2 = \frac{hb^2}{2m\omega(\underline{q}j)} \exp(-2W) \mid \underline{Q}\cdot\underline{e}(\underline{q}j) \mid^2 (n(\omega)+1) \quad , \tag{2}$$

where b is the coherent scattering length, m the mass, W the Debye-Waller factor, and $n(\omega)$ the Bose occupation number.

Thermal neutrons have energies and wavevectors which are similar to those of typical phonons so that both the energy ω and wavevector transfer $\underline{Q}$ may be directly measured to give the phonon frequencies through Eq. 1. Numerous experiments of this type have now been performed[1] and it is our intention to review neither the results nor the experimental techniques. It is sufficient that given a suitably large (preferably a few cc's) single crystal which does not contain isotopes which either scatter neutrons largely incoherently or absorb neutrons greatly, the phonon dispersion curves can be obtained in a routine manner. These dispersion curves can then be used either to test microscopic models of the interatomic forces or alternatively to provide accurate basic information which can be used in the theoretical interpretation of other measurements.

In most of these studies limited information has been made of the intensities, Eq. 2. This information enables one to readily distinguish between longitudinal and transverse polarizations, for example. Only very restricted attempts[2] have as yet been made to determine the eigenvectors $\underline{e}(\underline{q}j)$ in detail. This information is of importance in distinguishing between different microscopic models,[3] particularly of polyatomic crystals. It is, however, difficult to obtain partly because it is more difficult to measure intensities of scattered distributions than to measure centers of peaks, but also because it is necessary to solve a set of Eqs. 2 for the intensities at different scattering wavevectors $\underline{Q}$ to obtain the eigenvectors $\underline{e}(\underline{q}j)$. This problem is formally very similar to the problem of determining x-ray crystal structures from the intensities of Bragg reflections, except that now a crystal structure must be solved for each phonon wavevector $\underline{q}$.[4]

X-ray scattering is considerably less powerful because it is at present impossible to measure the frequency changes directly. The frequencies must then be deduced from the intensity of the scattering through Eq. 2. Unfortunately, the scattered intensity consists not only of the intensity scattered by a single phonon but is a superposition of the intensity scattered by all modes with the same wavevector, and also of the intensity scattered by multi-phonon scattering processes and from Compton scattering. It is clearly impossible to extract information about phonon dispersion curves unless a fairly reliable model is available initially with which to calculate the corrections and which may then be further improved by comparison with the measurements.

THE EFFECTS OF ANHARMONICITY

We expect anharmonicity to influence the scattering cross section by altering the frequencies of the normal modes of vibration and also by giving rise to a finite lifetime. This is indeed observed; the peaks in the cross section change in frequency and also broaden with increasing temperature. In detail, however, the behavior is more complex. The one-phonon cross section for an anharmonic crystal is given by

$$\sigma(\omega) = |F(\underline{Q}j)|^2 \, \mathrm{Im}[G(\underline{q}j,\omega)] \quad , \tag{3}$$

where the response function $G(\underline{q}j,\omega)$ is given by[5]

$$[\omega(\underline{q}j)^2 - \omega^2 + 2\omega(\underline{q}j)(\Delta(\underline{q}j,\omega) - i\Gamma(\underline{q}j,\omega))]G(\underline{q}j,\omega) = 2\omega(\underline{q}j) \quad , \tag{4}$$

where $\Delta(\underline{q}j,\omega)$ and $\Gamma(\underline{q}j,\omega)$ are the real and imaginary parts of the anharmonic self energy of the mode $(\underline{q}j)$ which depend on the frequency ω. In Fig. 1a we illustrate the two lowest order diagrams for the self energy and in Fig. 2 the calculated shape of Δ and Γ as a function of ω for the TO q=0 mode in KBr.[6] If Δ and Γ are sufficiently small then it is reasonable to approximate them by the simple forms $\Delta(\omega)=\Delta_0$, $\Gamma(\omega)=\omega\gamma_0$. These forms have been used to describe a large quantity of experimental results. In detail, however, the cross

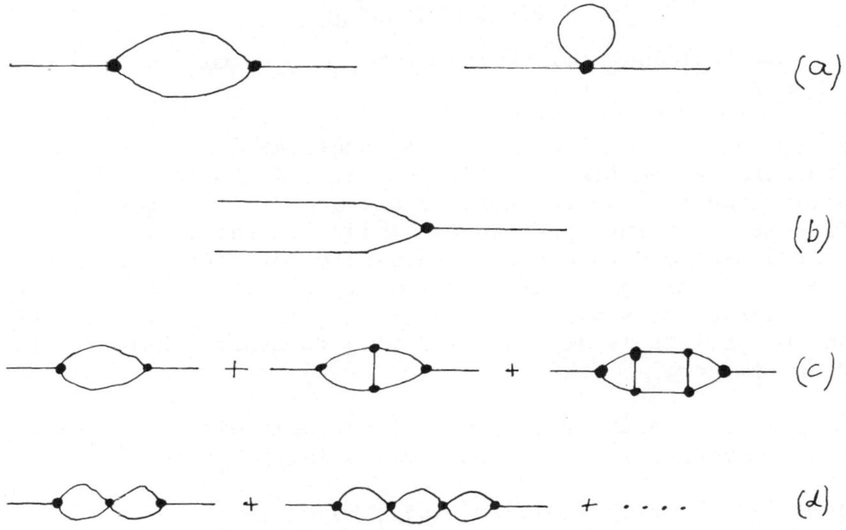

Fig. 1. Diagrams representing the different processes discussed in the text.

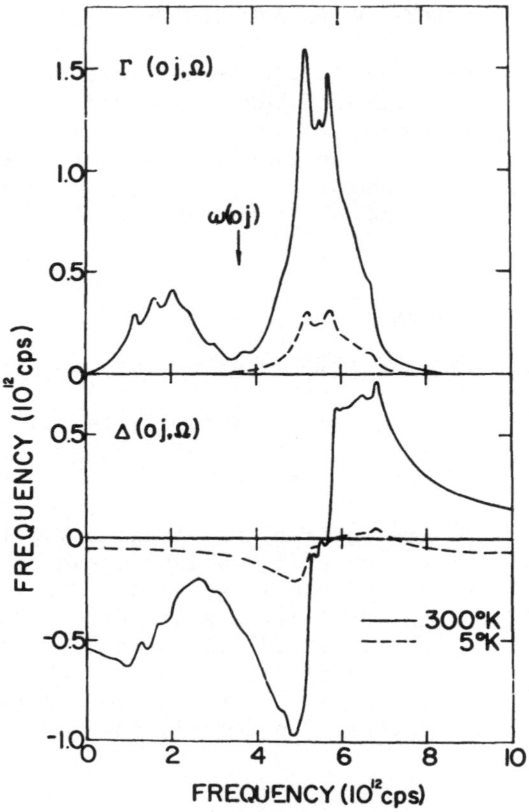

Fig. 2. The $\Delta(\omega)$ and $\Gamma(\omega)$ functions for the TO mode of KBr with $q = 0$.[6]

section may have a considerably more complicated shape because of the structure in the $\Delta(\omega)$ and $\Gamma(\omega)$ curves. A complex structure has been calculated for the LO modes of NaI and KBr.[6] These shapes lead to difficulties in interpreting the observed spectra in terms of a single frequency and intensity. In particular, it is necessary to use care in comparing the predictions of results because anharmonicity may introduce small but sharply varying features into the dispersion curves. It is at times tempting to assign these features to long-range forces.

In general, anharmonicity may also couple different normal modes of the same symmetry. The cross section is then given by[5]

$$\sigma(\omega) = \sum_{jj_1} F(\underline{Q}j)F(\underline{Q}j_1)\mathrm{Im}[G(\underline{q}jj_1,\omega)] \quad , \tag{5}$$

where Eq. 4 is now replaced by

$$\sum_{j} \left[(\omega(\underline{q}j)^2-\omega^2)\delta_{jj} + 2\omega(\underline{q}j)[\Delta(\underline{q}jj_1,\omega) - i\Gamma(\underline{q}jj_1,\omega)]G(\underline{q}j_1j_2,\omega)\right] = 2\omega(\underline{q}j)\delta_{jj_2} \quad .$$

If we solve for $G(\underline{q}j_1j_2,\omega)$ by iteration the leading term is

$$G(\underline{q}j_1j_2) = G(\underline{q}j_1,\omega)(\Delta(\underline{q}j_1j_2,\omega) - i\Gamma(\underline{q}j_1j_2,\omega))G(\underline{q}j_2,\omega) \ . \quad (6)$$

The imaginary part of this expression contains terms which depend on the imaginary part of $G(\underline{q}j,\omega)$ and other parts which depend on the real part. The former does not change the shape of the scattered intensity, but changes its magnitude; the latter term largely changes its shape. The effect of this term can be distinguished from the effects of $\Delta(\omega)$ and $\Gamma(\omega)$ by performing experiments for different $\underline{Q}$ but the same $\underline{q}$. The frequency-dependent self energy gives rise to the same shape for all $\underline{Q}$'s but the mode-mode interference terms may give rise to a different shape at each $\underline{Q}$. This is illustrated in the experiment of Harada et al.[7] on $BaTiO_3$, Fig. 3, where the shape of the cross section is quite different for the different wavevectors because of the interference of the TO and TA normal modes.

There is also the possibility of interference effects between the one-phonon scattering and the multi-phonon scattering. This is illustrated in Fig. 1b and the intensity can be written in the form

$$\sigma(\omega) = \mathrm{Im}[F(\underline{Q}j)G(\underline{q}j,\omega)V_3G_2(\omega)F_2(\underline{Q})] \quad ,$$

where V_3 is the cubic anharmonic interaction while $G_2(\omega)$ and $F_2(\underline{Q})$ are the two-phonon response functions and structure factors, respectively. As with the one-phonon interference, this term also gives rise to terms dependent upon $\mathrm{Im}[G(\underline{q}j,\omega)]$ which alter the intensity and $\mathrm{Re}[G(\underline{q}j,\omega)]$ which alter the shape. This type of interference can be distinguished by performing measurements at $\underline{Q}_1=\underline{\xi}+\underline{q}$ and $\underline{Q}_2=\underline{\xi}-\underline{q}$. Since the anharmonic coefficient is an odd function of $\underline{q}$, this interference changes sign in these two experiments unlike the mode-mode interference discussed above.

This effect has been observed in neutron scattering from the LO modes of KBr[10] as shown in Fig. 4 and results are in reasonable agreement with theoretical calculations. It also occurs in the x-ray scattering[9] as shown in Fig. 5 for NaCl. Since the magnitude of these effects is about 10% of the intensity, it is essential to perform some sort of correction before obtaining eigenvectors from neutron scattering intensities or frequencies from x-ray measurements.

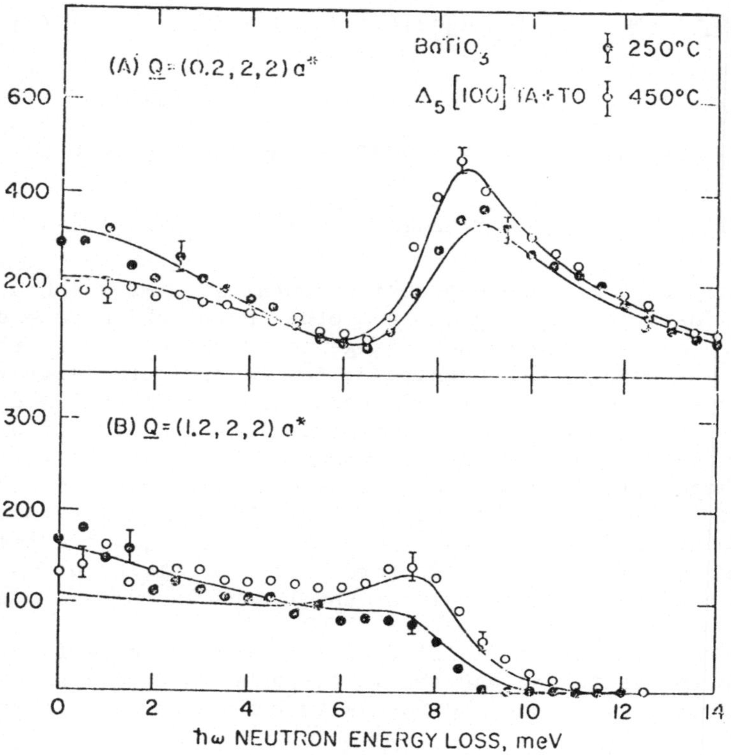

Fig. 3. The neutron scattering intensity in $BaTiO_3$ at different $\underline{Q}$ but the same $\underline{q}$. The difference in shape is due to interference between a TA and overdamped TO mode.[7]

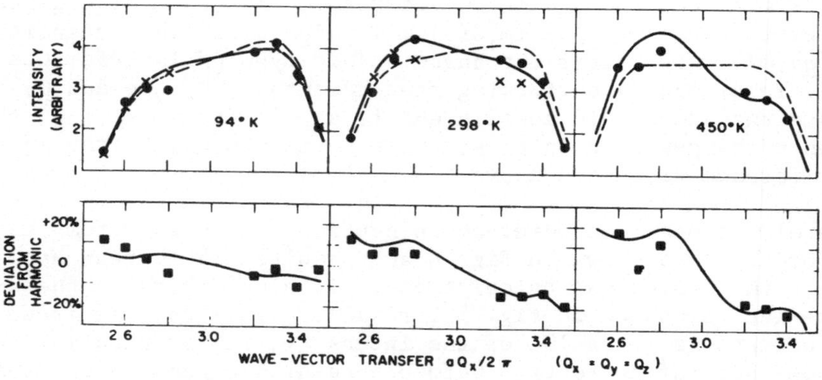

Fig. 4. The neutron scattering intensity for LO modes in KBr. In the upper curve the dotted line is the intensity calculated on a harmonic model and the full line the intensity including interference with the two-phonon background.[8]

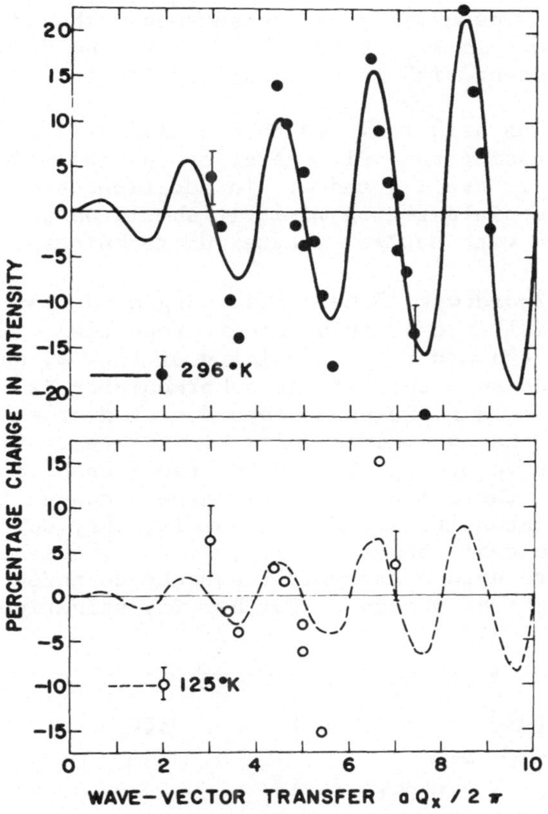

Fig. 5. The deviation in the x-ray scattered intensity for NaCl as a function of wavevector.[9]

MULTIPLE SCATTERING EFFECTS

The effects discussed in the previous section all arise from the application of low-order anharmonic perturbation theory. Recently there have been several developments that suggest this is inadequate. Initially Sham[11] pointed out that at small $\underline{q}$ and small ω the first diagram in the anharmonic self energy, Fig. 1a, was not a consistent solution. As $\underline{q}$ and ω become small the energy denominators become of the same size as the anharmonicity so that it is necessary to consider the infinite series of diagrams shown in Fig. 1c. When this is done it is necessary to distinguish between the results in the collisionless region $\omega\tau \gg 1$, and the collision-denominated or thermodynamic region $\omega\tau \ll 1$, where τ is the average lifetime of a thermal phonon. Ultrasonic elastic constant measurements are made in the $\omega\tau \ll 1$ region whereas neutron scattering measurements are made in the $\omega\tau \ll 1$. These different measurements should give different results as shown in Fig. 6 for KBr[12] where the difference is shown to be consistent with theoretical predictions.[13]

Recently it has been realized that a similar distinction is necessary in discussing the self energy of optical modes in piezoelectric crystals at small $\underline{q}$ and ω. In addition, theory predicts that in the intermediate region $\omega\tau \simeq 1$, it should be possible to observe second sound with neutron scattering techniques.

Ruvalds and Zawadowski[14] have also suggested that the series Fig. 1d is essential in describing other properties. They discuss the effect of the 4th order anharmonicity of Fig. 2, and the two-phonon spectrum and show that if the anharmonicity is large enough it is possible to obtain two-phonon bound states or at least considerable changes in the two-phonon density of states. Since this will change the $\Gamma(\omega)$ curve of Fig. 2, and the two-phonon response function $G_2(\omega)$, Eq. 6, these modifications cause a change in the shape of the scattered intensities. Unfortunately, they do not discuss the effects of the cubic anharmonicity, Fig. 1c, which for some modes gives rise to both a strongly frequency-dependent and lossy interaction between the phonons. Further investigation of both a

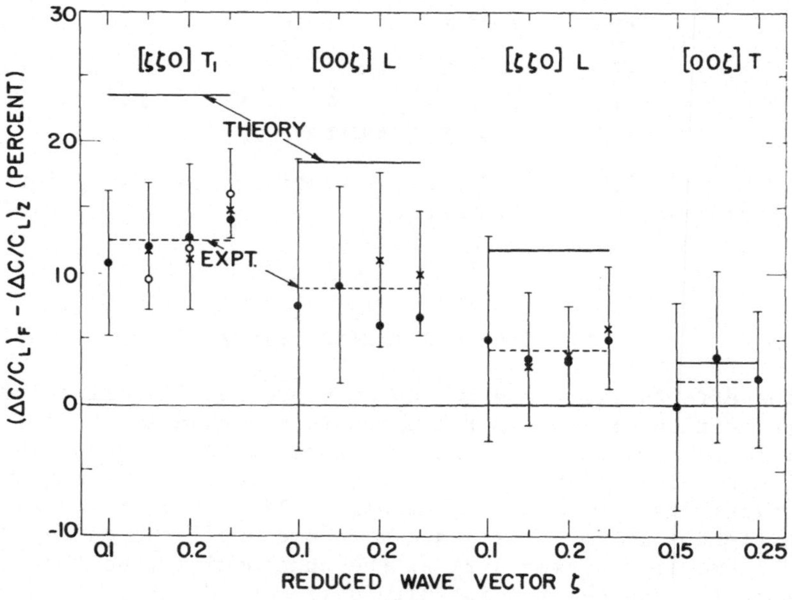

Fig. 6. The difference between the temperature dependence of the elastic constant for $\omega\tau \gg 1$ and $\omega\tau \ll 1$ for KBr and for different directions of propagation.[12]

theoretical and experimental nature is needed before these interactions between pairs of phonons are known to be as important as magnon-magnon intereactions in antiferromagnets.

In conclusion we have shown that anharmonicity may considerably influence the shape of the spectrum of scattered neutrons. It is essential to correct for these effects if the frequencies and eigenvectors appropriate to a harmonic model are to be obtained. These effects are, however, of considerable interest in themselves and provide very detailed information about the anharmonic process in crystals.

REFERENCES

1. See, for example, Neutron Inelastic Scattering, Vols. I and II, IAEA, Vienna (1968).

2. Skalyo, J., Frazer, B. C. and Shirane, G., Phys. Rev. B1, 278 (1970).

3. Leigh, R. S., Szigeti, B. and Tewary, U. K., Proc. Roy. Soc. A320, 505 (1971).

4. Cochran, W., Neutron Inelastic Scattering, Vol. I, IAEA, Vienna (1968).

5. Cowley, R. A., Advances in Physics 12, 421 (1963).

6. Cowley, E. R. and Cowley, R. A., Proc. Roy. Soc. A237, 259 (1965).

7. Harada, J., Axe, J. D. and Shirane, G., Phys. Rev. B4, 155 (1971).

8. Ambegaoker, U., Conway, J. and Baym, G., Lattice Dynamics, ed. by R. F. Wallis. (Pergamon: New York) p. 261 (1965).

9. Cowley, R. A. and Buyers, W.J.L., J. Phys. C. 2, 2262 (1969).

10. Cowley, R. A., Svensson, E. C. and Buyers, W.J.L., Phys. Rev. Letters 23, 525 (1971).

11. Sham, L., Phys. Rev. 156, 494 (1967).

12. Svensson, E. C. and Buyers, W.J.L., Phys. Rev. 165, 1063 (1968).

13. Cowley, R. A., Proc. Phys. Soc. 90, 1127 (1967).

14. Ruvalds, J. and Zawadowski, A., Phys. Rev. B2, 1172 (1970).

PHONON DISPERSION RELATIONS

H. Bilz

Technische Universität München, Germany

Physik-Department, 8046 Garching b. München

I. Introduction

The understanding of the basic properties of phonons has been increased very much during the last decade mainly due to inelastic neutron scattering. Successful models have been developed which allow a satisfactory description of simple insulators and metals with a few parameters.[1] Difficulties arise with crystals of highly polarizable ions and such of low symmetry. In addition, the explanation of the model parameters on a microscopic basis[2] is still to a large extent an open problem. This holds even more for phonon-determined crystal properties like infra-red absorption etc. There exist at the present time strong efforts to explain all these effects by rigorous quantum-mechanical treatments.

We shall concentrate in this paper on a critical review of mathematical methods and physical models. First we discuss the situation of phonon models mainly in insulators and secondly consider a few aspects of the microscopic treatments of the dynamical theory of lattice vibrations.[3]

II. Formal force constants

We use in this paper the quasi- or renormalized harmonic and adiabatic approximations.[1,3] The quasi-harmonic treatment seems to be sufficient for the majority of crystals at lower temperatures; the genuine anharmonic crystals are discussed by Koehler and Horner.

The adiabatic theorem is based on the large mass ratio of electrons and ions, and it means that electron-ion forces connected with lattice vibrations can be described as if they acted between the ions. This is valid not only in insulators, due to their energy gaps, but also in metals due to Pauli's exclusion principle,[4] and even in zero gap insulators like α-Sn[5] (with the exception of a very small energy-regime). With these approximations, the lattice potential ϕ of a crystal in thermal equilibrium can be derived from the free energy F(T,V)[3,6] leading to a bilinear form in the lattice displacements $\underline{u}$:

$$\phi \approx \phi^{(2)} = \frac{1}{2} \sum_{\lambda,\lambda'} \underline{u}(\lambda) \underline{\underline{\phi}}(\lambda\lambda') \underline{u}(\lambda') \qquad (2.1)$$

where $\lambda = (l,k)$ denotes the kth particle in the lth cell. The displacement vectors $\underline{u}(l,k)$ belong to the different ions localized at the equilibrium positions $\underline{x}(l,k)$ in the lth cell of the crystal. The 3 x 3 two-ion force constant matrices $\underline{\underline{\phi}}(\lambda\lambda')$ are subject to conservation laws as well as to the symmetry restrictions of the space group.[1a,c] They depend, in addition, on temperature and pressure due to the renormalization procedure with the anharmonic parts of the potential. From the equations of motion we obtain the eigenvalues ω of the system as the solutions $\omega_j^2(\underline{q})$ of the secular-equation

$$|\underline{\underline{D}}(\underline{q}) - \underline{\underline{m}}\,\omega^2 \underline{\underline{I}}| = 0 \;, \quad \underline{\underline{m}} = (m_k) \qquad (2.2)$$

$\underline{\underline{D}}(\underline{q})$ is the dynamical matrix built from the submatrices

$$\underline{\underline{D}}(\underline{q},kk') = \sum_{l'} \underline{\underline{\phi}}(\lambda\lambda') \exp\{i\underline{q}[\underline{x}(\lambda) - \underline{x}(\lambda')]\} \qquad (2.3)$$

with 3 x 3 cartesian elements $D_{\alpha\beta}$; $\underline{\underline{I}}$ means the corresponding unit matrix $\delta_{\alpha\beta}\delta_{kk'}$. The index j labels the different solutions for the ω^2's for every $\underline{q}$-value.

In practice, the problem is often to determine a set of force constants from a restricted number of eigenfrequencies $\omega_j(\underline{q})$, measured, say, by neutron spectroscopy. With the help of this set the complete spectrum of frequencies $\omega_j(\underline{q})$ and the one-phonon density of states $\sum_{(j,q)} \delta(\omega - \omega_j(\underline{q}))$ can be determined with the help of an interpolation scheme or model assumptions. This is sufficient to calculate crystal properties at low temperatures like the specific heat or, in a density approximation, the infrared and Raman cross sections of light. Very often, this approximation is poor and an explicite knowledge of the eigenvectors of the dynamical matrix is required. Unfortunately, the frequencies, even if completely known, determine the eigenvectors and the force constants not uniquely.[7] This shows that a set of (formal or model) force constants derived from experimental data has to be taken with caution in view of its physical significance. Additional arguments from microscopic theory etc. are necessary to prove that coupling parameters fitted to experimental data are more than a lucky simulation of the real mechanism. A measurement of eigenvectors would be helpful in this situation.

III. Rigid ion model

We obtain an instructive insight into the problems of phonons by discussing the situation in ionic crystals.

Kellermann[8] was the first to introduce a realistic description of lattice vibrations in alkali halides. His starting point was Born's theory of the cohesive energy of these crystals as an electrostatic interaction of pairs of free ions with closed and separated electronic shells, corrected by a Born-Mayer type of repulsion. The secular equation (2.5) can be derived from the equations:[1b]

$$\underline{\underline{m}}\,\omega^2 \underline{U} = (\underline{\underline{R}} + \underline{\underline{Z}}\underline{\underline{C}}\underline{\underline{Z}})\underline{U} \equiv \underline{\underline{D}}^{R I}\underline{U} \qquad (3.1)$$

with the matrix $\underline{R}$ for the repulsive short range forces and the matrix $\underline{ZCZ}$ for the Coulomb forces. Z means the static ("Born") charge which in alkali halides is near to $\pm e$. Assuming this value and n.n. central forces in the equilibrium condition $\phi^{(1)} = 0$, this model gives surprisingly good results with one parameter fitted to the elastic constants (Fig.1). The longitudinal optic branch, however, is described rather badly since the Lyddane-Sachs-Teller relation is not fulfilled. The high-frequency dielectric constant $\varepsilon_\infty = 1$ in this approximation.

IV. Dipole models

The distortion of the electronic charge by ion displacements induces dipole moments which give an additional contribution to the dynamical matrix. A convenient representation of these dipole forces is the shell model[9] (for ref. see [1b,c,d]). Here the electronic

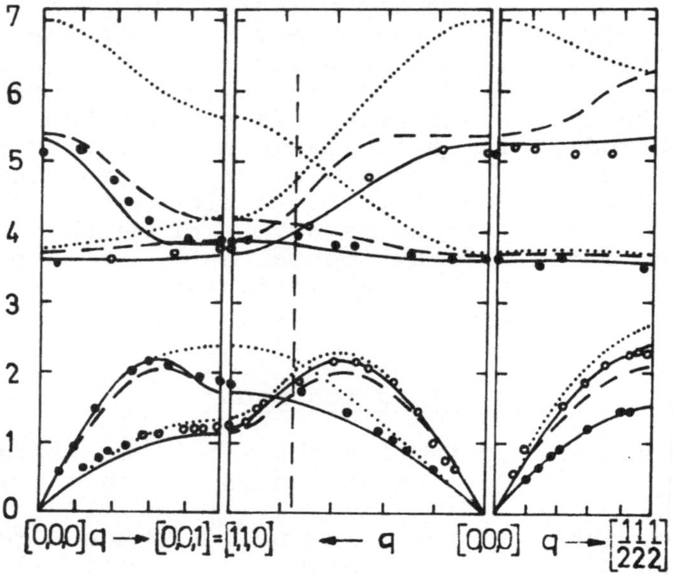

Fig 1: Dispersion curves for NaI derived from various models (...... rigid ion model; —— breathing shell model;--- shell model)

charge density is described by massless shells with charges Y_k coupled to the ion cores by harmonic coupling parameters. The model is very flexible with respect to certain generalizations (see sect.5,6,7). In this model, the dynamical matrix reads:

$$\underline{\underline{D}}(\underline{q}) = \underline{\underline{D}}^{RI} - (\underline{\underline{T}} + \underline{\underline{ZCY}})(\underline{\underline{S}} + \underline{\underline{k}} + \underline{\underline{YCY}})^{-1}(\underline{\underline{T}}^{+} + \underline{\underline{YCZ}})$$

$$\equiv \underline{\underline{D}}^{RI} + \underline{\underline{D}}^{DIP} \qquad (3.2)$$

If one assumes the short-range forces to be acting through the shells (which is sufficient for alkali halides), $\underline{T}^{+} = \underline{T} = \underline{S} = \underline{R}$. Furthermore, in crystals like NaI, the polarizability of the positive ions can be neglected against that of the negative ions. Using the parameters of the rigid-ion model for $\underline{D}^{RI}$ and fitting the shell charge Y and the polarization spring k of the negative ion to the dielectric constants ε_0 and ε_∞, we obtain the results of the three parameter simple shell model shown in Fig.1. The agreement between theory and experiment for the LO-mode is remarkably improved but considerable discrepancies still remain, especially at

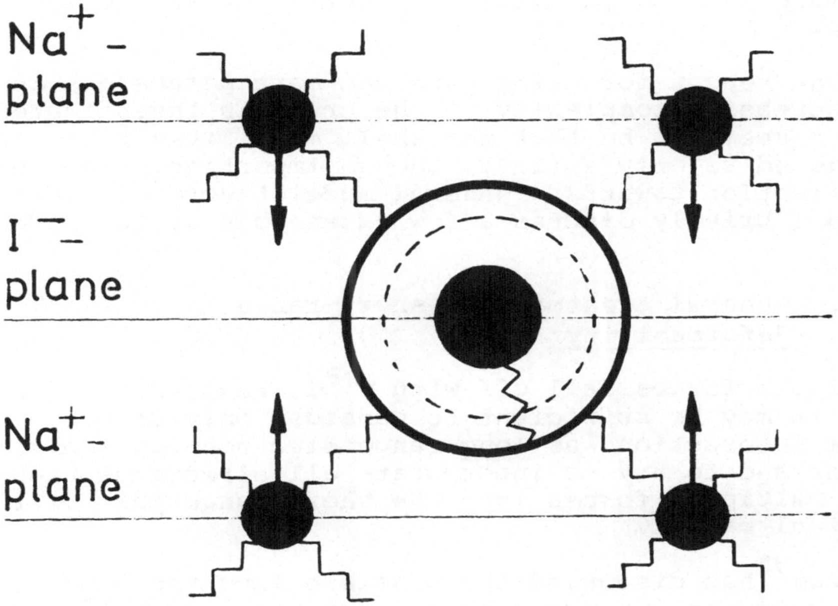

Fig 2: LO-mode at q = $\pi/2a$ (111) in NaI

$q = (1/2, 1/2, 1/2)$. At this point the shell model does not work since the planes with the polarizable I^- ions are at rest (Fig.2). This discrepancy can be removed by giving the positive ions pseudopolarizabilities which simulate the electronic deformation around the negative ions.[10] With the introduction of coupling constants between second-nearest neighbors, an extended shell model with at least nine parameters fits the experimental data within the limits of experimental error.

This extended shell model has been widely used during the last time as a useful tool for phonons in insulators,[1d] and it is about to replace the earlier formal force constant treatments completely. The reason for this success is based on the fact, that the shell model describes the long range dipole forces as the main part of the electron-ion interaction in phonons. Nevertheless, in highly polarizable crystals or those of low symmetry the number of parameters, required to describe the measured data, increases strongly. For TiO_2, e.g., which has a rutile structure, a shell model description with more than 20 parameters is still far from being correct.[11]

One reason for using more and more parameters is the increasing complexity of the crystals investigated, another seems to be that the shell model itself can be considered as only a first, though important, step in the direction towards a general model theory of phonons. We shall briefly discuss a few extensions of the model.

V. General treatment of short-range deformability

The dipole forces fall off with r^{-3} in real space. Therefore, it may be sufficient to consider only dipole-dipole interactions as long-range electron-ion contributions and to try to incorporate all effects of higher-order multipole forces into the short range part of the dynamical matrix.

Lax[12] has discussed the possible importance of long-range quadrupole - quadrupole interactions ($\propto r^{-5}$), but his results for germanium are inconclusive.

The dipole models contain short-range deformation effects of dipolar type only. The situation in NaI for $q = (1/2, 1/2, 1/2)$(Fig.3) indicates that short range deformations of the electronic charge density with even parity may be important.

The first attempt in this direction was the "breathing" shell model by Schröder.[13] This spherical degree of freedom describes a short-range radial electron-ion coupling. The dynamical matrix has to be extended to

$$\underline{\underline{D}} = \underline{\underline{D}}^{RI} + \underline{\underline{D}}^{DIP} + \underline{\underline{D}}^{SPH}, \quad \underline{\underline{D}}^{SPH} = -\underline{Q}H^{-1}\underline{Q}^{+\,13} \qquad (5.1)$$

The results of a 6 parameter calculation are shown in Fig. 1. Similar good agreements are obtained for all alkali halides.[1d,14] In this case, the "breathing" deformability can be assumed empirically to be equal to the short-range dipolar deformability. This keeps the number of adjustable parameters as small as the number of independent macroscopic parameters (three elastic constants, two dielectric constants and one optical frequency). The advantage is that the model can be used successfully without having neutron scattering data. The same is true for Hardy's[15] deformation dipole model, but results are less satisfactory since this

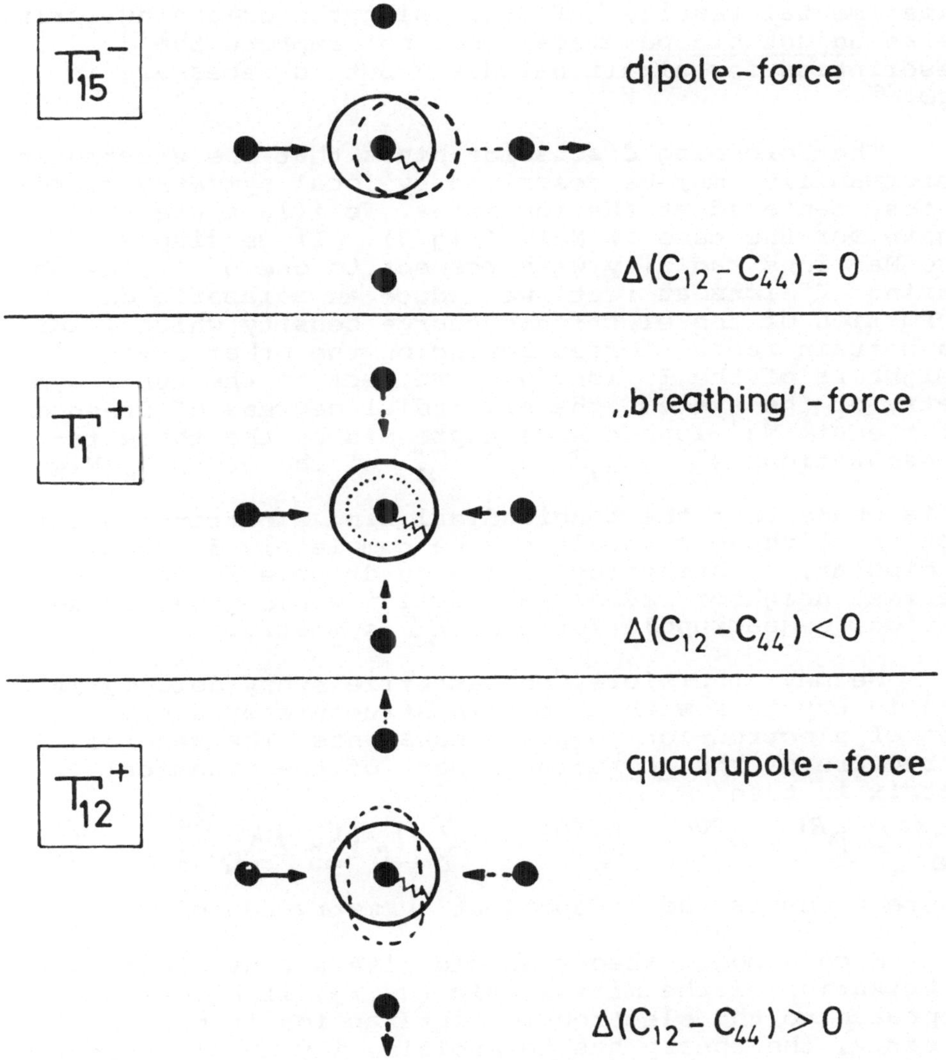

Fig. 3: Local symmetries of the displacement induced ion-electron-ion forces in a cubic crystal

model is missing the short-range deformability.

Difficulties remain also in this improved model. As in the shell model, the description is usually better with a static charge Z reduced from ±e to ±(0.8-0.9) e. Secondly, the breathing deformation affects the Cauchy relation without showing a reasonable connection with experimental results.[14] Third, using the breathing spring as an adjustable parameter does not improve the description for alkali halides [17] but is necessary in MgO.[16]

The foregoing discussion hints that the electronic deformability may be described by local symmetry coordinates, centered at the ion sites. We illustrate this again for the case of NaI. (Fig 3). If we displace one of the Na^+ ions radially with respect to one of its neighboring I^- ions at rest, we induce an adiabatic deformation of the electronic charge density which leads to certain radial forces acting on the other five Na^+ neighbors of the I^- ion. With respect to the cubic symmetry of the system, the six radial degrees of freedom of the six Na^+ ions can be expressed by the three representations Γ_{15}^-, Γ_1^+ and Γ_{12}^+ of the group O_h (Fig 3).

This means that the longitudinal n.n. electron-ion interaction in these crystals can be completely described by a dipolar, a "breathing" and a quadrupole force. Second-nearest neighbor radial interaction would give, in addition, a quadrupole force of Γ_{25}^+ symmetry.

We may, therefore, characterize every deformable ion in crystals with a certain structure by a symmetry set of electron-ion coupling constants. The general structure of the short-range part of the dynamical matrix is then

$$\underline{\underline{D}}^{SR} = \underline{\underline{D}}^{RI} + \underline{\underline{D}}^{POL}; \quad \underline{\underline{D}}^{POL} = - \sum_i \underline{\underline{T}}_{\Gamma_i} (\underline{\underline{S}}_{\Gamma_i} + \underline{\underline{K}}_{\Gamma_i})^{-1} \underline{\underline{T}}_{\Gamma_i}^{+} \qquad (5.2)$$

where i counts the independent symmetry coordinates.

Such a model theory should give a convenient parametrization of the microscopic theory, if a local approach to the electron-ion interaction is possible. Clearly, the theory has to explain, why an electron-ion coupling constant k_{Γ_i} is strong in a certain case or why not. From that point of view, the situation even in alkali halides is still challenging. As described above, the model theory gives for the anions in that case $(s_i \ll K_i)$ $K^{-1}(\Gamma_{15}^-) \approx K^{-1}(\Gamma_1^+)$ and $K^{-1}(\Gamma_{12}^+) \approx 0$

We should try to look for possible microscopic derivations of the model theories. Since the dielectric function approach is discussed in detail by Pick at this symposium, we focus on the treatment of overlap charges which lacks perhaps a certain rigour in the basic principles, but illuminates some aspect of the model theory very clearly.

VI. Overlap and effective charges

Lundquist[17a] has used Löwdin's[18] Heitler-London approach to the cohesive energy of ionic crystals for an investigation of the long-wavelength behaviour of alkali halides.

The classical theory of ionic crystals[3] with point charges $Z = \pm e$ leads to the Cauchy relation $C_{12} = C_{44}$. Empirically, the deviation $\Delta C \equiv C_{12} - C_{44}$ does not vanish, even not in alkali halides. The analysis of ΔC in the framework of a harmonic theory is difficult since the change of ΔC going from T = 0 to room temperature is of the order of ΔC itself, thus indicating the importance of anharmonic corrections. Nevertheless, one might hope to get the right sign and order of magnitude in a harmonic approximation at low temperatures. The overlap of the ions leads to long-range (at least) three-body corrections of the Coulomb interactions which give in the simplest approximation[17a]

$$\Delta C = \frac{e^2}{r_0^4} \cdot r_0 \left(\frac{\partial f}{\partial r}\right)_0 \cdot 2.3301 \qquad (6.1)$$

Here r_0 denotes the nearest neighbor distance in equilibrium, f(r) is a measure for the difference in size of neighboring ions and depends on their overlap. Therefore, $\partial f/\partial r$ should be generally less than zero or, $\Delta C = C_{12} - C_{44} < 0$. This is true for most of the alkali halides, but not for KBr and NaI, e.g.. Interestingly, a short-range "breathing" deformation always leads to a $\Delta^{B2} C < 0$[14] while a quadrupole force (Fig 3) gives $\Delta^{q4} C > 0$[19] It is therefore necessary in the model theories, to take into account (at least some part of) ΔC by a formal "non-central" force constant (see the discussion in[1d]) which overcompensates, e.g. in NaI, the overlap and/or breathing contributions. The long-range overlap corrections affect also the optical frequencies. At long wavelengths, but still in the non-retarded regime, the splitting of the longitudinal and transverse frequencies is given by the Lyddane-Sachs-Teller-relation[20]

$$\frac{\omega_{LO}^2}{\omega_{TO}^2} = \frac{\varepsilon_0}{\varepsilon_\infty} \qquad (6.3)$$

This gives

$$\Delta \omega_0^2 = \frac{1}{\varepsilon_\infty} (\varepsilon_0 - \varepsilon_\infty) \omega_{TO}^2 \qquad (6.4)$$

$\varepsilon_0 - \varepsilon_\infty$ measures the vibrational polarizability in a transverse mode[21]

$$4\pi\alpha_T^{VIB} = \varepsilon_0 - \varepsilon_\infty = 4\pi n \frac{Z_T^2}{m\omega_{TO}^2} \qquad (6.5)$$

with the reduced mass m, molecular density n, and the transverse effective charge Z_T , which is different from that in a longitudinal mode; $Z_T = \varepsilon_\infty Z_L$. So we have

$$m\Delta\omega_0^2 = 4\pi n e^2 \frac{1}{\varepsilon_\infty} Z_T^2 = 4\pi e^2 Z_T Z_L \qquad (6.6)$$

The charges Z_T and Z_L are related to Szigeti's[22] effective charge Z_S by $Z_T = Z_S(\varepsilon_\infty + 2)/3$. This charge describes the long-range (modified) Coulomb interactions between polarizable and deformable ions in a crystal. In the shell model (see Eqs. (3.1)(3.2)) the Szigeti charge matrix reads

$$\underline{\underline{Z}}_S = \underline{\underline{Z}} - \underline{\underline{T}}(\underline{\underline{S}} + \underline{\underline{K}})^{-1}\underline{\underline{Y}} \qquad (6.7)$$

clearly, $Z_L < Z_S < Z_T$ while in the rigid ion limit $|Z_L| = |Z_S| = |Z_T| = e$ .Empirically, the Szigeti charge in alkali halides is approximately 0.75e.

The overlap of the electronic charges contributes to the reduction of the static charge Z to smaller effective charges. With the same approximation as that used for eq. (6.1) one obtains for $\Delta\omega_0^2$ [17a]

$$m\Delta\omega_0^2 = 4\pi n e^2 (1 + 12f + 4r_0 f') \qquad (6.8)$$

Compared with Z_S, numerical calculations of f and f' with free ion wave functions give an effective ("Lundquist") charge near to 0.9e. This might explain why in shell model calculations, especially those with a least-square routine, the static charges turn out to be near to this value, since in the shell model the overlap effect is missing. The dynamical screening, inherent in the Szigeti charge, is not obtained by an overlap treatment in this approximation.

The importance of overlap effects in a model treatment was first taken into account by Dick[23] in his exchange charge model. While getting encouraging results at longwavelengths comparable with those discussed above, he found no improvement of the model for the critical LO-mode at $(\frac{1}{2}, \frac{1}{2}, \frac{1}{2})$ (Fig 2).

This result can be understood by using a generalized Heitler-London approach.[24] The overlap corrections then contain two parts: a "rigid" part corresponding to a change of the overlap, keeping the electrons in the (displaced) ground state; and a "polarization" part describing excitations of the overlap charge electrons during the ion motion. Only the first part has been taken into account by Lundquist. Even if one extends his treatment to all wavelengths by calculating rigorously a q-dependent (rigid overlap) charge tensor $\underline{\underline{Z}}(q)$, no improvement at higher q-values is obtained, as shown by Zeyher.[25] The same result can be obtained from the treatment by Verma and Singh[26] who used, unfortunately, some invalid[25] approximations but showed that an adjustable overlap-shell model gives good results (besides for ω_{LO} in NaI).

We conclude from this that the "polarization" part of the overlap charge is important in order to explain short-range effects as the "breathing" deformation. Qualitatively, the breathing effect should correspond to excitations of the electrons in the closed p^6-shells of the anions into excited (molecular orbital) states of spherical symmetry, but complicated correlations between the electrons are required to exclude quadrupole contributions.

On the other hand, deformations of quadrupole (Γ_{12}^+) symmetry would be possible if excitations from a closed d-shell into states with a s-symmetry via n.n. overlap polarization were possible. It seems that Ag^+ in AgCl[27] shows this effect[19] since here the TO($\frac{1}{2}$, $\frac{1}{2}$, $\frac{1}{2}$) mode is drastically lowered as compared with the (d-electronless) RbCl.[28]

VII. Homopolar crystals

The essential difference between ionic and homopolar crystals is the accumulation of the electronic charge into "bond charges"[29] between the ions paralleled by a strong (nearly metallic) crystal polarizability (Si, Ge, α-Sn). This situation makes it difficult if not impossible to find a unique model description in terms of polarizabilities and deformabilities of single ions. Cochran[30] showed that a five parameter shell model (putting Z = 0 in Eq.(3.2)) gives good results for Ge. Extending this model by a breathing deformation improves the description[31] and is comparable with a six-parameter "bond dipole" model[31] which might be considered as a screened "bond charge" model.[32] Valence forces treatments are also successful.[33] It seems to be not clear at the present time whether genuine long-range forces in these crystals are really important or not. Born's identity[3] ($4C_{11}(C_{11}-C_{44}) = (C_{11}+C_{12})^2$) and the validity of the sum rule ($\sum_j \omega_j^2(q)$ independent of q) are empirical facts indicating effective n.n.interactions, which are not understood in terms of a model or microscopic theory. Anharmonic corrections of the low-lying TA-branch seem to be important, too.[38]

A promising approach seems to be Sinha's[35] description of silicon which is a parametrized microscopic theory refining Martin's[32] ad-hoc approach to the dielectric function (see the paper by Pick).

Weakly ionic crystals[1d] like GaAs show a complicated mixture between properties of ionic and covalent crystals. For these crystals, an interpolation model would be very helpful. But a better understanding of homopolar crystals is still needed.

VIII. Resonant electronic polarization

The difficulties in describing phonons in highly polarizable crystals like TlBr, PbTe and β-Sn with simple and convincing models indicate that the treatments discussed in the foregoing sections might miss an important feature of the adiabatic electron-ion interaction. It is clear that e.g. in soft-mode transitions anharmonic effects play an important role. Nevertheless, there may exist long-range effects due to strong correlations in the electronic system not taken into account in, say, an extended shell model.

An interesting example for such a possibility are the recently observed anomalies in the phonon spectra of transition metal compounds[36] (Fig 4). These crystals behave in many respects like very hard, weakly ionic crystals with NaCl-structure but, in addition, they are metals with a small free-electron screening which removes the splitting of the optical modes at very long wavelengths.[37] The anomalies in the centre of the Brillouin-zone cannot be described within the framework of a shell model with metallic screening (without introducing a lot of additional parameters). It turns out that a type of resonant electronic coupling between the electron-ion dipole forces might explain the observed data.[37] In the language of the shell model this mechanism can be described by replacing the local electron-ion coupling constant $\underline{k}$ (eq.(3.2)) by some q-dependent entity $\underline{\underline{K}}(q)$ which approximately can be

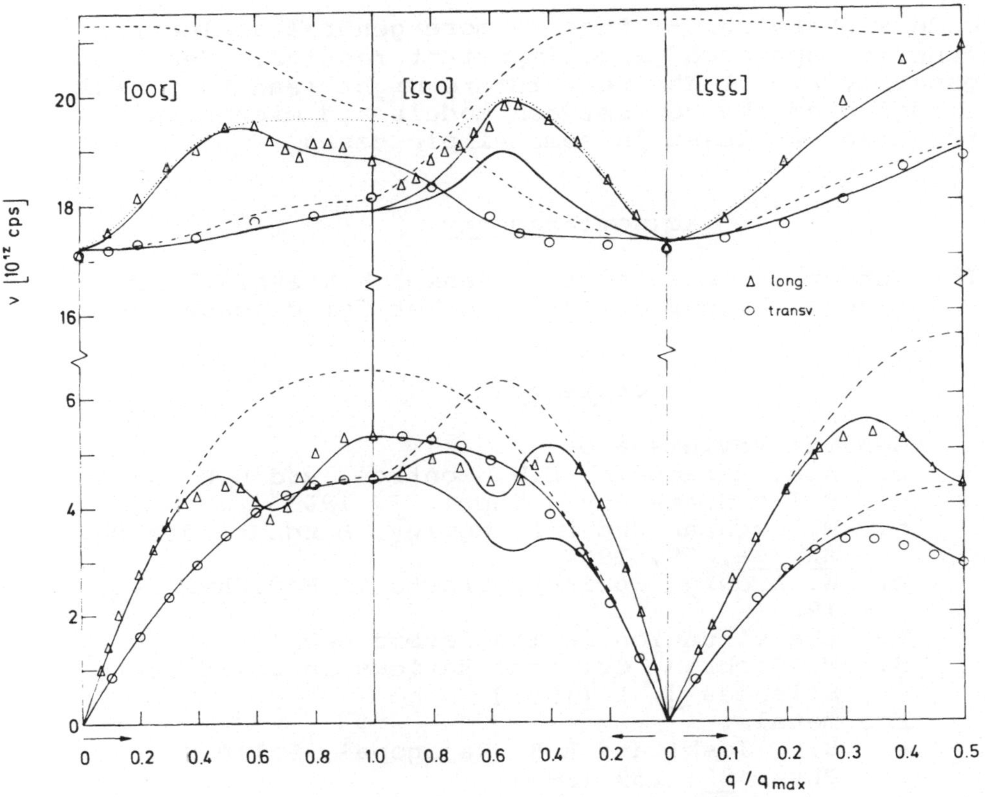

Fig 4: Phonon dispersion curves in TaC

interpreted as a q-dependent polarizability of the metal ions. If $\underline{K}(q)$ becomes small, soft-mode instabilities appear. One might hope that such a model could be able to describe similar effects in the pure transition metals etc. Furthermore, it may provide a useful basis for a microscopic treatment of crystals.

IX. Final remarks

The phonon models developed so far can be understood as the first steps towards a general model theory of phonons in terms of more or less localized charges, polarizabilities and deformabilities. The (semi-) microscopic treatment of the effects of the electronic polarization in terms of (rigid and polarizable) overlap charges illuminates some essential features of the models. Parallel to that, a more general dielectric function approach gives important results. (See the paper by Pick.) The near future might lead to a real bridging of the gap between models and microscopic treatment at least in some simple cases.

Acknowledgments

The author is indebted to W.Hanke, W.Kress,R.Pirc, D.Strauch , U.Schröder and W.Weber for discussions.

References

1. General reviews are
   a. A.A. Maradudin, E.W. Montroll and G.H. Weiss, Solid State Phys. Suppl. 3, 1963
   b. W. Cochran and R.A. Cowley, Handbuch der Physik XXV/2a, 59, 1967
   c. W. Ludwig, Springer Tracts in Mod.Phys. 43, 1967

   For the situation in insulators see
   d. W. Cochran, Critical Reviews in Solid State Sciences 2, 1 (1971)

   For metals:
   e. S.K. Joshi and A.K. Rajagopal, Solid State Phys. 22, 159 (1969)
2. A condensed review has been given by L.J. Sham in: The Simon Frazer Univ.Lect. 2, 143 (1969)
3. M. Born and K. Huang, Dynamical Theory of Crystal Lattices, 1954
4. E.G. Brovman and Yu.Kagan, JETP 52, 557 (1957)
5. D. Sherrington, to be published
6. P.E. Choquard, The Anharmonic Crystal, New York 1967 see also the paper in this sympos. by R.A. Cowley
7. B. Leigh, B. Szigeti and V.Tewary, Proc. Roy. Soc., A320, 505 (1971)
8. E.W. Kellermann, Phil. Trans. Roy. Soc. A238 , 513 (1940)
9. B.G. Dick and A.W. Overhauser, Phys. Rev. 112, 90 (1958)
10. R.A. Cowley, W. Cochran, B.N. Brockhouse and A.D.B. Woods, Phys.Rev. 131, 1030 (1963)
11. J.G. Traylor, H.G. Smith, R.M. Nicklow and M.K. Wilkinson, Phys. Rev. B 3, 3457 (1971)
12. M. Lax, Lattice Dynamics, Pergamon Press, Oxford, 179, (1965)

13. U. Schröder, Solid State Commun. 4, 347 (1966)
V. Nüsslein and U. Schröder, phys. stat. sol. 21, 309 (1967)
14. U. Schröder, private communication; G. Mahler and P. Engelhardt, phys. stat. sol.(b)45, 543 (1971)
15. J.R. Hardy, Phil. Mag. 4, 1278 (1959)
16. M.J.L. Sangster, G. Peckham, and D.H. Saunderson, J. Phys. C. (Solid State Phys.) 3, 1026 (1970)
17. J.S. Melvin, J.D. Pirie, and T. Smith, Phys. Rev., 175, 1082 (1968)
17a. S.O. Lundquist, Ark. Fysik, 6, 25 (1952), 9, 435 (1955), and 12, 263 (1957)
18. P.O. Löwdin, Thesis, Uppsala 1948; Adv. Phys. 5, 1 (1956)
19. K. Fischer, H. Bilz and W. Weber, to be published
20. R.H. Lyddane, R.G. Sachs, and E. Teller, Phys. Rev. 59, 673 (1941)
21. For a discussion see B. Szigeti, Enrico Fermi Summerschool, VARENNA, 1971
22. B. Szigeti, Trans. Faraday Soc., 45, 155 (1949); Proc. Roy. Soc., A204, 51 (1950)
23. B.G. Dick, Phys. Rev. 129, 1583 (1962)
24. B. Gliss, R. Zeyher and H. Bilz, phys. stat. sol. 44, 747 (1971)
25. R. Zeyher, phys. stat. sol., to be published
26. M.P. Verma and R.K. Singh, phys. stat. sol. 33, 769 (1969)
R.K. Singh and M.O. Verma, phys.stat. sol. 36, 335 (1969) and 38, 851 (1970)
27. P.R. Vijayaraghavan, R. M. Nicklow, H.G. Smith and M.K. Wilkinson, Phys. Rec. B 1, 4819 (1970)
28. G. Raunio and S. Rolandson, phys. stat. sol. 40, 749 (1970)
29. J.C. Phillips, Rev. Mod. Phys. 42, 317 (1970)
30. G. Dolling, R.A. Cowley, C. Schittenhelm, and I.M. Thorson, Phys. Rev., 147,577, 1966
31. W. Kress, to be published in phys.stat. sol.(b), 49 (1972)
32. R. Martin, Phys. Rev. Lett. 21, 536 (1968)
33. A.W. Solbrig, Jr., J. Phys. Chem. Solids 32, 1761 (1971)
34. R. Brout,Phys. Rev. 113, 43 (1958)
35. S.K. Sinha, R.P. Gupta and D.L. Price, Phys. Rev. Lett. 26, 1324 (1971)
36. H.G. Smith and W. Gläser, Phys. Rev. Letters 25, 1611 (1970)
37. W. Weber, H. Bilz and U. Schröder, to be published
38. H. Jex, phys. stat. sol. (b), 45, 343 (1971)

MICROSCOPIC THEORY OF PHONONS IN SOLIDS

Robert M. Pick

DPh-G/PSRM - C.E.N. Saclay - 91, Gif-sur-Yvette and

Departement de Recherches Physique, Université Paris,
Paris, FRANCE

I. INTRODUCTION

This paper is a short review of the microscopic theory of phonons, its achievements and its present difficulties.

Let us first recall that the aim of such a theory is to obtain the dispersion curves of a solid directly from the many-body Hamiltonian, which takes into account all the electron-electron and electron-nucleus interactions, of a given solid.

In principle, a microscopic calculation of such dispersion curves is a static as well as a dynamic problem. Indeed, the determination of the equilibrium position of the nuclei of a crystal is a necessary preliminary to the computation of the excitation spectrum; in this sense, the existence of non-imaginary phonon frequencies is a proof that an equilibrium situation has actually been found. Such a condition is, nevertheless, not a sufficient one, because the possibility of internal stresses cannot be completely ruled out within the framework of lattice dynamics. One should then make sure that one starts with an equilibirum position in the absence of any externally applied pressure.

In practice, such verifications of the self consistency of a calculation are rarely done. One usually satisfies himself by using a more or less approximate Hamiltonian function of the position of the nuclei,for the description of the electronic states of the crystal, and by assuming that the binding energy which may be computed from this Hamiltonian is indeed a minimum when the nuclei stand at their experimentally measured positions. In order to shorten our discussion we shall also adopt this attitude and concentrate ourselves on the problem of obtaining the phonon spectrum of a crystal, its equilibrium being assumed.

We shall start by giving a brief summary of the essential aspects of the microscopic theory. Then we shall describe the simplifications which are usually made in such calculations and discuss their validity. Finally we shall review very quickly what has actually been achieved in the case of simple metals and semi-conductors, and point out which difficulties have rendered these calculations unfeasible up to now in transition metals and insulators.

II. SUMMARY OF THE MICROSCOPIC THEORY

Some Limitations of the Present Theory

A large amount of literature has been devoted to the microscopic theory of lattice dynamics. After the pioneer work of G. Baym[1] who gave a detailed description of the problem in a metal, P. N. Keating[2] was the first to attack the problem of a general system, irrespective of its metallic or insulating properties. Unfortunately, his paper overlooked the macroscopic differences existing between the two classes of materials. Within the framework of the harmonic approximation, the proof that the same theory leads, for the two classes of materials, to different results was independently given by L. Sham[3] and R. M. Pick, M. H. Cohen and R. M. Martin,[4] and this section is just a brief summary of the physical content of the two above-quoted papers.

Let us note that such an approach, based on the harmonic approximation, is not a self-consistent one from the point of view of the compressibility sum rule. Even at zero temperature, elastic constants obtained as the second derivative of the energy of the crystal with respect to a strain, differ from those deduced from an analysis of the speed of sound. This discrepancy is inherent to the harmonic approximation. As shown in metals by E. G. Brovman and Y. Kagan,[5] and a more general context by G. Meissner,[6] only an anharmonic theory of phonons is able to reproduce the apparently harmonic result of the first method. This casts some doubts on the possibility of comparing the calculations within the harmonic approximation to experiments.

We shall also admit here that we can consider the nuclei as classical particles, so that the equilibrium position of every nucleus can be defined. As we deal with crystals, we shall label with $\underline{R}_L$ the center of a lattice cell, and a nucleus of mass M_s and charge Z_s has a position $\underline{R}_s$ inside this cell.

For the sake of simplicity, we shall decide for the time being to treat all the electrons on the same footing. This is physically unrealistic because, obviously, some of them (the core electrons) rigidly follow the nuclei in any of their movements, while others (the valence electrons) are mainly responsible for the binding energy and the forces which show up in the dispersion curves. But this approach will help the theory to be simple.

Finally, for the same reason, we shall make the adiabatic approximation that the electrons are always in their ground state, whatever is the position of the nuclei.

Physical Approach to the Harmonic Approximation

Let us admit that one can define an inverse dielectric function, $\varepsilon^{-1}(\underline{r},\underline{r}')$ which describes the potential felt by a test charge at point $\underline{r}$ when an external potential ΔV_{ext} is applied at point $\underline{r}'$. The variation of the potential at the position of a nucleus $\underline{R}_L^s = \underline{R}_s + \underline{R}_L$ if the nucleus at $\underline{R}_{L'}^{s'}$ is moved by an amount $U_{s' L'}^{\beta}$ in the direction β is

$$\Delta V_{tot}(\underline{R}_L^s) = \sum_\beta \int \varepsilon^{-1}(\underline{R}_L^s,\underline{r}') \frac{\partial}{\partial R_{L'}^{s'\beta}} \frac{Z_{s'}}{|\underline{r}'-\underline{R}_{L'}^{s'}|} U_{L'}^{s'\beta}\, d^3r' \quad . \tag{1}$$

The force constant between the two nuclei being the gradient of the potential energy of the nucleus which stands ar $\underline{R}_{L'}^{s'}$, one has

$$C_{L\ L'}^{\alpha\ \beta}{}_{s\ s'} = \frac{\partial^2}{\partial R_L^{s\alpha}\, \partial R_{L'}^{s'\beta}} Z_s\, Z_{s'} \int \varepsilon^{-1}(\underline{R}_L^s,\underline{r}') \frac{1}{|\underline{r}'-\underline{R}_{L'}^{s'}|}\, d^3r' \quad . \tag{2}$$

The dynamical matrix of the crystal has the form

$$\| C_{s\ s'}^{\alpha\ \beta}(\underline{q}) - M_s\, \delta_{ss'}\, \delta_{\alpha\beta}\omega^2 \| = 0 \quad , \tag{3}$$

where, due to Eq. (2) and the translational invariance of the lattice, which reflects itself into $\varepsilon^{-1}(\underline{r},\underline{r}')$,

$$C_{a\ a'}^{\alpha\ \beta}(\underline{q}) = 4\pi \sum_{\underline{k},\underline{k}'} Z_s e^{i(\underline{k}\cdot\underline{R}_s)} (\underline{q}+\underline{k})_\alpha \varepsilon^{-1}(\underline{q}+\underline{k},\underline{q}+\underline{k}') \times \frac{(\underline{q}+\underline{k}')_\beta}{|\underline{q}+\underline{k}'|^2} e^{-i(\underline{k}'\cdot\underline{R}_s')} Z_{s'} \quad , \tag{4}$$

where $\underline{q}$ is inside the first Brillouin zone, and $\underline{k},\underline{k}'$ are reciprocal lattice vectors.

From a macroscopic point of view, it is preferable that the usual symmetry property of the force constants, $C^{\alpha\ \beta}_{s\ s'}(\underline{q}) = C^{\beta,\ \alpha}_{s'\ s}(-\underline{q})$, is clearly fulfilled. This can be achieved by defining

$$S^{-1}(\underline{q}+\underline{k},\underline{q}+\underline{k}') = \varepsilon^{-1}(\underline{q}+\underline{k},\underline{q}+\underline{k}') \times \frac{1}{|\underline{q}+\underline{k}'|^2} , \tag{5}$$

and it can be shown that $S^{-1}(\underline{q}+\underline{k},\underline{q}+\underline{k}')$, which contains all the information about the electrons, possesses the correctly symmetry, namely,

$$S^{-1}(\underline{q}+\underline{k},\underline{q}+\underline{k}')^* = S^{-1}(\underline{q}+\underline{k}',\underline{q}+\underline{k}) . \tag{6}$$

Remark: A more detailed analysis[4,7] shows that Eq. (2) contains an additional term of the form $\overline{C}^{\alpha\beta}_{\substack{s\\L}}\ \delta_{ss'}\delta_{LL'}$. This term simply adds a $\underline{q}$ independent, $\delta_{ss'}$ term to Eq. (4), and it can be shown that its role is just to insure the translation invariance of the crystal, i.e., the existence of acoustical branches. Those properties allow it to be computed once Eq. (2) is known, so that there is no need to discuss it any longer.

Susceptibility and Inverse Dielectric Functions

Let us briefly review the important properties of $S^{-1}(\underline{q}+\underline{k},\underline{q}+\underline{k}')$. This function can readily be obtained in closed form at 0°K, from the eigenfunctions and eigenvalues of the many-body Hamiltonian,

$$\mathcal{H} = \sum_i \frac{P_i^2}{2m} - \sum_{i,s,L} \frac{Z_s}{|\underline{r}_i-\underline{R}^s_L|} + \tfrac{1}{2}\sum_{i,j} \frac{1}{|\underline{r}_i-\underline{r}_j|} . \tag{7}$$

On the other hand, its analytical properties, especially in the vicinity of $\underline{q} = 0$ are less simple to obtain. An analysis of the latter has been done by V. Ambegaokar and W. Kohn,[8] and P. C. Martin,[9] and later a comprehensive discussion of the properties which are important in the phonon problem has been given in Ref. 4. We shall summarize here the most important points.

Let us first introduce the self-consistent, one-electron Hamiltonian for the same crystal,

$$\mathcal{H}_{SCF} = \frac{P^2}{2m} - \sum_{s,L} \frac{Z_s}{|\underline{r}-\underline{R}^s_L|} + \int \frac{1}{|\underline{r}-\underline{r}'|} \langle\rho_{SCF}(\underline{r}')\rangle\, d^3r' , \tag{8}$$

where $<\rho_{SCF}(\underline{r}')>$ is the self-consistent electronic density. The eigenfunctions of Eq. (8) can be taken as a complete set of functions with which one can express $S^{-1}(\underline{q}+\underline{k},\underline{q}+\underline{k}')$. Then one can show (see, in particular, Ref. 9) that one can write

$$S^{-1}(\underline{q}+\underline{k},\underline{q}+\underline{k}') = \delta_{\underline{k},\underline{k}'}|\underline{q}+\underline{k}|^2 + \sum_{\underline{k}''} \chi(\underline{q}+\underline{k},\underline{q}+\underline{k}'')\, S^{-1}(\underline{q}+\underline{k}'',\underline{q}+\underline{k}'), \quad (9)$$

where $\chi(\underline{q}+\underline{k},\underline{q}+\underline{k}\ )$ is the sum of all the irreducible interaction diagrams between the approximate ground state electronic eigenfunction and the electronic density operator,

$$\sum_{i,j} \delta(\underline{r}_i-\underline{r}_j) \quad ,$$

each diagram starting with a line $\underline{q}+\underline{k}$, and ending with a line $\underline{q}+\underline{k}'$.

As Eq. (9) can be written under the form

$$\sum_{\underline{k}''} S^{-1}(\underline{q}+\underline{k},\underline{q}+\underline{k}'')\, [\delta_{\underline{k}''\underline{k}'}|\underline{q}+\underline{k}'|^2 - 4\pi\, \chi(\underline{q}+\underline{k}'',\underline{q}+\underline{k}')] = \delta_{\underline{k},\underline{k}'} \quad , \quad (10)$$

it means that $S^{-1}(\underline{q}+\underline{k},\underline{q}+\underline{k}')$ may be viewed as the inverse of the matrix

$$S(\underline{q}+\underline{k},\underline{q}+\underline{k}') \equiv \delta_{\underline{k},\underline{k}'}|\underline{q}+\underline{k}|^2 - 4\pi\, \chi(\underline{q}+\underline{k},\underline{q}+\underline{k}') \quad (11)$$

in which $\underline{k}$ is the row index and $\underline{k}'$ the column index.

Finally, it is clear that the lowest diagram entering into $\chi(\underline{q}+\underline{k},\underline{q}+\underline{k}')$ contains only the creation and destruction of one electron-hole pair, and is just the expression of the susceptibility in the self-consistent field approximation. More explicitly, one has in this approximation:

$$\chi_{SCF}(\underline{q}+\underline{k},\underline{q}+\underline{k}') = \sum_{1,2} \frac{f(E_1) - f(E_2)}{E_1 - E_2} <1|e^{-i(\underline{q}+\underline{k})\underline{r}}|2><2|e^{-i(\underline{q}+\underline{k}')\underline{r}}|1> \quad , \quad (12)$$

where E_1 and E_2 are the eigenvalues of Eq. (8) corresponding respectively to the eigenfunctions $|1>$ and $|2>$, and $f(E)$ is the usual Fermi function

$$f(E) \begin{cases} = 0 & E > E_F \\ = 1 & E \leqslant E_F \end{cases} .$$

Metals Versus Insulators

Using Eqs. (10), (11) and (12), the analytical properties of $S^{-1}(\underline{q}+\underline{k},\underline{q}+\underline{k}')$ in the SCF approximation are quite obvious, and the paper of L. Sham[3] shows very clearly that such properties persist when diagrams of higher order than Eq. (12) are taken into account in Eq. (11). Let us briefly recall them.

In metals, due to the existence of a Fermi surface, $\chi(\underline{q}+\underline{k},\underline{q}+\underline{k}')$ is finite for any value of $\underline{q},\underline{k}$ and $\underline{k}'$ (except for accidental cancellation of matrix elements for off-diagonal terms). Then, $S^{-1}(\underline{q}+\underline{k},\underline{q}+\underline{k}')$ is also analytic for any value of $\underline{q},\underline{k}$ and $\underline{k}'$ and so is $C^{\alpha\ \beta}_{s\ s'}(\underline{q})$.

In insulators, on the other hand, for $\underline{k}$ or $\underline{k}' = 0$, $\chi(\underline{q}+\underline{k},\underline{q}+\underline{k}')$ systematically tends to zero with $\underline{q}$. This shows that $S^{-1}(\underline{q}+\underline{k},\underline{q}+\underline{k}')$ must have a non-analytic behaviour in the vicinity of $\underline{q} = 0$. In this region, its analytic and non-analytic parts may be separated out and one can show[4] that $C^{\alpha\ \beta}_{s\ s'}(\underline{q})$ may be written under the form

$$C^{\alpha\ \beta}_{s\ s'}(\underline{q}) = C^{\alpha\ \beta}_{1\ s\ s'}(\underline{q}) + \frac{\left(\sum\limits_{\gamma} Z^{\alpha\ \gamma}_{s}(\underline{q})q_{\gamma}\right)\left(\sum\limits_{\delta} q_{\delta}\, Z^{\beta\ \delta *}_{s}(\underline{q})\right)}{\sum\limits_{\gamma\delta} q_{\gamma}\ \varepsilon^{\gamma\delta}(\underline{q})\ q_{\delta}} , \quad (13)$$

where the three quantities $C^{\alpha\ \beta}_{1\ s\ s'}(\underline{q})$, $Z^{\alpha\ \beta}_{s}(\underline{q})$, and $\varepsilon^{\gamma\delta}(\underline{q})$ have an analytic behaviour in the vicinity of $\underline{q} = 0$.

The form of Eq. (13) could have been anticipated from a macroscopic point of view. Its origin may be traced to the dipolar-dipolar character of the force constants between two charges in an insulator. The anisotropic character of the crystal changes this charge into a charge tensor, $[Z^{\alpha\ \gamma}_{s}(\underline{q})]_{\underline{q}\,=\,0}$ for each atom, and this interaction is screened by an anisotropic but finite dielectric constant $[\varepsilon^{\gamma\ \delta}(\underline{q})]_{\underline{q}\,=\,0}$.

Two more points need to be made in this respect: (a) The charge tensors must satisfy the sum rule,

$$\sum_{s} [Z^{\alpha\ \gamma}_{s}(\underline{q})]_{\underline{q}\,=\,0} = 0 \qquad \text{all } \alpha,\gamma \qquad , \quad (14)$$

which comes from, and is strictly equivalent to, the charge neutrality of the crystal, electrons and nuclei being included; and (b) In actual cases, the absolute value of any charge tensor is always $\leqslant 1$. On the other hand, one can show that

$$Z_s^{\alpha\,\gamma}(\underline{q}) = Z_s\Big(\sum_{\underline{k}}\Big(\delta_{\alpha\gamma}\delta_{\underline{0k}} + \sum_{\underline{k}'}{}' \, 4\pi \frac{\partial}{\partial q_\alpha} \chi(\underline{q},\underline{q}+\underline{k}')$$

$$\times\ \overline{S}^{-1}(\underline{q}+\underline{k}',\underline{q}+\underline{k})(\underline{q}+\underline{k})_\gamma\Big)\, e^{-i\underline{k}\cdot\underline{R}_s}\Big)^* \quad , \qquad (15)$$

where $\overline{S}^{-1}(\underline{q}+\underline{k},\underline{q}+\underline{k}')$ is the inverse of $S(\underline{q}+\underline{k},\underline{q}+\underline{k}')$, once the line $\underline{k} = 0$ and the corresponding column $\underline{k}' = 0$ have been deleted.

As Z_s is always positive and much larger than unity, those two remarks show the fundamental role of the off-diagonal terms of $\chi(\underline{q}+\underline{k},\underline{q}+\underline{k}')$ in the lattice dynamics of insulators; neither could the magnitude of the charge tensors be obtained, nor, _a fortiori_, could the sum rule, Eq. (14), be fulfilled without taking those terms into account. On the other hand, they do not play such fundamental roles in metals, at least in simple ones; this explains why the computational situation is rather different in the two cases.

III. THE USUAL APPROXIMATIONS AND THEIR LIMITATIONS

We must now reduce our problem to a manageable form by introducing physically realistic approximations. They are usually of two different types and are done either independently or simultaneously.

The first one consists in using a finite number of electronic bands in the computation of $(\underline{q}+\underline{k},\underline{q}+\underline{k}')$. The basic idea here is that, when taking into account all the electrons on the same footing, we have to describe the motion of the core electrons which actually follow their nucleus in its own motion. This is physically unwise, and at the same time mathematically awkward because, in order to reproduce the movement of a well-localized charge density, one needs to use a large number of wave functions; this means that one should properly take into account highly excited states and large $\underline{q}+\underline{k}$ Fourier transform of the electronic wave functions. One usually overcomes this difficulty by defining a local pseudopotential, $z_s(\underline{r})$, considered as the one seen by the outer valence electrons. Then its effective charge $[z_s(\underline{q})]_{\underline{q}=0}$ is equal to Z_s minus the number of core electrons, and $z_s(\underline{q})$ is such that it rapidly decreases with increasing $\underline{q}$. The SCF Hamiltonian in Eq. (6) is then replaced by

$$\mathcal{H}_{P.SCF} = \frac{p^2}{2m} - \sum_{s,L} \int \frac{1}{|\underline{r}-\underline{r}'|}\, z_s(\underline{r}'-\underline{R}_L^s)\, d^3r'$$

$$+ \int \frac{1}{|\underline{r}-\underline{r}'|} \langle\rho_{SCF}(\underline{r}')\rangle\, d^3r' \quad , \qquad (16)$$

and $\chi(\underline{q}+\underline{k},\underline{q}+\underline{k}')$ will eventually be computed with the help of the sole eigenfunctions and eigenvalues of Eq. (16). It is straightforward to realize that the phonon spectrum will be given by the same formulae as in the preceding section, except for the change of Z_s into $z_s(\underline{q}+\underline{k})$ in Eq. (4) and (15), and the fact that the SCF unperturbed electronic ground state wave functions now contain only pseudo-wavefunctions, solutions of Eq. (16), the energy of which is below the Fermi energy.

One must nevertheless realize that this method may be inconsistent in two respects: (a) The basic idea underlying the use of a pseudo-potential is that the wiggles of the real wave functions in the vicinity of the core have no relationship with the binding energy of the crystal and that one should replace in that region the real wave functions by their smoothed part. It turns out that practical computations do not coverge quickly, i.e., large determinants must be used which take many plane waves into account. Then the pseudo wavefunction is no longer smooth in the vicinity of the core where it was assumed to be so.

(b) In principle, the actual electronic wavefunction may be obtained from the pseudo wavefunction by orthogonalizing it to the core electrons.[10] In fact, the orthogonalization procedure expels some electrons from the core region so that the actual wave function represents an integrated charge density smaller than unity. It means that in the real crystal, the electronic charge density far from the core is larger than the one computed with a pseudo-potential techniques. As those regions are the most important for the binding energy, the use of such pseudo wavefunctions can lead to large errors.

The second approximation consists in retaining only the SCF part of the susceptibility function. In fact, the only case where such an approximation has not been made is that of simple metals to which we shall return in the next section.

Let us finally note that, even if such approximations are done, one is left with the double numerical problem of (a) computing with a reasonable accuracy the susceptibility function $\chi(\underline{q}+\underline{k},\underline{q}+\underline{k}')$; and (b) inverting the $S(\underline{q}+\underline{k},\underline{q}+\underline{k}')$ matrix and verifying that, in an insulator, the effective charge tensor sum rule is actually fulfilled.

Clearly, except for alkali metals, the charge of the pseudopotential is larger than unity. For example, in ClNa, the six electrons of the 3p shell will be taken into account, in the preceding approach, as valence electrons, and an effective charge of +5 will be given to the pseudopotential of the chlorine ion. Large cancellation effects must then take place because the effective charge tensor is approximately 0.9. In some sense, it means that only the outer electrons of the 3p shell play a role in the effective forces. One could expect to avoid such cancellations effects, which transform into computational difficulties, by considering that a large amount of the valence electrons rigidly move with the nucleus as do the core electrons. This approach has been used by S. K. Sinha[11] but is too complicated to be discussed here. To our knowledge it has not yet been used in actual calculations.

IV. THE PRESENT SITUATION - SIMPLE METALS

Simple metals are the field where the microscopic theory of phonons has been applied for the longest time, as it was used in fact since 1963, i.e., before the theory was fully understood.

In fact, since that time the lattice dynamics of most simple metals, i.e., metals in which the d or f electrons do not play any role, has been explored, and for many of them more than ten calculations now exist (see References, e.g., Ref. 12). It is not our purpose to discuss and compare these various works. We shall only point out here why such calculations are easy to perform, what kind of information has been gained, and what seem to be, for the time being, the most reliable methods.

The basic simplification with simple metals is that one can show (see, e.g., M. H. Cohen and V. Heine[13] or W. Harrison[10a]) that their pseudo potential is so weak that it perturbs very slightly the electronic wavefunction of the free electron gas. One is then justified in using a free electron gas susceptibility function $\chi(\underline{r},\underline{r}')$. As this susceptibility is only a function of $(\underline{r}-\underline{r}')$, the matrix $S(\underline{q}+\underline{k},\underline{q}+\underline{k}')$ is a diagonal one and no problem exists for its inversion. One can then concentrate on two problems mentioned in the preceding sections, namely (a) what is the best local potential which must be used; and (b) what is the influence of the correlation and exchange corrections to $S(\underline{q}+\underline{k},\underline{q}+\underline{k}')$ on the phonon spectrum; by systematically varying one of the terms while leaving the second constant. It turns out that, for one given local pseudopotential, exchange and correlation give a contribution which may give a thirty per-cent correction to the dispersion curves. As could be anticipated, this correction is mostly important for longitudinal phonons which involve a local change of the electronic density, while its role is much weaker for transverse phonons. These results appear clearly in the work of D. L. Price *et al.*[14] on alkali metals, where four different susceptibilities have been used, or that of E. R. Floyd *et al.*[15] on magnesium. Nevertheless, one must note that the sign of the effect is not *a priori* predictable and might depend on the assumed pseudopotential. For instance, D. L. Price *et al.*[14] find a decrease of all the phonon frequencies when going from the random phase to the J. Hubbard[16] susceptibility function and so does M. A. Coulthard *et al.*,[17] while E. R. Floyd *et al.*[15] obtain an increase for the same change. Nevertheless, they all find a decrease with respect to the random phase approximation if they use a still more sophisticated susceptibility function.

On the other hand, it is found that one can always adjust or slightly modify the few parameters of a local pseudopotential in order to achieve a good fit to experimental data. For instance, W. Hartmann *et al.*[12] obtain equally good fits for aluminum with three different pseudopotentials, two of which use the same susceptibility function. In this last case, those pseudopotentials nevertheless differ, e.g., by 20% at the second reciprocal lattice vector. This shows that the use of local pseudopotentials is largely questionable, even in the case of simple metals, and that the amount of information gained by applying the microscopic theory of phonons to such a system might be rather poor.

Let us finally point out that there exists a more fundamental approach to the pseudopotential, where its role of orthogonalizing valence or conduction electrons to the core ones is seriously taken into account. Those methods are not easily amenable to the formalism summaried in Section II. On the other hand, for simple metals, one can again use the idea that these *non-local* pseudopotentials weakly perturb a uniform electron gas, and obtain the total energy of the metal up to second order in pseudopotentials. The dependence of this energy on the position of the atoms can directly be obtained.[18] Such methods put all the stress on the direct calculation of energies without using the intermediate step of wave-function computations, and, in this respect, they are more reliable. Those methods have recently been improved, and many aspects of their latest developments may be found in Ref. 19. They give good results when, once again, the exchange and correlation properties of the free electron gas are taken into account and extended to the actual electronic charge density.[17,20,21] On the other hand, as the electronic wavefunctions are not direct ingredients of the calculation, such methods are not easily extendable, e.g., to the computation of frequency-dependent susceptibilities.

V. THE PRESENT SITUATION - SEMICONDUCTORS

The application of the microscopic theory of phonons has been up to now limited to elemental semiconductors of the zinc blend structure Si and Ge. The corresponding calculations are still crude ones where both the local potential and the random phase approximations are made. The basic simplification expected with these materials was the following: symmetry imposes that the effective charge tensor $[Z_s^{\alpha\beta}(\underline{q})]_{\underline{q}=0}$ must be equal to zero. It means that no dipole-dipole interaction exists and it could be expected that a treatment based on local pseudopotentials and a free electron susceptibility function $\chi(\underline{q}+\underline{k},\underline{q}+\underline{k}')\delta_{\underline{k},\underline{k}'}$ would give a first-order approximation to the dispersion curves. Unfortunately, in this calculation, the transverse acoustic frequencies turned out to be imaginary which meant that the off-diagonal elements of the susceptibility function could no longer be neglected.

Two proposals have been made to overcome this difficulty without actually computing those off-diagonal terms. Let us assume that the diagonal part of $S(\underline{q}+\underline{k},\underline{q}+\underline{k}')$ has been computed for the actual semiconductors. As

$$\mathrm{Lim}_{q \to 0}[S(\underline{q}+\underline{k},\underline{q}+\underline{k}')]_{\substack{\underline{k}=0\\ \underline{k}'=0}} = 0 \quad , \tag{17}$$

it means that the pseudopotential is not completely screened at long wave length by the diagonal part of $S(\underline{q}+\underline{k},\underline{q}+\underline{k}')$. J. C. Phillips[22] and R. M. Martin[23] proposed that the absence of neutrality of the screened pseudopotential could be seen as a reflection of the existence of a covalent bond charge which lies always between the two nearest atoms at the same distance from both, and has a charge such that any lattice cell is neutral. R. M. Martin[23] has calculated a dispersion curve which does not differ from the measured one by more than 20% by assuming that the pseudopotentials interact via the diagonal part of $S(\underline{q}+\underline{k},\underline{q}+\underline{k}')$ while the covalent charges interact between themselves and with the unscreened part of the pseudopotentials via a direct dipolar interaction.

On the other hand, the S. K. Sinha <u>et al</u>.[24] ansatz has been that one can write

$$S(\underline{q}+\underline{k},\underline{q}+\underline{k}') = \tilde{S}(\underline{q}+\underline{k})\delta_{\underline{k},\underline{k}'} + \sum_{P,P'} A_P^*(\underline{q}+\underline{k})\alpha_{PP'}(\underline{q})A_{P'}(\underline{q}+\underline{k}') \quad , \tag{18}$$

where $A_P(\underline{q}+\underline{k})$ represents the rigid motion of a muffin-tin charge density centered on some lattice point. An analysis of their model shows that it differs from J. C. Phillips' proposal[22] in two ways: (a) the pseudo-potentials and the potential created by the muffin-tin charge density now interact between themselves and one with the other through the same diagonal function $S(\underline{q}+\underline{k})\delta_{\underline{k},\underline{k}'}$; and (b) the muffin tins also interact through a "shell-shell" interaction $\alpha^-_{PP'}(\underline{q})$ which is adjusted so that the effective charge tensor goes to zero with $\underline{q}$. If this method is more consistent, the agreement obtained is not substantially improved over Martin's. Let us note that none of those calculations have been self-consistent, in the sense that the pseudopotentials used in the phonon calculations have no relationship with the susceptibility function used in the same calculation. Hopefully, a more consistent calculation in this respect, as well as a local approximation of the Slater[25] or Hohenberg, Kohn and Sham[26] type, will improve these results. Also, an actual calculation of the off-diagonal term of $S(\underline{q}+\underline{k},\underline{q}+\underline{k}')$ using the same wave functions which lead to the susceptibility function used by S. K. Sinha et al.[24] would be a good test of their ansatz.

VI. THE PRESENT SITUATION - TRANSITION METALS AND INSULATORS

The situation here is still worse than for semiconductors. One can more or less safely state that no calculation is available up to now.

In the case of transition metals where the Kohn anomaly[27] is large, it is clear that its importance is due to the large density of states of the d electrons at the Fermi level. Those d electrons which are fairly localized must be taken into account in the computation of $\chi(\underline{q}+\underline{k},\underline{q}+\underline{k}')$ which has no large off-diagonal terms. No such calculations have been made up to now. The role of the d electrons have just been theoretically explored by S. Barisic[28] in a model which unfortunately is still too crude to be applicable in actual cases. One is then faced with an extremely bad situation where no calculation is able to interpret the large experimental information actually available because the phenomenological models are unable to describe the Kohn anomaly, due to its intrinsic microscopic origin.

In insulators the situation is scarcely better. The only calculation, done before the microscopic theory was really understood, is that of B. Gliss and H. Bilz[29] and its agreement with experiment was rather poor. The present efforts are preliminary ones, and tend to overcome the difficulty of inverting a large $S(\underline{q}+\underline{k},\underline{q}+\underline{k}')$ matrix. S. K. Sinha's proposal[30] is to use the same ansatz as that used for silicon, with a presumably very small diagonal part. R. M. Pick's suggestion[31] is to reformulate the problem in terms of the charge transfers associated with the electronic movements of the pseudopotentials. Both approaches lead to a reformulation of the phonon problem which now looks quite analogous to a breathing shell model, as developed from semi-empirical arguments by U. Schroeder[32] and co-workers, and analyzed from a symmetry point of view by H. Bilz[33] in this volume. But actual computations still have to be made in order to see if any progress has been done in the microscopic understanding of these phonon spectra, which are very well described by a few parameters in the phenomenological descriptions.

VII. CONCLUSIONS

The microscopic theory of phonons is certainly a wide-open field for computational problems. It is clear that a large amount of physical information can be obtained from such calculations in the cases where the computations have not yet started. On the other hand, simple metals might be a less attractive field in the sense that the latest developments have not really improved our knowledge of the electronic wave functions. At the outset of the theory, they seemed to be the important quantities that could be obtained by a careful comparison between calculations and experiments. Because a phonon spectrum is only in the adiabatic approximation, a difference in binding energy between two different nucleus configurations, it is quite possible that very good fits may be obtained without having achieved this intermediary goal.

REFERENCES

1. G. Baym, Ann. Phys. 14, 1 (1961).

2. P. N. Keating, Phys. Rev. 175, 1171 (1968).

3. L. J. Sham, Phys. Rev. 188, 1431 (1969).

4. R. M. Pick, M. H. Cohen and R. M. Martin, Phys. Rev. B1, 910 (1970).

5. E. G. Brovman and Y. Kagan, Soviet Physics JETP 30, 721 (1970).

6. G. Meissner, to be published.

7. L. J. Sham, Proc. Roy. Soc. A283, 33 (1965).

8. V. Ambegaokar and W. Kohn, Phys. Rev. 117, 423 (1960).

9. P. C. Martin and J. Schwinger, Phys. Rev. 115, 1392 (1950).

10a W. A. Harrison, Pseudo-Potentials in the Theory of Metals. (Benjamin, New York, 1966).

10b R. Pick and G. Sarma, Phys. Rev. 135, A1363 (1964).

11. S. K. Sinha, Phys. Rev. 169, 477 (1968).

12. W. M. Hartmann and T. O. Milbrodt, Phys. Rev. B3, 4133 (1971).

13. M. H. Cohen and V. Heine, Phys. Rev. 122, 1821 (1961).

14. D. L. Price, K. S. Singwi and M. P. Tosi, Phys. Rev. B2, 2983 (1970).

15. E. R. Floyd and L. Kleinman, Phys. Rev. B2, 3947 (1970).

16. J. Hubbard, Proc. Roy. Soc. A243, 336 (1957).

17. M. A. Coulthard, J. Phys. C. Sol. St. Phys. C3, 821 (1970).

18. W. Harrison, Phys. Rev. 136, A1107 (1964); and
R. Pick, Journal de Physique, 28, 539 (1967).

19. R. W. Shaw, Jr., J. Phys. C. Sol. St. Phys. C2, 2350 (1969).

20. R. W. Shaw, Jr. and R. Pynn, J. Phys. C. Sol. St. Phys. C2, 2071 (1969).

21. T. Schneider, this volume.

22. J. C. Phillips, Phys. Rev. 166, 832 (1968).

23. R. M. Martin, Phys. Rev. 186, 871 (1969).

24. S. K. Sinha, R. P. Gupta and D. L. Price, Phys. Rev. Letters (1971).

25. J. C. Slater, Phys. Rev. 81, 385 (1951).

26. P. Hohenberg and W. Kohn, Phys. Rev. 136, B864 (1964); and
W. Kohn and L. J. Sham, Phys. Rev. 140, A1138 (1965).

27. W. Kohn, Phys. Rev. Letters 2, 393 (1959); and
W. Kohn and S. J. Nettel, Phys. Rev. Letters 5, 8 (1960).

28. S. Barisic, Ph. D. thesis, University of Paris XI (unpublished); and Phys. Rev. (to be published).

29. B. Gliss and H. Bilz, Phys. Rev. Letters 21, 884 (1968).

30. S. K. Sinha, Phys. Rev. 177, 1256 (1969).

31. R. M. Pick, International Conference on Phonons. Rennes, France (ed. by M. Nusimovici) 1971 (to be published).

32. U. Schroeder, Sol. St. Comm. 4, 347 (1966).
V. Nusslein,and U. Schroeder, Phys. Stat. Solid. 21, 309 (1967).

33. H. Bilz, this volume.

COMPUTATIONAL ASPECTS OF ANHARMONIC LATTICE DYNAMICS

T. R. Koehler

IBM Research Laboratory, San Jose, California 95114

I. INTRODUCTION

One major subfield of lattice dynamics is concerned with the evaluation of the physical properties of a defectless or ideal crystal for which the adiabatic approximation should be valid. In this approximation one assumes that the solid is well modeled by a collection of atoms which interact through an interatomic potential and that electronic effects, except as they contribute to the potential, are negligible. This approximation is appropriate for insulating crystals and should be especially good for the solid isotopes of helium, commonly called the quantum crystals, and the rare gas solids. In practice, it is also found to work even for the lattice dynamics of metals.

In this contribution, I will try to give some perspective on the computational aspects of theories which are particularly well-suited to the treatment of a system for which the adiabatic approximation is valid. Only methods which are essentially perturbative in nature will be discussed. This qualification is used to denote an approach in which the calculation of the properties of the system starts with an exactly soluble approximation which yields a zeroth order result for the free energy of the crystal and a non-interacting set of elementary excitations (phonons). The true properties of the crystal are then obtained as successive corrections to the lowest order results by evaluating expressions which involve only the bare properties. This method can be contrasted with an important new approach to lattice dynamics in which renormalized properties are calculated in as self-consistent a manner as possible by using expressions in which only renormalized properties are present. The latter ideas are discussed by Heintz Horner in this volume.

In Section II some ideas of the traditional approach to lattice dynamics--the harmonic approximation-- will be reviewed and the computational implications of the essential formulae will be discussed. In Section III, a newer theory, self-consistent harmonic approximation, will be treated in the same manner. In Section IV, a few conclusions and suggestions will be given.

There will be no original work presented in this paper. However, the computational implications of formulae used in anharmonic lattice theory are not generally discussed in the original papers and so should be of use to someone entering the field. In addition, much older work in lattice dynamics is concerned with techniques for the approximate evaluation of expressions which can now be easily evaluated exactly. Thus the older literature tends to provide a misleading focus on areas of computational difficulty. The list of references will not be exhaustive; however, references which themselves provide a guide to the literature will be used whenever possible.

II. HARMONIC APPROXIMATION

In the notation which will be used, $u_\alpha\binom{i}{\mu}$ represents the Catesian of the displacement from equilibrium of the μth atom, of mass m_μ, in the ith unit cell. Similarly, the equilbrium position is denoted by $R_\alpha\binom{i}{\mu}$ and the true position by $x_\alpha\binom{i}{\mu}$; then $u_\alpha\binom{i}{\mu} = x_\alpha\binom{i}{\mu} - R_\alpha\binom{i}{\mu}$. The gradient with respect to $x_\alpha\binom{i}{\mu}$ is $\nabla_\alpha\binom{i}{\mu}$. The vector to the point of origin of the unit cell is $\underline{R}(i)$.

The Hamiltonian for the crystal in the adiabatic approximation is

$$\mathcal{H} = K + V \quad , \tag{1}$$

where

$$K = -\Sigma(i,\mu,\alpha)\nabla_\alpha\binom{i}{\mu}$$

and V is a function only of the $u_\alpha\binom{i}{\mu}$.

The traditional approach to lattice dynamics is the harmonic (HA) or quasi-harmonic[1] (QHA) approximation as described in the classic book of Born and Huang[2] or in the later volume of Maradudin, Montroll and Weiss.[3] In this method, the potential is expanded in a Taylor series about the equilibrium position of the atoms and one writes, schematically,

$$H = K + V^{(0)} + V^{(2)} + V^{(3)} + V^{(4)} + \ldots \tag{2}$$

where

$$V^{(n)} = \frac{1}{n!}\,\Sigma^{(A)}\Phi_{\alpha_1\ldots\alpha_n}\begin{pmatrix}i_1\ldots i_n\\ \mu_1\ldots\mu_n\end{pmatrix}u_{\alpha_1}\begin{pmatrix}i_1\\ \mu_1\end{pmatrix}\ldots u_{\alpha_n}\begin{pmatrix}i_n\\ \mu_n\end{pmatrix} \tag{3}$$

with

$$\Phi_{\alpha_1\ldots\alpha_n}\begin{pmatrix} i_1\ldots i_n \\ \mu_1\ldots\mu_n\end{pmatrix} = \nabla_{\alpha_1}\begin{pmatrix} i_1 \\ \mu_1\end{pmatrix}\ldots\nabla_{\alpha_n}\begin{pmatrix} i_n \\ \mu_n\end{pmatrix} V \Big|_{\text{equilibrium}} \qquad (4)$$

and $\Sigma^{(A)}$ used to denote a sum over all indices. The $V^{(1)}$ term vanishes if the lattice is in equilibrium.

Since the displacements from equilibrium are expected to be small, the relationship $V^{(n+1)} < V^{(n)}$ should hold and one can write $\mathcal{H} \approx H$ where

$$H = K + V^{(0)} + \tfrac{1}{2}\Sigma^{(A)} u_\alpha\binom{i}{\mu}\Phi_{\alpha\beta}\binom{ij}{\mu\nu}u_\beta\binom{j}{\nu} \qquad (5)$$

is a harmonic Hamiltonian which serves as an exactly soluble model Hamiltonian for the crystal. The $\Phi_{\alpha\beta}\binom{ij}{\mu\nu}$ are customarily called force constants.

The properties of H are well known; however, a few points essential for a computational perspective will be reviewed. The eigenstates are states in which a number of non-interacting, elementary excitations (bare phonons) are present. The squares of the phonon frequencies $\omega_j(\underline{k})$ and the polarization vectors $e_\alpha(\mu|{\underline{k} \atop j})$ are the roots and eigenvectors of the 3n x 3n dynamical matrix $D(\underline{k})$ whose components are

$$D_{\alpha\beta}\binom{\underline{k}}{\mu\nu} = \frac{1}{N}\frac{1}{\sqrt{m_\mu m_\nu}}\sum_{i,j}\Phi_{\alpha\beta}\binom{ij}{\mu\nu}\exp(-i\underline{k}\cdot[\underline{R}(i)-\underline{R}(j)]) \qquad (6)$$

In matrix notation, $\underline{De} = \omega^2\underline{e}$. For each wave vector $\underline{k}$ there are 3n phonon branches where n is the number of atoms per unit cell. The index j in $\omega_j(\underline{k})$ and $e_\alpha(\mu|{\underline{k} \atop j})$ runs from 1 to 3n.

The free energy of H, which is the lowest order approximation to the free energy of the true crystal, is

$$F^{(0)} = \frac{1}{\beta}\sum_{j,\underline{k}} \ln\,[2\sinh(\tfrac{1}{2}\beta\hbar\omega_j(\underline{k}))] \qquad (7)$$

and the energy is

$$E^{(0)} = \sum_{j,k}(n_j(\underline{k}) + \tfrac{1}{2})\hbar\omega_j(\underline{k}) \qquad (8)$$

where $\beta = 1/kT$ and $n_j(\underline{k}) = (e^{\beta\hbar\omega_j(k)} - 1)^{-1}$ is the phonon occupation number.

'The evaluation of $F^{(0)}$ or $E^{(0)}$ requires a sum over a mesh of points in $\underline{k}$-space of simple functions of the $\omega_j(\underline{k})$. Thus, whether one is interested in phonon dispersion curves or in the thermal properties of crystals, the computation of the $\omega_j(\underline{k})$ is necessary. The only such calculations that are practical by hand are for $\underline{k}$ values along symmetry directions especially if contributions from only one or two shells of nearest neighbors are used. With the aid of a computer of modest power, any expression that involves only a single scan over the Brillouin Zone can be evaluated easily. One might suspect that the diagonalization of $\underline{D}(\underline{k})$ would be the most time consuming step, but in fact its construction takes the most time, especially if contributions from many shells of neighbors are used. However, neither of these steps is prohibitively time consuming even for a crystal with several atoms per unit cell. In the case of more complicated structures, the greatest problem is the programming effort.

The modifications which must be made to the properties of H to obtain the properties of are called anharmonic corrections. Anharmonicity gives corrections to the free energy of the crystal and induces interactions among the phonons which change their energy and introduce damping.

Perturbative analyses[4,5] of anharmonic effects show that the two leading contributions to the free energy should be roughly of equal importance and should involve $V^{(4)}$ to first order and $V^{(3)}$ to second order. These are known as quartic and cubic contributions, respectively, and will be denoted by $\delta F^{(4)}$ and $\delta F^{(3)}$. Thus, $F \simeq F^{(0)} + \delta F^{(3)} + \delta F^{(4)}$. The specific form of the terms are

$$\delta F^{(3)} = -\frac{\hbar^2}{48N}\Sigma_{(1,2,3)}\frac{\Delta(k_1+k_2+k_3)|\Phi(1,2,3)|^2}{\omega_1\omega_2\omega_3} \times \frac{(n_1+1)(n_2+1)(n_3+1) - n_1n_2n_3}{\omega_1+\omega_2+\omega_3} + \frac{(n_1+1)(n_2+1)n_3 - n_1n_2(n_3+1)}{\omega_1+\omega_2-\omega_3} \tag{9}$$

and

$$\delta F^{(4)} = \frac{\hbar^2}{32N}\Sigma_{(1,2)}\frac{\Phi(1,-1,2,-2)}{\omega_1\omega_2}(2n_1+1)(2n_2+1) \tag{10}$$

where $\Delta(\underline{k})=0$ unless $\underline{k}=0$ or a reciprocal lattice vector.

Two notational changes are made in the above equations: First, since all indices k_i and j_i are summed over, i alone is used to denote both indices. And, second, a transform of the derivatives of the potential is used in which $\Phi(1...)$ is defined in a similar way to $\Phi_{\mu...}\binom{i...}{\mu...}$ but with

$$\nabla_1 \equiv \nabla_j(k) = \Sigma_{(i\mu\alpha)} \frac{1}{\sqrt{m_\mu}} e_\alpha(\mu|\tfrac{k}{j}) \nabla_\alpha(\tfrac{i}{\mu}) e^{i\underline{k}\cdot\underline{R}(i)} \tag{11}$$

replacing $\nabla_\alpha(\tfrac{i}{\mu})$ in Eq. (4).

The effect of anharmonicity on the phonon frequencies is generally obtained by evaluating the quantities Δ and Γ which enter into the single phonon Green's function G which is obtained from

$$\Sigma_{j'} \{(\omega_j(\underline{k})^2 - \Omega^2)\delta_{jj'} + 2\omega_j(\underline{k})(\Delta_{jj'}(\underline{k},\Omega) - i\Gamma_{jj'}(\underline{k},\Omega)\}$$

$$G_{j'j''}(\underline{k},\Omega) = 2\omega_j(\underline{k})\delta_{jj''}/\beta\hbar \quad . \tag{12}$$

In many cases, an inelastic neutron scattering experiment measures the imaginary part of G. Deviations from this concept are discussed by Cowley in another paper in this volume, and special difficulties that one might expect in the case of solid helium have recently been discussed by Werthamer.[6] The cubic anharmonic contributions to Δ and Γ are

$$\begin{matrix}\Delta^3_{ij}(\underline{k},\Omega)\\ \Gamma^3_{ij}(\underline{k},\Omega)\end{matrix} = -\frac{\hbar^2}{16\sqrt{\omega_i(\underline{k})\omega_j(\underline{k})}} \sum_{(1,2)} \Phi(\tfrac{-\underline{k}}{i},1,2)\Phi(\tfrac{\underline{k}}{j},1,2)$$

$$\times \quad (n_1+n_2+1)\begin{matrix} P[(\omega_1+\omega_2+\Omega)^{-1}] + P[(\omega_1+\omega_2-\Omega)^{-1}] \\ \pi\delta(\omega_1+\omega_2+\Omega) \quad - \quad \pi\delta(\omega_1+\omega_2-\Omega)\end{matrix}$$

$$+ \quad (n_2 - n_1)\begin{matrix} P[(\omega_1-\omega_2+\Omega)^{-1}] + P[(\omega_1-\omega_2-\Omega)^{-1}] \\ \pi\delta(\omega_1-\omega_2+\Omega) \quad - \quad \pi\delta(\omega_1-\omega_2-\Omega)\end{matrix} \quad . \tag{13}$$

There is no quartic contribution to Γ; the contribution to Δ is

$$\Delta^{(4)}_{ij}(k,\Omega) = \frac{\hbar^2}{8N\sqrt{\omega_i(\underline{k})\omega_j(\underline{k})}} \sum_1 \frac{\Phi(\tfrac{\underline{k}}{i},\tfrac{-\underline{k}}{j},1,-1)}{\omega_1} (2n_1+1) \quad . \tag{14}$$

If the off-diagonal elements of G are small because Δ and Γ are small, or zero because of symmetry, G is then approximately diagonal, Eq. (12) simplifies, and there is a $G_j(\underline{k},\Omega)$ for each mode. Furthermore, if Δ and Γ are nearly constant in the vicinity of $\omega_j(\underline{k})$, $\mathrm{Im}[G_j(k,\Omega)]$ is a Lorentzian line of width $\Gamma_{jj}[\underline{k},\omega_j(\underline{k})]$ whose peak has shifted to $\Omega = \omega_j(\underline{k}) + \Delta_{jj}[\underline{k},\omega_j(\underline{k})]$. Thus, one need only evaluate Δ and Γ at a single frequency. However, if the anharmonicity is large, the frequency dependence of Δ and Γ causes appreciable departures from a Lorentzian shape and these quantities must be evaluated as a function of Ω, and if G is not diagonal from symmetry, the full matrix algebra implicit in Eq. (12) must be performed.

The quartic contribution $\delta F^{(4)}$ and $\Delta^{(4)}$ are easy to evaluate. Both expressions involve only quantities whose computation requires a single scan over the Brillouin Zone. Additionally, $\Delta^{(4)}$ is frequency independent.

However, $\delta F^{(3)}$ involves a double scan over the Brillouin Zone and is about two orders of magnitude more time-consuming than is the evaluation of $F^{(0)}$. The calculation of $\Delta^{(3)}$ and $\Gamma^{(3)}$ involves only a single scan, but needs to be obtained at a variety of $\underline{k}$ vectors if one is calculating dispersion curves along the symmetry directions, and needs to be evaluated at a number of Ω values if the anharmonicity is large. The evaluation of all of the cubic corrections can be quite time-consuming.

In addition to the machine time requirements, the programming effort involved in calculating anharmonic effects is considerably greater than is required for the harmonic properties, especially for more complicated structures.

However, with the latest generation of computers, we can perform anharmonic calculations with adequately fine scans in k space quite routinely. As examples, in fcc or bcc structures, a $\delta F^{(3)}$ calculation employing a grid of 1000 points in the first Brillouin Zone takes less than 30 seconds and a $\Delta^{(3)}$ and $\Gamma^{(3)}$ calculation using a grid of 8000 points in k space and 200 values of Ω takes about 3 minutes.

Many of the anharmonic calculations that were performed prior to 1968 were those of Cowley and co-workers. The work was primarily concerned with anharmonic effects in alkali halides and is described or referred to in a review article by Cowley.[7]

Calculations of the thermal properties of rare gas solids using QHA with anharmonic corrections have been performed by Klein, Horton and Feldman.[8] An important result of these calculations was the indication that the quasi-harmonic approximation with anharmonic corrections cannot be extended reliably to higher temperatures. Their results agreed quite well with experiment at temperatures up to about 1/3 of the melting temperature, but above that agreed very poorly.

In work similar to that reported in this conference by T. Schneider, the lattice dynamics of several metals has been investigated in QHA using a pseudopotential for V. Recently, calculation of anharmonic effects in potassium[9] and aluminum[10] have extended the QHA results. Here, investigation of the temperature dependence of the phonon energies and widths provides a good confrontation between theory and the results of inelastic neutron scattering experiments.

III. SELF-CONSISTENT HARMONIC APPROXIMATION

Several alternative and closely related approaches to lattice dynamics have become increasingly popular in the last few years. These are generaloy called self-consistent phonon theories. One particular form of the self-consistent theory is structurally similar to QHA, and I call it the self-consistent harmonic approximation (SCHA). Here again, one models the crystal Hamiltonian with a harmonic Hamiltonian, but the recipe for choosing the force constants is changed. Only this particular approach will be discussed in this section because virtually all of the calculations that have been performed in the self-consistent spirit have used the equations pertinent to this approach even though the equations may have been derived as limiting cases of those obtained in more fully self-consistent theories.

SCHA was originally suggested by Born[11] and a closely following paper by Born and Hooton[12] attempted to exploit the idea. However, after this original effort, the approach lay dormant for a number of years until it was independently re-discovered by several workers around 1966 and 1967.[13] There was good reason for not following this approach immediately--the results in lowest order in SCHA require several times as much computation to obtain as does the evaluation of $F^{(0)}$ in QHA, and the original proposal by Born came at a time when a single summation of $\omega_j(k)$ over the first Brillouin Zone was virtually impossible.

The ideas of SCHA are most easily introduced if one first notes that the ground state wavefunction of an arbitrary harmonic Hamiltonian whose force constant matrix is Φ may be written as

$$|0\rangle = A \exp\left[-\tfrac{1}{2}\Sigma^{(A)} \sqrt{m_\mu m_\nu}\, u_\alpha\binom{i}{\mu} G_{\alpha\beta}\binom{ij}{\mu\nu} u_\beta\binom{j}{\nu}/\hbar^2\right] \quad , \tag{15}$$

where A is a normalization constant and

$$G^2_{\alpha\beta}\binom{ij}{\mu\nu}\sqrt{m_\mu m_\nu} = \hbar^2 \Phi_{\alpha\beta}\binom{ij}{\mu\nu} \quad . \tag{16}$$

The wavefunction $|0\rangle$ can now be used as a variational wavefunction with the $G_{\alpha\beta}\binom{ij}{\mu\nu}$ as variational parameters. One can then define a trial ground state energy as $E_0 = A^2\langle 0|\mathcal{H}|0\rangle$ where

$$E_0 = \frac{1}{4}\,\Sigma\, G_{\alpha\alpha}\binom{ii}{\mu\mu} + A^2\langle 0|V|0\rangle \quad . \tag{17}$$

One can then show that $\partial E_0/\partial G_{\alpha\beta}\binom{ij}{\mu\nu} = 0$ implies

$$\Phi_{\alpha\beta}\binom{ij}{\mu\nu} = A^2\langle 0|\nabla_\alpha\binom{i}{\mu}\nabla_\beta\binom{j}{\nu}V|\rangle \quad , \tag{18}$$

where the relation between G and Φ has already been given in Eq. (16). Equation (18) is self-consistent because Φ (or G) enters explicitly on the lhs and implicitly on the rhs.

Thus SCHA uses the ground state average rather than the equilibrium value of the second derivative of the potential to determine the force constants. In the infinite temperature version of the theory, which for reasons of brevity has not been discussed, the ensemble average is used.[16] It has been shown[15] that expansion of the potential in terms of the eigenfunctions of the self-consistent model Hamiltonian introduces a series like that described in Eq. (2) and following, but with averages of all derivatives the potential replacing the related equilibrium value. As has been pointed out,[17] one can then construct a perturbative theory of lattice dynamics in a manner similar to the approach of Ref. 4. This theory resembles QHA with anharmonic effects except that ensemble averages of the derivatives of the potential replace the equilibrium values at all places, for example in Eqs. (4) and (11) and therefore in Eqs. (9), (10), (13) and (14). However, it is found that the contribution from Eqs. (10) and (14), the quartic contribution, vanishes identically; this part of the renormalization has already been accounted for in the lowest order expressions. Further details can be found in the references.

The computational implementation of SCHA is inherently more difficult than for QHA. First of all, the ensemble averages of derivatives of the potential must be obtained by numerical integration. It has been shown[18] that the many-body integral implied by the averaging can be avoided; however, the avoidence requires a calculational effort equivalent to obtaining the phonon spectrum throughout the first Brillouin Zone (as well as several three-dimensional numerical integrals). And second, obtaining the self-consistent solution to Eq. (18) requires an iterative procedure each cycle of which is roughly equivalent to two free-energy calculations in QHA.

However, when cubic anharmonic corrections are also obtained, this additional overhead is not particularly important. Nor is the time saved because the quartic contribution need not be evaluated particularly significant.

In summary, then, the lowest order calculations with SCHA are almost half an order of magnitude more time-consuming than with QHA; however, with the addition of cubic anharmonic corrections, both are comparable.

Since SCHA uses phonon energies in all expressions which are corrected for quartic anharmonicity in lowest order, one would expect that improvement over QHA would be found in cases where quartic renormalization was the most important. The systems which have been analyzed within the framework of SCHA have met this criterion. They are the rare gas solids, the quantum crystals, and a recent study of a model para-electric.

In the rare gas solids, even though cubic and quartic contributions to the free energy are expected to be approximately equal, numerically, at T = 0°K, the quartic contribution is about ten times larger. Calculations of the thermal properties of various rare gas solids using SCHA and cubic anharmonic corrections have been performed with a one-neighbor model[20] and an all-neighbor model.[21] Considerable improvement over the results of Ref. 8 was found.

In the quantum crystals, the use of some form of self-consistent theory is not merely a theoretical nicety; it is an essential. In these substances the expansion of the solid due to zero-point energy is such that the equilibirum position of each atom is at a place where the second derivative of the potential is negative and all of the frequencies in QHA are imaginary.[22] However, the phonon renormalization in lowest order provided by SCHA remedies this. Unfortunately, the large zero-point motion in the quantum crystals leads to another problem--hard-core overlap. The integrals required to evaluate the matrix elements $\langle 0|\nabla^n V|0\rangle$ do not converge and some means to introduce additional short-range correlation is imperative. Those which have been tried to accomplish this fall into three general groups. One way is for one to introduce a short-range correlation function of the Jastrow type. Theories using this approach which have a perturbative framework similar to SCHA have recently been described.[23] Unfortunately, the additional computational complexity of exactly evaluating (by Monte-Carlo integration, for example) the many-body matrix elements which now include a Jastrow function in addition to a correlated Gaussian has become apparent in recent calculations.[24] Thus two areas of major computational effort have to be merged for one proper calculation to be performed. A second approach to the hard-core problem is described at this conference by Horner. A third approach is the t-matrix method as described, for example, by Glyde[25] and Brandow.[26] A good overall reference to the theory of solid helium is the review article of Guyer[27] although this article slightly predates the recent emphasis on anharmonic effects.

Another recent application[28] of SCHA is to a model para-electric in the NaCl structure. Here one notes that the TO mode (and many other modes as well) for the Coulomb lattice alone is unstable. However, with the addition of a quartic interaction alone between nearest neighbors, all the frequencies become real in the lowest order of SCHA. But, because of the thermal averaging of the second derivative of the potential, the stabilizing effect of the quartic term is very temperature dependent and the calculation leads, in a very natural way, to the well-known falling mode phenomena.

IV. DISCUSSION

Two related but somewhat different methods for calculating the properties of crystal lattices in the adiabatic approximation have been presented. Both methods model the true crystal Hamiltonian by an exactly soluble harmonic Hamiltonian and then use the eigenfunctions and eigenvalues of the model Hamiltonian as a basis for perturbative calculations.

The current computational perspective on both approaches is that, with the aid of the latest generation of computers, the lowest order calculations can be accomplished with ease although the programming effort may be substantial. The computer requirements and programming effort for the evaluation of cubic anharmonic corrections is more difficult but by now can be performed routinely.

One fact I particularly want to emphasize here is the current routine nature of cubic anharmonic calculation, at least for lattices which are either cubic or have only one atom per unit cell. The practicability of performing such calculations in hcp substances has recently been established[29] and any lattice with two atoms per unit cell would probably be no more difficult.

One thing that is lacking in the current scene is the general availability of lattice dynamical programs, a problem which is shared with other computational areas in which programming effort is substantial. A possible remedy for this situation would be research directed at systematizing the results of anharmonic calculations in such a way that reliable estimates of anharmonic effects could be made without performing the actual calculations.

Finally, if one takes the perturbative approach seriously, it is now probably feasible to evaluate numerically all of the terms in the next order. These terms have been written down in a recent paper by Shuhla and Cowley[30] and estimates of their contributions were made.

REFERENCES

1. HA treats effects caused by changes of the lattice constant as perturbations whereas QHA uses a different set of bare quantities at each lattice constant. Since the latter approach only requires the change of one data element in a computer program, we see little need to consider the former approach at all. Thus, in Section III what could be called SCQHA is called SCHA.

2. M. Born and K. Huang, Dynamical Theory of Crystal Lattices. (Oxford University Press, London, 1954).

3. A. A. Maradudin, E. W. Montroll and G. H. Weiss, Solid State Physics (ed. by F. Seitz and D. Turnbull) (Academic Press, New York, 1963), Suppl. 3.

4. A. A. Maradudin, P. A. Flinn and R. A. Coldwell-Horsfull, Ann. Phys. (N.Y.) 15, 337 (1961).

5. R. A. Cowley, Advan. Phys. 12, 421 (1963).

6. N. R. Werthamer, Phys. Rev. A2, 2050 (1970).

7. R. A. Cowley, Rept. Progr. Phys. 31, 123 (1968).

8. M. L. Klein, G. K. Horton and J. L. Feldman, Phys. Rev. 184, 968 (1969).

9. W.J.L. Buyers and R. A. Cowley, Phys. Rev. 180, 755 (1969).

10. T. Högberg and R. Sandström, Phys. Stat. Solidi 33, 169 (1969); T. R. Koehler, N. S. Gillis and D. C. Wallace, Phys. Rev. B1, 4521 (1970); and T. R. Koehler and N. S. Gillis, Phys. Rev. B3, 3568 (1971).

11. M. Born, Fest. d. Akad. Wiss. Göttingen (1951). A translation of this article has been issued as Bell Laboratories TR.70-14.

12. D. J. Hooton, Phil. Mag. 46, 422 and 433 (1955).

13. The appropriate references can be found in Ref. 14 and 15. Of these two, Ref. 14 has the simplest exposition of the theory.

14. N. R. Werthamer, Am. J. Phys. 37, 763 (1969).

15. N. R. Werthamer, Phys. Rev. B1, 572 (1970).

16. N. S. Gillis, N. R. Werthamer and T. R. Koehler, Phys. Rev. 165, 951 (1968).

17. T. R. Koehler, Phys. Rev. 165, 942 (1968).

18. T. R. Koehler, Phys. Rev. 144, 789 (1966).

19. T. R. Koehler, Phys. Rev. Letters 17, 589 (1966).

20. M. L. Klein, V. V. Goldman and G. K. Horton, J. Phys. Chem. Solids 31, 2441 (1970). References to other calculations by these authors may be found in this reference.

21. T. R. Koehler, unpublished.

22. F. W. de Wette and B.R.A. Nijboer, Phys. Letters 18, 19 (1965).

23. T. R. Koehler and N. R. Werthamer, Phys. Rev. A2074 (1971); and P. Gillissen and W. Biem, Z. Phys. 242, 250 (1971).

24. T. R. Koehler and N. R. Werthamer, to be published.

25. H. R. Glyde and F. C. Khanna, to be published.

26. B. H. Brandow, Phys. Rev. A4, 422 (1971).

27. R. A. Guyer in Solid State Physics, Vol. 23 (ed. by F. Seitz D. Turnball and H. Ehrenreich) (Academic Press, New York, 1969).

28. N. S. Gillis and T. R. Koehler, Phys. Rev. to be published.

29. T. R. Koehler and R. L. Gray, Bull. Am. Phys. Soc. 16, 439 (1971).

30. R. C. Shukla and E. R. Cowley, Phys. Rev. B3, 4055 (1971).

ANHARMONIC LATTICE DYNAMICS: RENORMALIZED THEORY

Heinz Horner

Institut für Festkörperforschung, Kernforschungsanlage

Jülich, WEST GERMANY

ABSTRACT

An anharmonic perturbation theory for phonons is discussed using partially renormalized coupling constants. Criteria are given which allow calculation of these in a self-consistent way, and the similarity to conventional anharmonic perturbation theory is outlined. The theory can be applied to strongly anharmonic crystals including hard-core interactions. As examples, calculations on quantum crystals and rare gas crystals are discussed.

INTRODUCTION

Rare gas crystals have served for considerable time as model substances for lattice dynamics, mainly because they form simple lattices and because the interaction among the rare gas atoms is radial symmetric and sufficiently well known. The heavier rare gas solids are, at least at low temperatures, well-described by the harmonic interaction or low-order anharmonic perturbation theory. At higher temperatures, however, and at all temperatures for the light rare gas solids, this approach fails (1). This situation is most obvious for solid helium where a harmonic approximation yields not even real phonon frequencies (2).

Attempts to overcome these difficulties have lead to a theory using partially renormalized harmonic and anharmonic coupling constants. Its simplest form, the self-consistent harmonic approximation, was introduced by Born (3) and applied first to solid helium by Koehler (4). Since, it has been extended to include residual anharmonicities (5,6) and applied to various problems.

An additional problem arises from the strongly repulsive nature of the interaction among the rare gas atoms at short distances. This difficulty was recognized for a long time for solid helium (7), but even for the heavier rare gas crystals a careful treatment of short-range correlations seems necessary at higher temperatures.(8)

In the following a formulation of a renormalized anharmonic phonon theory will be outlined which allows us to discuss the questions mentioned above in a relatively simple fashion. (9)

EQUILIBRIUM CONDITION

Let us consider a crystal under the influence of some external forces, for instance external pressure, described by the Hamiltonian

$$H = \sum_i \frac{p_i^2}{2m} + \sum_{i>j} V(x_i - x_j) + \sum_i u_i x_i \quad , \tag{1}$$

where p_i and x_i are momentum and position operators of the particle at lattice site i, respectively.* The equation of motion for the position operators is found to be

$$-m \frac{\partial^2}{\partial t^2} x_i = \sum_j{}' \nabla_i V(s_i - x_j) + u_i \quad , \tag{2}$$

where the external force could be time dependent eventually.

We introduce expectation values of the position of each particle,

$$d_i(t) = \langle x_i(t) \rangle \quad , \tag{3}$$

where the expectation value in the presence of time-dependent forces is defined in the usual way (10),

$$\langle \sigma(t) \rangle = \mathrm{Tr}[T\sigma(t) \exp\{-i \int_0^{-i\beta} d\tau H(\tau)\}]/\mathrm{Tr}\ldots \tag{4}$$

where T is the time-ordering operator for imaginary times. In the absence of time-dependent forces this expression goes over into the usual expectation value in statistical mechanics.

Taking the average of Eq. (3) analogous to (4) we find the equation of motion for the average positions in the external field,

---

* Throughout this paper we use a vector and tensor notation to reduce the number of indices.

$$-m \frac{\partial^2}{\partial t^2} d_i(t) = K_i(t) + U_i(t) \quad , \tag{5}$$

where the average internal force on particle i due to the presence of the remaining particles is

$$K_i(t) = \sum_j{}' \int d^3r g_{ij}(rt) \nabla V(r) \quad . \tag{6}$$

Here,

$$g_{ij}(rt) = \langle \delta(x_i(t) - x_j(t) - r) \rangle \tag{7}$$

is the relative part of the equal time distribution function of the pair of particles i and j. It gives the probability of finding the particles i and j separated by r. The time dependence is due to the presence of the external force.

Especially in the absence of time-dependent external forces, the left hand side of Eq. (5) vanishes and the right hand side just represents the equilibrium condition stating that the sum of external and internal forces has to vanish for each lattice site.

So far the problem has just been shifted to the pair distribution function.

PHONONS

To obtain information about phonons we have to investigate the response of the crystal to changes of the external forces. More precisely we observe shifts in the average positions of a particular particle i at a time t provided the force on some other particle j has been changed at some time t'. The response is described by the so-called phonon propagator or phonon Green's function,

$$i \frac{\delta d_i(t)}{\delta u_j(t')} = D_{ij}(tt') \quad . \tag{8}$$

The physical picture which we have to have in mind is not that of a vibrating classical particle but rather a particle distributed over a cloud due to its zero point and thermal motion. The response is a vibration of the center of mass of this cloud where its shape might eventually change during a period of this oscillation. Such a situation is pictured for the two-particle distribution functions for next neighbors in Fig. 1.

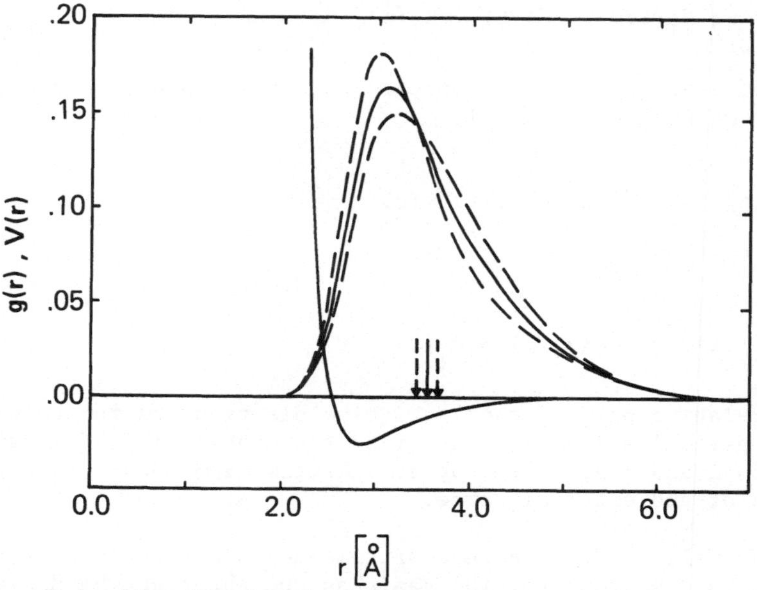

Fig. 1. Pair distribution function for next neighbors in bcc ^3He for three adjacent average distances.

To calculate the phonon propagator we need its equation of motion which is easily derived from Eq. (5). So far the external forces were the independent variables for our problem, especially in the correlation function Eq. (7). It is convenient to consider $g_{ij}(rt)$ and $K_i(t)$ as a function of the average positions $d_i(t)$ which themselves depend on the external forces. Differentiating Eq. (5), we obtain

$$- m \frac{\partial^2}{\partial t^2} D_{ij}(tt') - \sum_{\ell} \int_0^{-i\beta} d\tau M_{i\ell}(t\tau) D_{\ell j}(\tau t') = i\delta_{ij}\delta(t-t'), \quad (9)$$

where

$$M_{ij}(tt') = \frac{\delta K_i(t)}{\delta d_j(t')} = \sum_{\ell}' d^3r \frac{\delta g_{i\ell}(rt)}{\delta d_j(t')} \nabla V(r) \quad . \quad (10)$$

This equation of motion resembles very much the harmonic approximation where here $M_{ij}(tt')$ plays the role of a time-dependent generalized dynamical matrix.

As in the discussion of the equilibrium condition, the problem has been shifted to the pair distribution function.

PAIR DISTRIBUTION FUNCTION

So far we have expressed the equilibrium condition and the generalized dynamical matrix in terms of the pair distribution function. The approximations now will be made in choosing an appropriate form of it. The simplest choice would be to take a δ function at the average distance,

$$g_{ij}(rt) = \delta(r - d_i(t) + d_j(t)) \quad , \tag{11}$$

which would just reproduce the quasi-harmonic approximation.

In situations where the zero point or thermal vibrations are not small this choice is insufficient and we have to find a more general form. This choice is, however, not arbitrary and especially some moments of this function have to be determined self-consistently. The first condition is obtained from the definition of the average position Eq. (3),

$$d_i(t) - d_j(t) = \int d^3 r \; r \; g_{ij}(rt) \quad . \tag{12}$$

From the definition of the expectation value, Eq. (4), and from the phonon propagator, Eq. (8), we find

$$D_{ij}(tt') = \langle T x_i(t) x_j(t') \rangle - \langle x_i(t) \rangle \langle x_j(t') \rangle \quad , \tag{13}$$

which is the fluctuation-dissipation theorem relating the response function to the fluctuations in equilibrium. Especially for the equal-time phonon propagator we find

$$D_{ii}(tt) - D_{ij}(tt) - D_{ji}(tt) + D_{jj}(tt)$$
$$= \int d^3 r \left(r - d_i(t) + d_j(t)\right)^2 g_{ij}(rt) \quad , \tag{14}$$

which is the second consistency condition. Relations for the higher moments of the pair distribution function can be obtained in a similar way, but might be unnecessary for practical calculations.

For hard-core interactions we expect that the pair distribution function behaves for short distances as the one for an isolated pair of particles. Especially at absolute zero,

$$g_{ij}(r) \xrightarrow[r \to 0]{} \psi_0^2(r) \quad , \tag{15}$$

where $\psi_0(r)$ is the s-wave scattering solution for a pair of particles. At high enough temperatures, in the classical limit, we expect

$$g_{ij}(r) \xrightarrow[r \to 0]{} e^{-\beta V(r)} \quad . \tag{16}$$

This asymptotic behavour together with the first two-moment relations, Eqs. (12) and (14), and the normalization condition,

$$\int d^3 r g_{ij}(rt) = 1 \quad , \tag{17}$$

determine the pair-distribution function generally with sufficient accuracy. As a special choice we can use

$$g_{ij}(r) = af_0(r)e^{-A(r-R)^2} \quad , \tag{18}$$

where $f_0(r)$ is a short-range correlation function being equal to Eqs. (15) and (16). The normalization constant a, the vector R, and the width parameter A which is in general a tensor have to be determined such that the consistency conditions, Eqs. (12), (14) and (17) are fulfilled.

For soft-core interactions where $f_0(r)$ might be replaced by unity, we find for the ansatz Eq. (18),

$$R = d_i(t) - d_j(t) \quad ,$$

$$A^{-1} = -\tfrac{1}{2}\{D_{ii}(tt) - D_{ij}(tt) - D_{ji}(tt) + D_{jj}(tt)\} \quad , \tag{19}$$

$$\alpha = \sqrt{1 \det A^{-1}/ 3} \quad .$$

For general $f_0(r)$, however, the parameters in Eq. (18) depend in a more complicated way on the equilibrium distances and phonon propagators and have to be determined numerically.

RENORMALIZED COUPLING CONSTANTS

Having found a special form for the pair distribution function, we can calculate the internal forces K_i entering the equilibrium conditions, Eq. (5). To calculate the generalized dynamical matrix $M_{ij}(tt')$ given by Eq. (10) we have to know how $g_{ij}(r)$ changes with the average positions d_i. The consistency conditions, Eqs. (12), (14) and (17) have to hold for any configuration of the $d_i(t)$ and we have to use this fact to determine M_{ij}.

First we have to find how the parameters in $g_{ij}(r)$, Eq. (18), depend on the d_i and on the equal-time phonon propagators D_{ij}.

From this we determine the "renormalized coupling parameters" considering d_i and D_{ij} as independent for the moment:

$$C_{ij} = \sum_{\ell} \int d^3r \left.\frac{\delta g_{i\ell}(r)}{\delta d_j}\right|_D \nabla V(r)\{\delta_{ij} - \delta_{\ell j}\} \quad , \tag{20}$$

and

$$C_{ijk} = \sum_{\ell} \int d^3r \left.\frac{\delta g_{i\ell}(r)}{\delta D_{i\ell}}\right|_d \nabla V(r)\{\delta_{ij} - \delta_{\ell j}\}\{\delta_{ih} - \delta_{\ell h}\} \quad . \tag{21}$$

This again has to be done numerically in general. In the absence of the short-range correlation function we find

$$C_{ij} = -\sum_{\ell} \int d^3r g_{i\ell}(r)\nabla^2 V(r)\{\delta_{ij} - \delta_{\ell j}\} \quad , \tag{22}$$

and

$$C_{ijk} = \sum_{\ell} \int d^3r g_{i\ell}(r)\nabla^3 V(r)\{\delta_{ij} - \delta_{\ell j}\}\{\delta_{ih} - \delta_{\ell h}\} \quad . \tag{23}$$

In this special case the renormalized parameters are equal to the ordinary coupling parameters averaged over the pair distribution function.

The generalized dynamical matrix is now easily obtained and we find

$$M_{ij}(tt') = C_{ij}\delta(t-t')$$

$$- i \sum_{\substack{h\ell \\ mn}} C_{ih\ell} \int d\tau d\tau' D_{hm}(t\tau) D_{\ell n}(t\tau') Q_{mnj}(\tau\tau' t') \quad , \tag{24}$$

where we have used Eq. (9) and

$$Q_{ijk}(tt'\tau) = \frac{\delta M_{ij}(tt')}{\delta d_h(\tau)} \quad . \tag{25}$$

In lowest order we find

$$Q_{ijk}(tt'\tau) = C_{ijk}\delta(t - t')\delta(t - \tau) \quad . \tag{26}$$

In this case the second term in Eq. (24) has the same structure as the cubic anharmonic correction in ordinary perturbation theory. (11) Due to its explicit time dependence it is also responsible for damping.

DISCUSSION

Putting all pieces together we have a self-consistent set of equations which can be solved using the following iteration procedure. We start from a given set of lattice vectors d_i and width parameters D_{ij}. Then: (a) calculate the renormalized vertices, Eqs. (22) and (23); (b) calculate phonon frequencies from Eqs. (24) and (9); and (c) calculate new width parameters and return to (a). With reasonable initial values this iteration procedure is rapidly converging and the computational problems are not too much more involved than for ordinary anharmonic perturbation theory.

We finally have to discuss the question under which conditions the present formulation might be superior to conventional anharmonic theory. For quantum crystals this is obviously the case since the conventional theory is not applicable. For the heavier rare gas solids the situation is different.

Figure 2 shows the relative changes of the renormalized second-order and third-order coupling constants as a function of the root mean square displacements Δ. The bars at the bottom of the figure show the region of existence of the rare gas crystals between absolute zero and melting. The upper set of curves gives values obtained from a Gaussian average without short-range correlation functions, the lower curves with. As we see, near melting, the renormalized coupling constants are drastically changed from their values at $\Delta=0$. The effect of short-range correlations is also marked.

In ordinary perturbation theory there is an addition to the terms in Eq. (24) a contribution linear in $C_4^{(0)}$ in the generalized dynamical matrix. This is essentially the first-order contribution to the renormalization of the harmonic coupling parameter. Near melting its contribution is approximately equal to $C_2^{(0)}$ and it gives about 90% of the total renormalization. The contribution due to cubic anharmonicities, especially damping of phonons, is, however, underestimated by a factor of approximately four in the conventional perturbation theory near melting since $C_3 \simeq 2\times C_3^{(0)}$ and since C_3 enters quadratic. It should also be pointed out that a calculation without short-range correlations overestimates those effects by approximately a factor of two near melting.

Concluding we can say: The questions raised in this paper are of great importance for the theory of quantum crystals. For the heavier rare gas crystals they affect only some quantitative results. Nevertheless they are of increasing interest since calculations including even higher order anharmonicities have been done (12) or are under way and since neutron experiments are being carried out (13) to determine phonon damping, for which the differences to ordinary perturbation theory are most pronounced.

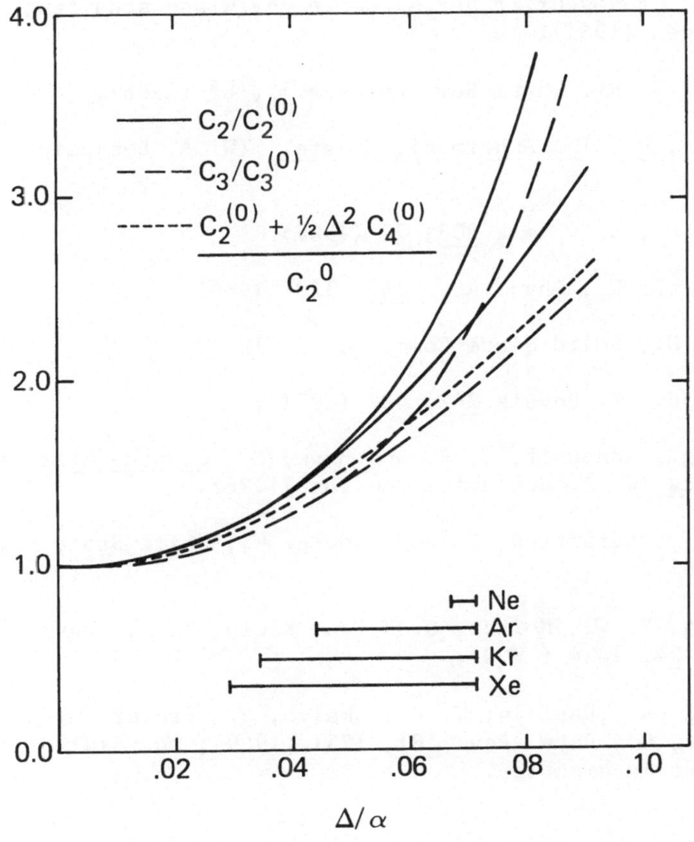

Fig. 2. Second- and third-order renormalized coupling constants as functions of the root mean square displacement Δ. Upper curves without short-range correlations, lower curves with.

REFERENCES

1. Klein, M. L., Horton, G. K. and Feldmann, J. L., Bull. Am. Phys. Soc. 13, 689 (1968); and Klein, M. L. and Horton, G. K., Proc. L. T. 11, St. Andrews, Scotland I, 553 (1968).

2. De Wette, F. W. and Nijboer, B.R.A., Phys. Letters 18, 19 (1965).

3. Born, M., Festschrift der Akademie der Wissenschaften, Goettingen (1951).

4. Koehler, T. R., Phys. Rev. Letters 17, 89 (1966).

5. Choquard, P., The Anharmonic Crystal (W. A. Benjamin, New York) 1967.

6. Horner, H., Z. Physik 205, 72 (1967).

7. Nosanow, L. H., Phys. Rev. 146, 120 (1966).

8. Horner, H., Solid State Comm. 9, 79 (1971).

9. Horner, H., Z. Physik 242, 432 (1971).

10. See, e.g., Kadanoff, L. P. and Baym, G., Quantum Statistical Mechanics (W. A. Benjamin, New York) 1962.

11. See, e.g., Leibfried, G. and Ludwig, W., Solid State Phys. 12, 275 (1961).

12. Goldmann, V. V., Horton, G. K. and Klein, M. L., Phys. Rev. Letters 24, 1424 (1970).

13. Leake, J. A., Daniels, W. B., Skalyo, J., Frazer, B. C. and Shirlane, G., Phys. Rev. 181, 1251 (1969); and Skalyo, J., private communication.

VI

Localized Imperfections and Dislocations

INTRODUCTORY REMARKS

A. B. Lidiard

Theoretical Physics Division, A.E.R.E., Harwell, Berks., U.K.

The papers in this section represent four different aspects of the theory of point defects and dislocations in crystalline solids: (i) static or, more accurately, thermodynamic properties of point defects (Lidiard and Norgett and Bullough) (ii) dynamic and optical properties of point defects (Balkanski) (iii) properties of the dislocation core (Bullough) and (iv) dislocation properties dependent on their long-range elastic strain fields (Lothe). These papers are representative of important aspects of the theory of the defect solid state. An apparent asymmetry in the coverage of topics provided by the four papers comes about (a) because point defects in contrast to dislocations, are all 'core' in the sense that their most important properties are determined by what happens in the immediate vicinity of the defect where continuum elasticity theory cannot be used, and (b) because the study of localised and perturbed electronic and vibrational states in the vicinity of dislocation cores has proved more difficult than the corresponding studies for point defects, although some studies of the perturbation by dislocations of localised or more accurately resonant modes due to point defects show promise (Busse and Haider[1]). The most obvious gaps in the coverage offered by these are two. The first is the absence of a detailed discussion of the theory of localised electronic states, e.g. colour centres, where a variety of approaches are in use; some of these are very analogous to those used in molecular theory. The second is the absence of a review of radiation damage cascades and other high energy events in solids.

In this brief introduction to the section we aim to set the scene by providing some brief discussion of the questions posed to the conference by the organisers. It is worth repeating these (slightly modified). "Which mathematical methods and physical models should we develop and which abandon? What impact will the availability of bigger and faster computers, etc. have on the theory of localised imperfections and dislocations? What are the really important calculations that should be done in these areas over the next few years".

The variety of theory in this area is enormously wide. At one extreme there may be little formal mathematical analysis at all as for example in the numerical solution of Newton's equations of motion for a many-particle assembly. Such an approach is the basis of 'molecular dynamics' studies of liquids[2] and of analogous studies of high energy radiation damage cascades in solids[3]. Such simulation studies can be immensely valuable in several ways. Firstly they can point to new physical effects, for example, the prediction of the occurrance of the phenomenon of the channelling of energetic ions down open crystal directions (Robinson and Oen[4]). Secondly, they may allow the verification of an approximate analysis of a complex situation or the determination of its limitations. Examples are provided by the simulation of radiation damage displacement cascades in comparison to approximate analyses such as that of Kinchin and Pease[5]. Another interesting example of this kind is the simulation of the diffusive motion of defects in a lattice in thermal equilibrium in comparison to the rate-process theory which lies at the basis of almost all descriptions of thermally activated atomic migrations[6]. In these examples the comparisons not only test the analytical statistical theories, they also allow the determination of the physical parameters arising in them, e.g. displacement energies, activation energies etc. Thirdly, in so far as these simulations allow the rigorous description of a system otherwise only incompletely or approximately analysed, they permit more rigorous tests of the basic atomic model e.g. of the interatomic potential by which the atoms interact.

It seems clear that this class of calculation is one likely to be greatly stimulated by the increasing availability of large computers. However, the benefits will not come automatically. The need for good atomic models and for imaginative analysis is as great as the need in experimental solid state work for well characterised systems and imaginative experiments.

Moving away from the most computational kind of theory one may consider localised electronic and vibrational states. Here the basic problem of solving the Schrödinger equation is initially reduced by the use of methods often closely analogous to those used in molecular theory, e.g. molecular orbital schemes, normal mode analysis etc. Similar problems may arise e.g. the difficulty of integrals (multicentre, penetration etc.) arising with centres containing several electrons although the different nature of the atomic orbitals sometimes appropriate to defects (e.g. to electron excess vacancy centres in ionic crystals, F-centres) means that sometimes these difficulties are reduced[7]. In other cases, however, e.g. defects in covalent solids, these difficulties arise in a very analogous form[8]. One suspects that despite the ability to do large configuration interaction calculations the correct handling of correlation effects is as serious as in band theory. Possibly those current developments in band theory which go beyond the Hartree-Fock approximation will have an impact upon the calculation of localised states as well. In the meantime it should be said that little more than a start has been made in applying known methods to many of the centres of interest; for example in ionic crystals the F-aggregate centres and the interstitial centres have been little studied. In calculating vibrational states the correct analogy is to a molecule coupled to a crystal lattice so that one requires the response function of the lattice, the calculation of which requires a knowledge of the vibrational states of the lattice as well as its eigenfrequencies. Again to make realistic calculations requires a knowledge of the basic interatomic forces, although of course only over more restricted ranges of interatomic separation than in the molecular dynamics class of calculation. Such requirements may be met by fundamental approaches such as those of Pick described in another section of the conference, or more commonly by the use of semi-empirical models such as those described by Bilz. The development of stronger links between the two approaches would obviously do much to remove uncertainties in the significance of many empirically determined shell parameters. For the coupling of the 'molecule' to this lattice one may either calculate from the Schrödinger equation, as when photon absorption changes the electronic state, or again use potential models as for example for the absorption or inelastic scattering of a photon by coupling to a localised vibrational mode.

Turning now to defect properties in conditions of or near to thermodynamic equilibrium, (e.g. diffusion, dislocation core structure etc.) statistical mechanics and approximations such as the quasi-harmonic approximation allow us often to reduce the problem to an equivalent static lattice problem e.g. for energies of defect formation, interaction and migration. When this is so the mathematical problem is that of minimising the energy function to determine the equilibrium configuration around the defect. Minimisation has of course been extensively studied by numerical analysts in recent years but some of their achievements have yet to be fully applied in this field. However, generally speaking, one is now in a situation where the predictions can be made accurately for a given potential model. One problem in matter transport theory which has so far defied successful statistical mechanical analysis is that of transport in a temperature gradient, i.e. of calculating the heat-of-transport parameter. Extension of the molecular dynamical calculations of Tsai et al[6] would probably be very valuable here.

In conclusion we observe that although this survey of the present state of defect theory is only superficial there emerges clearly the need for more soundly based models of interatomic forces. The greatest service which could be rendered to the defect studies by those whose primary interest is electronic theory would be the provision of improved potential functions valid for arbitrary configurations of atoms. The difficulties are already quite apparent. Thus in metals pseudopotential theory supplies some guidance, but for relatively few metals. Bullough for example does not trust his models of Cu and α-Fe to give absolute defect energies, yet ultimately can we be confident of the inferred structure if we cannot trust the minimum value of the function we are minimising to obtain that structure? In covalent materials bond models which appear satisfactory for lattice dynamics are far too soft when tested against the forces produced by vacancies in the lattice[9]. Only in the ionic crystals do we appear to have models (shell models) which can make valid predictions of defect energies, and even here very few examples have been accurately evaluated.

REFERENCES

1. G. Busse and G. Haider, Optics Commun. 2, 45 (1970).

2. For a general review see e.g. A Rahman in The Growth Points of Physics (European Physical Society 1969) p. 315.

3. See e.g. C. Erginsoy, G.H. Vineyard and A. Englert, Phys. Rev. 133, A595 (1964); C. Erginsoy, G.H. Vineyard and A. Shimizu, Ibid 139, A118 (1965); I.M. Torrens and M. T. Robinson in Proceedings of the Battelle Colloquium on Interatomic Potentials and Simulation of Lattice Defects (June 1971), to be published.

4. M. T. Robinson and O. S. Oen, Phys. Rev. 132, 2385 (1963).

5. G. H. Kinchin and R. S. Pease, Rep. Prog. Phys. 18, 1 (1955).

6. D. H. Tsai, R. Bullough and R. C. Perrin, J. Phys. C 3, 2022, (1970).

7. See e.g. M. J. Norgett, J. Phys. C. 4, 1289 (1971).

8. See e.g. C. A. Coulson and M.J. Kearsley Proc. Roy. Soc. A241, 433 (1957); C. A. Coulson and F. P. Larkins, J. Phys. Chem. Solids, 30, 1963 (1969); F. P. Larkins, Ibid, 32, 965 (1971).

9. F. P. Larkins and A. M. Stoneham, J. Phys. C, 4, 143 and 154 (1971).

LOCALIZED VIBRATIONAL MODES IN CRYSTALS

M. Balkanski

Laboratoire de Physique des Solides associe au C.N.R.S.,

Universite Paris VI, Paris, FRANCE

INTRODUCTION

The presence of imperfections in a crystal lattice create electronic or vibrational localized states which destroy the translational symmetry and give rise to a certain number of observable effects. Two reasons generally make possible the spectroscopic observations of these effects: (a) The lattice defects destroy the overall symmetry and relax the perfect crystal selection rules, thus rendering observable optical transitions which would otherwise be forbidden, for example, on normal modes at critical points of the Brillouin zone; and (b) The newly created localized state is characterized by a spectrum of electronic and vibrational levels which can be excited by interaction with the radiation field.

In the first case, the induced polarizability in homopolar crystals, for example by high defect concentration, provides the possibility of direction observation of the phonon spectrum at high density of states points throughout the Brillouin zone either by infrared spectroscopy or by light scattering. Recent observations of these phenomena in ion implanted and amorphous semiconductors show how far-reaching this effect is. The calculation of the phonon distribution spectra and its comparison with experimental results may lead to a better understanding of the vibrational properties of heavily disturbed lattices.

In the second case, the defect itself is more directly concerned, and the created localized state depends specifically on the nature of the defect. When the vibrations of defects are considered, one realizes that the normal modes of the perturbed crystals differ from those in the perfect case.

The modification of the normal modes is naturally larger in the immediate vicinity of the defect. Essentially, two different types of modes may occur. According to the nature of the imperfection, there may be modes with frequencies which cover the range of allowed frequencies for the perfect crystal with only a slight modification near the defect (although there may be an enhancement of the amplitude over a narrow range of frequencies which gives the characteristics of a resonance) or there may be modes which are localized in the vicinity of the defect and have isolated frequencies outside the range of those of the perfect lattice.

The theory of the effects of impurities on the vibrational spectrum has been recently reviewed by Elliott[1] and Maradudin.[2]

Infrared spectroscopy[3] is particularly appropriate for the study of these impurity modes. The absorption spectrum gives directly the frequency response of the lattice. In the case of homopolar crystals, where no other phonon transitions are allowed in first order, the defects carrying an electric charge produce electric dipole moments in their displacement which couple directly with the radiation field. Thus the only optical transitions to be observed in first order are those due to impurity phonons.

There has been extensive work on defects in silicon[4,5,6,7,8] in particular.

In ionic crystals there is, of course, large absorption in the infrared by the perfect lattice which occurs at the restrahl frequencies. However, optical absorption associated with defects may be directly measured if the frequency is sufficiently far removed from the strong, perfect lattice absorptions.[9]

Detailed calculations were first made on the simplest models of such defects assuming a single substitutional impurity with changed mass.[10]

More complicated defects arise from the pairing of impurities in silicon. In particular, lithium can be interstitially associated with, for example, substitutional boron.[4] Then changes in force constants have to be considered. P. Pfeuty[11] has developed calculations for a certain number of models related to complicated defects.

More recently, analogous calculations have been carried out[12] for the complex $V_{As}Cu_{Ga}$ with mass and force constant changes due to Cu in GaAs.

We shall review here the main features of these calculations and discuss experimental results which create new situations and needs for further theoretical investigations.

REVIEW OF BASIC THEORY

The theory of the dynamics of perturbed lattices has been developed on the basis of the Green's function formalism first proposed by Lifshitz.[13] The perturbed crystal vibrations are deduced from the perfect crystal ones. The interaction of the radiation field with the phonons in the perturbed lattice can then be easily found.

The Defect Vibrations

Let $u_\alpha\binom{\ell}{k}$ be the displacement in the α direction of the kth atom in the ℓth cell which is centered at $\vec{x}(\ell)$. The mass of the atom is $M\binom{\ell}{k}$ and $\Phi_{\alpha\beta}\binom{\ell\ell'}{kk'}$ are the force constants of the crystal in the harmonic approximation.

If the lattice is perfect, i.e., possessing translational invariance, M is independent of ℓ and Φ depends only on $x(\vec{\ell}) - \vec{x}(\ell')$. The normal coordinates $\sigma_\alpha(k|{\vec{q} \atop j})$ are the eigenvectors and the squares of the characteristic frequencies $\omega\binom{\vec{q}}{j}$ are the eigenvalues of the dynamical matrix:

$$\Psi_{\alpha\beta}\left({\vec{q} \atop k\ k'}\right) = \frac{1}{N}\sum_{\ell'} \frac{\Phi^0_{\alpha\beta}\binom{\ell\ell'}{kk'}}{M^{\frac{1}{2}}_{\alpha k} M^{\frac{1}{2}}_{\beta k'}} \exp\left[i\vec{q}.\left(\vec{x}(\ell)-\vec{x}(\ell')\right)\right]. \tag{1}$$

In the perturbed lattice the equations of motion are conveniently written in matrix form. Φ is a general 3sN x 3sN matrix for a crystal of N cells, s atoms per cell, and 3 dimensions. M is diagonal and u a column vector. Then,

$$M\frac{d^2u}{dt^2} + \Phi u = 0 \quad . \tag{2}$$

The new characteristic frequencies and normal modes are the eigenvalues and eigenvectors of $-M\omega^2+\Phi$. It is convenient to consider the deviation from the perfect case by defining

$$C_{\alpha\beta}\binom{\ell\ell'}{kk'} = \Phi_{\alpha\beta}\binom{\ell\ell'}{kk'} - \Phi^0_{\alpha\beta}\binom{\ell\ell'}{kk'} + \varepsilon\binom{\ell}{k}M_k\omega^2\delta_{kk'}\delta_{\alpha\beta}\delta(\ell\ell'). \tag{3}$$

We define the inverse of the perfect matrix as the Green's function matrix,

$$g = (M_0\omega^2 - \Phi^0)^{-1} \quad , \tag{4}$$

where, because of the form of (1),

$$g_{\alpha\beta}\binom{\ell\ell'}{kk'}\omega^2 = \frac{1}{3sN}\sum_{jq}\frac{\sigma_\alpha(k|{}^{\vec{q}}_{j})\sigma_\beta(k|{}^{\vec{q}}_{j})\exp\left[i\vec{q}.(\vec{x}(\ell)-\vec{x}(\ell'))\right]}{\omega^2-\omega_j^2({}^{\vec{q}}_{j})}$$

$$\times\frac{1}{M_{\alpha k}^{\frac{1}{2}}M_{\beta k'}^{\frac{1}{2}}} \quad . \tag{5}$$

Then,

$$-M\omega^2+\Phi = -M_0\omega^2+\Phi^0+C = g^{-1}(gC-1) \quad . \tag{6}$$

Now C is a small 3r x 3r matrix covering only r sites specified by i, which are perturbed by the defect. The characteristic frequencies can be determined from 1-gC taken only over the same sites. This matrix can usually be considerably simplified by using the symmetry of the defect; g has the point symmetry of the lattice which is normally greater than that of C. Transforming to coordinates belonging to particular irreducible representations of the group of C will reduce the matrix to blocks for each representation which can be solved separately. For the localized modes of interest here, it is entirely equivalent to a molecular problem.

Optical Absorption

The probability of absorption of light by the defect depends on the atomic charges and on the correlation function of the displacements u. It has been shown[1,10] that the absorption coefficient is

$$\alpha(\omega) = \frac{4\pi\omega}{nc}p\Lambda\sum_{ii'}e_ie_{i'}G^I_{ii'}(\omega) \quad , \tag{7}$$

where p is the concentration of defects, n the refractive index, and Λ the local field correction. G^I is the imaginary part of the imperfect Green's function defined as the inverse of (6),

$$G^I + \operatorname{Im}\operatorname*{Lt}_{\psi\to 0} G(\omega+i\psi); \quad G = -(gC-1)^{-1}g \quad . \tag{8}$$

This expression, when weighted with the e, will also have the point symmetry of defect. Only those representations which transform like a vector will give a non-zero result.

For frequencies in the perfect lattice band, $g(\omega+i\varepsilon)$ has a finite imaginary part from which G^I may be obtained by writing

$$\underset{\psi\to 0}{\mathrm{Lt}}\ g(\omega+i\psi) = g'(\omega) + \frac{i\pi k(\omega)}{2\omega M_0} \quad , \tag{9}$$

k is related to the density of modes; in fact, for $\alpha=\beta$, $k=k'$, $\ell=\ell'$, it is precisely that. For isolated, localized mode frequencies outside the band, G^I takes the form of a set of delta functions at the zeros ω_e of

$$D = \det(1 - gC) \quad . \tag{10}$$

Then if A_{ij} is the cofactor in D,

$$G^I_{ii'} = \sum_{\omega_e} \frac{\sum_j A_{ij} g_{ji'} \pi\delta(\omega-\omega_e)}{\left.\frac{dD}{d\omega}\right|_{\omega=\omega_e}} \quad . \tag{11}$$

If D is block diagonalized by the transformation to coordinates with appropriate point symmetry, the parts of D belonging to each irreducible representation may be considered separately, say,

$$D = \prod_\tau D_\tau \quad .$$

CALCULATIONS FOR VARIOUS PAIR CONFIGURATIONS IN SILICON

To treat pair defects it is necessary to know g for the perfect lattice only over a short range, in fact between the two atoms labelled k=1 or 2 in one cell and in two adjacent cells where $x(\ell)-x(\ell')=\frac{1}{2}a(1,1,0)$. Because of the tetrahedral point symmetry and cubic overall symmetry, this involves only six independent variables. All others can be obtained from them by using the rotational symmetry of the lattice.

Without going into any of the computational details we shall only briefly outline the different models considered for pair defects in Si.

Two Nearest-Neighbor Substitutional Impurities (Model 1)

The defect is assumed to consists of an atom of mass defect $\varepsilon_1 = (M-M_1)/M$ at site k=1 and an atom of mass defect $\varepsilon_2=(M-M_2)/M$ at site k=2 of the same unit cell $\ell=0$. The force constant between these two nearest neighbors is changed by ϕ for displacements along the (1,1,1) bond joining them, and by ϕ' for displacements perpendicular to this. The point symmetry of this defect is C_{3v} with axis along the bond. The mechanical representation for displacement vectors of the two defect atoms is

$$2\Gamma_1 + 2\Gamma_3 \quad . \tag{12}$$

(The notation of Koster et al.[14] will be used throughout.) The 6 x 6 matrix 1-gC will therefore be reducible to a 2 x 2 matrix corresponding to the Γ_1 vibrations along the bond and two identical 2 x 2 matrices corresponding to the Γ_3 vibrations perpendicular to this.

There are two special cases of particular interest. If the atoms are the same $\varepsilon_1=\varepsilon_2=\varepsilon$ the defect has inversion symmetry as well and the point group is D_{3d}. The vibrational representations are

$$\Gamma_1^+ + \Gamma_1^- + \Gamma_3^+ + \Gamma_3^- \quad . \tag{13}$$

The solutions are now

$$\varepsilon M\omega^2 = \frac{1}{g_1 \pm g_u} - \phi(1\pm 1) \quad , \tag{14}$$

the signs corresponding to $\Gamma_i^{\pm}$. For a single impurity the threefold degenerate localized modes are given by the equation $\varepsilon M\omega^2 = 1/g_\ell$. We see that the six degenerate modes of two uncoupled impurities are split into two singlets and two doublets by the interaction. The high frequency modes are Γ_i^+, i.e., with atoms vibrating in opposite directions. The Γ_i^- modes where the atoms vibrate in the same direction are seen to be independent of ϕ. If the impurity atoms have the same charge, only the Γ_i mode is optically active.

If one defect becomes similar to the host $\varepsilon_2 \to 0$ (for example, a phosphorus defect which would still be charged), there is only one equation each, for a singlet and a doublet of the form,

$$\varepsilon_1 M\omega^2 = \frac{1-2\phi(g_1-g_u)}{g_1-\phi(g_1^2-g_u^2)} \quad . \tag{15}$$

Interstitial Defect (Model 2)

We consider an interstitial atom of mass M' on the tetrahedral site of the diamond lattice as shown in Fig. 1. The defect is limited to the interstititał and its first four neighbors by neglecting the coupling to more distant atoms and leaving unchanged all the forces in the host lattice. The point symmetry is then that of the tetrahedral group T_d. The 15 dimensional subspace spanned by the displacement vectors on all five atoms belongs to representations[12]

$$\Gamma_1 + \Gamma_3 + \Gamma_4 + 3\Gamma_5 \quad , \tag{16}$$

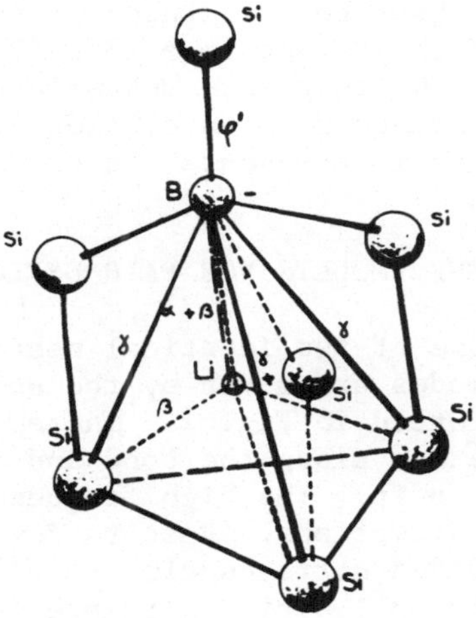

Fig. 1. Structure of the boron lithium defect showing the different force constant parameters.

so that by choosing suitable symmetry coordinates the part of the matrix corresponding to the first three is diagonal and the Γ_5 modes give 3 identical 3 x 3 matrices. These are the only infrared active modes.

Substitutional-Interstitial Pair (Model 3)

This defect consists of an interstitial on a tetrahedral site and a substitutional impurity on one of the neighboring sites, chosen as site 5 of the last problem. If we restrict changes of force constant to those between these five atoms, the problem has the same 15 x 15 dimension as Model 2, but the symmetry is reduced to C_{3v}. The representations are[12]

$$4\Gamma_1 + \Gamma_2 + 5\Gamma_3 \quad , \tag{17}$$

and both Γ_1 and Γ_3 are infrared active.

A Six-Atom Model (Model 4)

In Model 3 we neglected any change of force constant between the substitutional boron and its own nearest neighbors on the lattice. Experiments on isolated boron[5] when compared with the theory of a mass defect only[10] show that the force constants are decreased by about 10 per cent in that case. We shall find that we need larger changes than this in the pair defect. Since none of the nearest neighbors of the boron have been included in the defect so far, it is necessary to extend the defect to a 27 x 27 problem to include them all (see Fig. 1). We therefore assume a change only in the force constant to the neighbor (atom 6) along the B-Li direction t_5. This adds one further representation of Γ_1 and Γ_3 type.

DISCUSSION OF THE RESULTS FOR PAIR DEFECTS IN SILICON

In the special case of two identical nearest nieghbor impurities, the localized modes are given by the solution of (14). The results for $\phi=0$ are plotted in Fig. 2. There are two singlet modes corresponding to vibration along the bond and two doublets for vibration perpendicular to it. The high frequency pair corresponds to motion in opposite directions. Results for $\phi=0$ as a function of ε_2 when $\varepsilon_1=0.644$ and 0.608 corresponding to B^{10} and B^{11}, respectively, are shown in Fig. 3. Again there are four modes corresponding to the splitting of the triplets associated with the two isolated impurities provided ε_2 is large enough (>0.15) to produce a localized mode. The splitting of the B triplets are largest when $\varepsilon_2=\varepsilon_1$. The effect on the Γ_1 modes of a change in ϕ is shown in Fig. 4. The effect on the Γ_3 modes is very similar. The changes

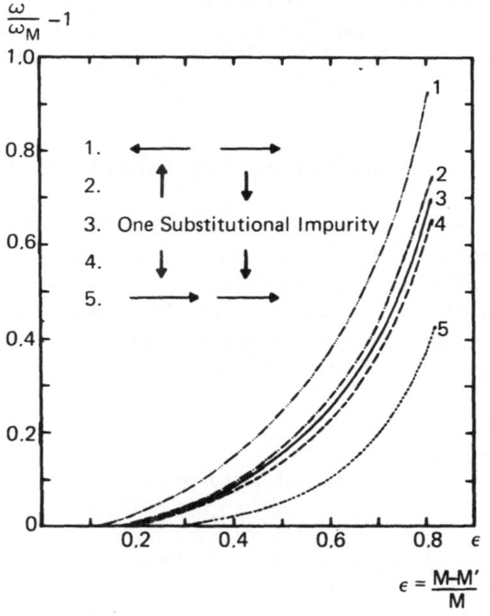

Fig. 2. Localized mode frequencies for a pair of identical substitutional impurities without change of force constant between them ($\phi=0$) $(\omega/\omega_M)-1$ is plotted vs. $\varepsilon_1=\varepsilon_2=\varepsilon$, pair modes Γ_1^+ curve 1, Γ_3^+ curve 2, Γ_1^- curve 4, Γ_1^- curve 5. Single impurity Γ_5 mode curve 3.

include one which is actually as great as the Si force constant in the perfect lattice on the nearest neighbor model.

As expected, the modes vibrating along the axis are shifted to lower frequencies by lowering the force constant.

The intensities of the lines for impurities of the same and opposite charges have also been calculated.

These calculations provide only a general framework for the understanding of actual experimental data. Boron pairing in Silicon seems to be a good example where, in order to quantiatively interpret the experimental data,[5] one finds it necessary to employ P atoms close by the defect to reduce the symmetry and relax the selection rule.

To fully understand all the available experimental data, more complex defect models involving other changes in force constants should probably be calculated.

The localized mode due to B impurities in Si observed recently[15] by Raman scattering has frequencies which compare well with that calculated by this model.

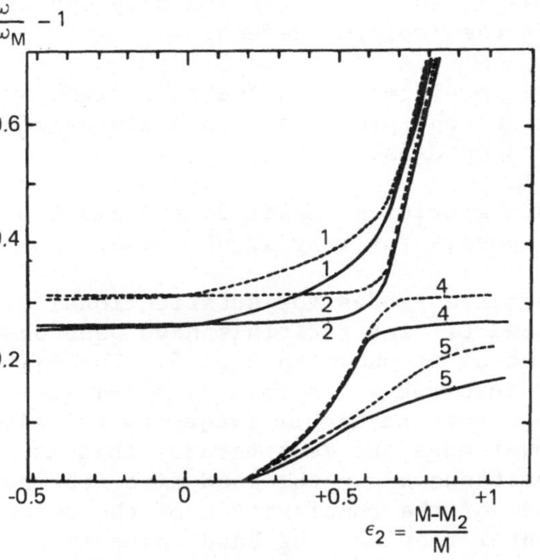

Fig. 3. Localized mode frequencies for a pair of two different impurities without change of force constant. Full lines B_{11}, ε_1=0.608; Dashed lines B_{10}, ε_1=0.644. Numbers indicate symmetry types as in Fig. 2.

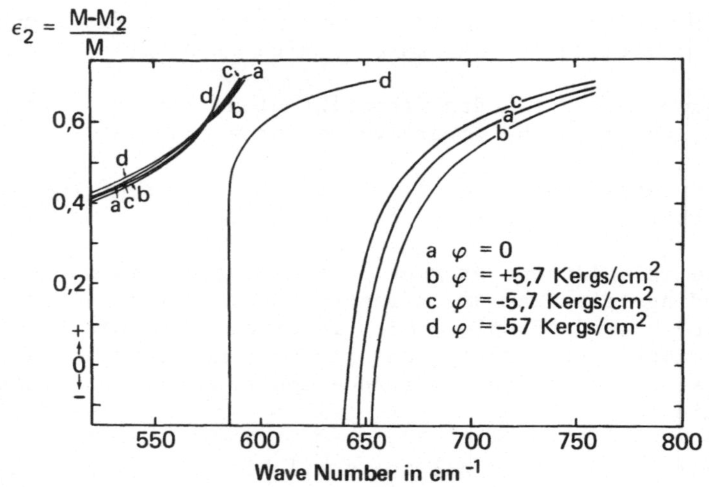

Fig. 4. Frequencies of localized modes of Γ_1 type for a pair of two different impurities with a positive or negative change of force constant ϕ when one impurity is $B_{11}(\varepsilon_1=0.608)(\omega/\omega_M)-1$ is plotted vs. ε_2 for various values of ϕ.

Raman scattering[16] from local modes due to Si defects in Ge gives two lines (475 and 450 cm^{-1}) at high concentration attributed to pairing whereas at low concentration only one line (390 cm^{-1}) is observed for the isolated defect.

The splitting of localized modes due to a light substitutional impurity (Li^-) due to the presence of a heavy neighbor (Te^+) has also been observed in GaAs.

For interstitial impurities (Model 2) and small β values, a localized mode only appears for very light atoms.

Well-defined localized modes for substitutional B_{10} and B_{11} associated with Li^6 and Li^7 interstitials have been observed.[4,6,7] The experimental spectrum is shown in Fig. 5. The B_{10} and B_{11} lines are each split into two. A single line for each Li isotope is observed near ω_M. Those at higher frequency are almost twice as intense as the lower ones and we interpret this as indicating that the singlet vibrations along the bond direction have the lower frequency. In view of the complexities of the calculation and the lack of experimental results, the band absorption will not be considered.

The inclusion of the B nearest neighbors in Model 4 produced too complicated a defect. However, a similar effect may be produced by weakening the next-nearest neighbor force constants and these bonds are certainly stretched by the inclusion of Li.

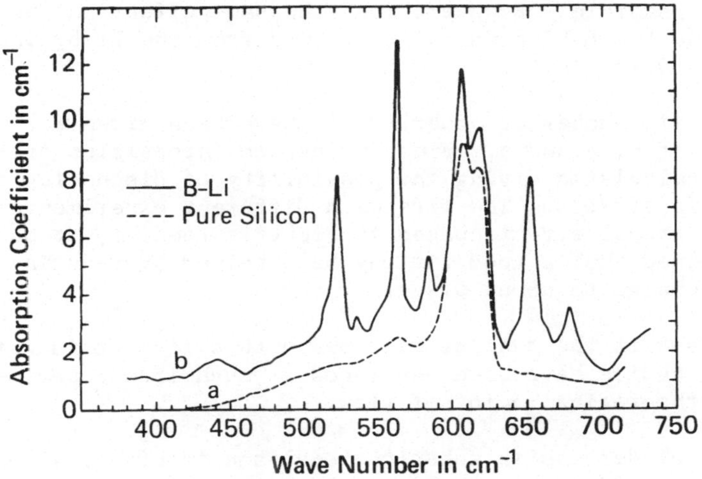

Fig. 5. Experimental absorption spectrum due to boron and lithium in silicon (full line). The absorption spectrum of pure silicon is shown for comparison (dotted line).

This reduces the frequency of both Γ_1 and Γ_3 modes and causes only a small splitting between them. This splitting may be mainly accounted for by weakening the B-Li force constant by α. A good fit may be obtained for both B modes assuming α=-7.5 kerg/cm^2 and α= -69 kerg/cm^2. Of course, α<0 leads to a negative force constant

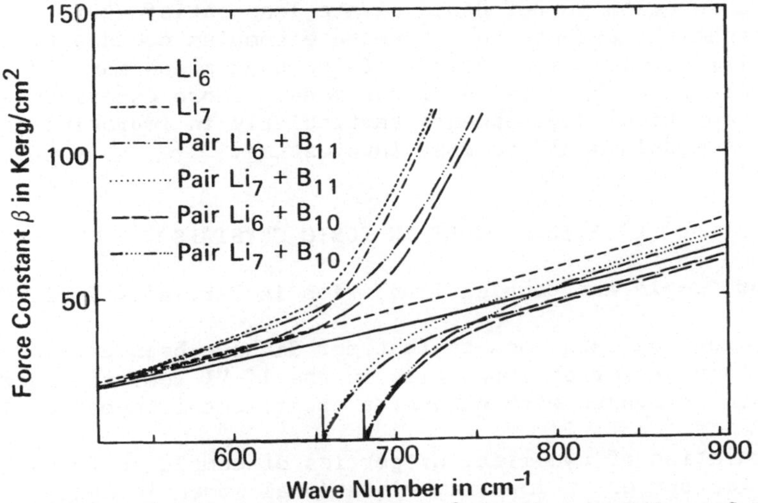

Fig. 6. Effect of the force constant β on the position of the triplets due to boron and lithium isotopes (by considering Model 3).

which is impossible. However, much the same effect may be produced by weakening the B-Si bond pointing away from the Li by ϕ' such that $\phi'+\alpha=-69$ kerg/cm^2.

A certain number of other trials have been given for different values of β, λ and α. Frequencies and intensities of the bands have been calculated giving the possibility of discussing different isotopic effects. The fits with different experimental results have been extensively discussed in Pfeuty's thesis. It has also been mentioned that a good fit may be obtained by reducing the Li effective charge to about 0.7 e.

At least in the case of Si doped with different impurities extensive studies have been developed as much from experimental as from a theoretical point of view.

Pairs of defects in substitutional and interstitial sites have been considered using the local point symmetry to simplify the calculations. The general effect of such pairing is to split and shift the localized and resonant modes associated with the individual impurities. Detailed comparison between theory and experiment has been achieved for the localized modes of B-Li centers in Si, and a satisfactory model of these centers has been found. It suggests that the oscillations along the B-Li bond have lower frequency than those perpendicular to it, mainly because the B-Si bond in this same direction is weakened, possibly by the attraction of its electrons towards the Li. It would be interesting to check the degeneracy of the lines by reducing the local symmetry with strain[18] or other impurities.[19] The detailed predictions of the band absorption in B-P doped Si are less satisfactory. It is possible that the defects here have more complex coordination than simple pairing. The calculations of the band modes may also be more dependent on the details of the model. More experiments and further theoretical developments particularly in improving the complex defect models would be most interesting.

LOCALIZED MODES IN IONIC CRYSTALS

Models for Simple and Complex Impurities in Partially Ionic Crystals

Analogous calculations to that for Si have been carried out for a certain number of impurities in the II-VI semiconductor compounds[20] and compared with extensive experimental investigations.[21]

Calculation of dynamical properties of complex defects has been carried out on Cu impurity vibrational modes in GaAs.[12]

It has been recognized that Cu impurity situated next to a vacancy can form a complex with weakened force constants which can give rise to resonant modes. The complex $V_{As}Cu_{Ga}$ with mass and force constant change due to vacancies creates a "defect space" of eight atoms; hence a dynamical problem of 24 x 24 had to be solved. In order to reach a good fit, the detailed structure of the complex defect has to be discussed and a possible model considered.

Raman Scattering

As for infrared absorption, the presence of imperfections changes the selection rules and leads to the possibility of observing Raman scattering transitions which would be forbidden for the perfect crystal. The symmetry group of the system is reduced from the crystal space group to the point group of the lattice site of impurity. Heavy substitutional impurities activate the whole range of normal mode frequencies and render observable the normal modes at all critical points of the Brillouin Zone where the density of states generally have their maxima. It is then possible to measure by Raman spectroscopy the normal mode frequencies at high symmetry points of the Brillouin Zone when these modes are activated by the presence of imperfections which give rise to resonant modes.

Raman scattering measurements and group theory analysis have been recently carried out in the case of Cu_2O[22] and ZnS[23] containing Mn.

A particularly interesting point in the determination of the resonant mode frequencies is the demonstration[24] of the fact that resonant modes exhibit the same resonant Raman effect as the normal modes like LO, for example, whereas the localized modes exhibit a weak resonant effect at the frequency of the impurity electronic states.

TEMPERATURE DEPENDENCE OF THE INFRARED ABSORPTION BY LOCALIZED AND RESONANCE MODES IN IONIC CRYSTALS

For some impurities such as substitutional Ag^+ in KI which give rise to resonance modes as well as for $^7Li^+$ in KBr, it has been observed[25] that the integrated absorption under the peak shows a strong decrease with increasing temperature. Whereas in the case of localized modes associated with H^- ions substitutional in alkali-halide crystals, the experimental results are more contradictory. Evidence has been given for a strong decrease[26] of the integrated absorption as well as for an increase[27] with temperature.

Numerous theoretical investigations have been carried out.[28,29] Maradudin[30] has discussed recently the role anharmonicity can play in lattice dynamical defect problems, and I shall try to briefly summarize his arguments.

In the case of absorption by localized modes it is shown that the imaginary part of the dielectric response function can be put into the form

$$\varepsilon(\omega) = \varepsilon_{\ell\cdot m\cdot}(\omega) + \varepsilon_{s\cdot b\cdot}(\omega) + \ldots \quad .$$

The first term describes the absorption by the localized mode, while the remaining terms describe the side bands to the localized mode peak.

The one-phonon absorption by the localized mode consists of a delta function peak centered at the harmonic approximation value of the localized mode frequency,

$$\varepsilon_{\ell\cdot m\cdot}(\omega) = 2\pi^2 \frac{n_d}{V} \frac{(e^*)^2}{M'\omega_0} e^{-2M} \delta(\omega-\omega_0) \quad .$$

In obtaining this expression, the effective Hamiltonian which leads to a temperature-dependent broadening of the peak and to a shift in its position from $\omega=\omega_0$ has been consistently neglected. The temperature dependence of the contribution to $\varepsilon(\omega)$ arises entirely from the factor exp(-2M) which resembles the Debye-Weller factor. Here,

$$2M = \frac{12}{\hbar^2} \sum_{ps\cdot s_2} V^2_{ps\cdot s_2} \frac{2n_p+1}{\omega_p^2} \quad .$$

The subscript p refers to the 3rN-3 band modes and s denotes the 3 localized modes; ω_p is the frequency of the p the band mode, and ω_0 is the frequency of the triple degenerate localized mode, e* is the transverse effective charge of the impurity ion, and M' is its mass.

The results of numerical calculations[31] of 2M(T) in the low temperature limit for various models of an H^- U-center in KCl show that values of 2M(T) large enough to explain the observations of an increase of the integrated absorption[26] with temperature can be obtained for quite reasonable models of the impurity host crystal interaction.

In the case of resonance modes which are not true normal modes of the perturbed crystal, the procedure is different. The fact that resonance modes associated with substitutional impurities in crystals of the rocksalt structure are infrared active means that the impurity ion is vibrating in a resonance mode with the frequency of the mode. On the other hand, the low frequency of the resonance mode associated with a light impurity, such as Li^+ in KBr, indicates that the harmonic force constants of the interaction between the impurities and the host lattice are much weaker than the corresponding force constants in the perfect crystal. Inasmuch as experimental results show that the resonance mode peak in infrared absorption spectra are sharp, it is reasonable to make the assumption that the vibrational spectrum of the crystal contains a well-defined resonance mode which, in the harmonic approximation, can be described by a three-dimensional harmonic oscillator with the frequency Ω of the resonance mode, which is weakly coupled to the remaining modes of the crystal.

In the theoretical treatment of Maradudin[30] weak harmonic coupling of the impurity ions to the host lattice is assumed. The anharmonic interactions between the impurity ion and the ions of the host crystal are not regarded as weak, however, and in fact are considered to be responsible for the strong temperature effect.

In the low temperatures at which the experiments[25] are carried out ($T < \hbar\Omega/k_B$) only the transitions from the ground state to the first three excited states of the impurity oscillator contribute appreciably to the absorption spectrum. At a site of O_h symmetry, these first three excited states of an isotropic oscillator remain degenerate in the presence of anharmonic terms in the potential energy of the oscillator.

Detailed calculation[30] also leads to the conclusion that the absorption spectrum depends strongly on temperature through a factor $\exp(-2M(T))$ with

$$2M(T) = \frac{1}{N}\sum_{P}[\gamma_P(\ell j;\ell j) - \gamma_P(0;0)]^2(2n_P+1) \quad .$$

Thus, by removing the large anharmonic terms from the interaction Hamiltonian by means of a canonical transformation, an expression for the absorption spectrum is obtained involving multiphonon processes which already in the lowest order of perturbation theory displays the observed strong temperature dependence for absorption by resonant modes.

CONCLUSION

Although a general understanding of the physics of defect vibrations in a lattice is available, detailed calculations for specific models, in particular with regard to complex defects, will be extremely useful in the future.

Investigation of the vibrational properties of a defect is of a nature to furnish a picture of the structure of the local defect site and to eventually even become an analytical tool. The practical implications of such an eventuality are obvious.

The study of extended defects and of solids with high defect concentrations opens the way to a more fundamental understanding of materials such as mixed crystals, alloys, and amorphous solids. There is a vast area for development of computational physics.

REFERENCES

1. Elliott, R. G., "Phonons," Aberdeen Summer School Lectures. (Oliver & Boyd) 1965.

2. Maradudin, A. A., Rep. on Prog. in Phys. 19, 331.

3. Balkanski, M., Optics in Solids, ed. by Abeles (North Holland).

4. Balkanski, M. and Nazarewicz, W., J. Phys. Chem. Solids 25, 457 (1964); and 27, 671 (1966).

5. Angress, J. F., Goodwin, A. R. and Smith, S. D., Proc. R. Soc. A287, 64 (1965).

6. Waldner, M. Hiller, M. A. and Spitzer, W. A., Phys. Rev. 140, A172 (1965).

7. Chrenko, R. M., McDonald, R. S. and Pell, E. M., Phys. Rev. 138, A1775 (1965).

8. Newman, R. C. and Willis, J. B., J. Phys. Chem. Solids 26, 373 (1965).

9. Schafer, G., J. Phys. Chem. Solids 12, 233 (1960); Sievers, A. J., Maradudin, A. A. and Jaswal, S. S., Phys. Rev. 138A, 272 (1965); Renk, K. F., Phys. Letters 14, 281 (1965); and Hayes, W., Phys. Rev. 138A, 1272 (1965).

10. Elliott, R. J. and Dawber, P. G., Proc. R. Soc. A273, 222 (1963); and Proc. Phys. Soc. London 81, 521 (1963).

11. Balkanski, M., Elliott, R. J., Nazarewicz, W. and Pfeuty, P., International Symposium on Lattice Defects in Semiconductors. Kyoto, Japan (1966); Pfeuty, P., Thesis, Paris, 1968; and Elliott, R. J. and Pfeuty, P., J. Phys. Chem. Solids 28, 1789 (1967).

12. Grimm, A., Maradudin, A. A., Ipatova, I. P. and Subashiev, A. V., J. Phys. Chem. Solids 32 (1971); and Ipatova, I. P., International Conference on Phonons, Rennes, France, 1971.

13. Lifshitz, Nuevo Cimento Suppl. 3 716 (1956).

14. Koster, G. F., Dimmoch, J. O., Wheeler, R. G. and Statz, H., Properties of 32 Point Groups. M.I.T. Press.

15. Nazarewicz, W., Balkanski, M., Morhange, J. F. and Sebenne, C., Sol. Stat. Comm. (to be published).

16. Feldman, D. W., Ashkin, M. and Parker, I. H., Phys. Rev. Letters 17, 1209 (1966).

17. Hayes, W., Phys. Rev. Letters 12, 373 (1964).

18. Elliott, R. J., Hayes, W. and McDonald, H. F., Phys. Rev. Letters 14, 961 (1965).

19. Mirlin, D. N. and Reshina, H., Soviet Phys. Solid State 8, 116 (1966).

20. Pfeuty, P., Birman, J. L., Nusimovici, M. A. and Balkanski, M., Localized Excitations in Solids. (Plenum Press) 1968.

21. Beserman, R., Thesis, Paris, 1968; Beserman, R. and Balkanski, M., Phys. Rev. B1, 608 (1970); and Beserman, R. and Balkanski, M., J. Phys. Chem. Solids 31, 355 (1970).

22. Reydellet, J., Thesis, Paris, 1971.

23. Beserman, R., Zigone, M. and Balkanski, M., International Conference on Phonons, Rennes, France, 1971.

24. Zigone, M., Beserman, R. and Balkanski, M., Proc. International Conference on Light Scattering in Solids, Paris, 1971.

25. Takeno, I. S. and Sievers, A. J., Phys. Rev. Letters 15, 1020 (1965).

26. Schaefer, G., J. Phys. Chem. Solids 12, 233 (1960); Takeno, I. S. and Sievers, A. J., Phys. Rev. Letters 15, 1020 (1965); and Mitra, S. S. and Singh, R. S., Phys. Rev. Letters 16, 694 (1966).

27. Mirlin, D. N. and Reshina, I. I., Fiz. Tver. Tela 6, 945 (1964); (English translation: Soviety Physics--Solid State 6, 728 (1964)); Fritz, B., Gross, U. and Bauerle, D., Phys. Stat. Solidi 11, 231 (1965); and Kleinman, L., Pryce, M.H.L. and Spitzer, W. G., Phys. Rev. Letters 17, 304 (1966).

28. Mitra, S. S. and Singh, R. S., Phys. Rev. Letters 16, 694 (1966); Highes, E. A. Phys. Rev. 173, 860 (1968); and Ipatova, I. P., Subashiev, A. V. and Maradudin, A. A., Annals of Physics (N.Y.) 53, 376 (1969).

29. Ipatova, I. P., Maradudin, A. A. and Subashiev, A. V., Fiz. Tver. Tela 11, 2271 (1969)(English translation: Soviety Physics--Solid State 11, 1834 (1970)).

30. Maradudin, A. A., International Conference on Phonons, Rennes, France, 1971.

31. Ipatova, I. P., Subashiev, A. V. and Maradudin, A. A., Annals of Physics (N.Y.) 53, 376 (1969).

POINT DEFECTS IN IONIC CRYSTALS

A. B. Lidiard and M. J. Norgett

Theoretical Physics Division, A.E.R.E., Harwell, Berks., U.K.

I. INTRODUCTION

In recent years there have been many calculations of the properties of point defects in ionic crystals, particularly calculations of the energies of vacancies, interstitials and impurity ions. Apart from those on colour centres where one is obliged to solve for the quantum mechanical states and energy levels of the centre, these calculations have been almost entirely made using classical ionic models derived from the original work of Born. Even in the case of colour centres questions of the relaxation of the crystal about the centre and of electron-lattice coupling are answered by using such models to describe the lattice response. The importance of the long-range Coulomb forces in all these models clearly differentiates them from those for other types of solid. It is this fact also which in the long run allows us to make quantitative predictions for ionic crystals with greater confidence than for other materials.

It is our purpose here to review critically the various physical models and the mathematical methods used to describe the response of an ionic lattice to the presence of a defect, but not to survey the many applications in detail - in any case there have been two recent surveys of these applications to which the reader may turn[1,2]. We will be largely concerned with the static response of the lattice to those defects of principal interest in the understanding of atomic and ionic transport processes, although some of what we have to say is applicable more widely. While a wider ranging review might be of greater value, it seems necessary in an article of this length to limit the discussion to a well defined part of the subject. However we believe that our limited examination of this class of problems is timely since the rate at which calculations are appearing is increasing and likely to grow even faster before long. Viewed in the light of recent developments it seems clear that many previous calculations have been made using relatively inefficient methods while some of the physical models can now be seen to be rather bad once accurate deductions from them are possible. The demonstration of these assertions is in keeping with the aims of this conference.

The next section reviews the physical models used. It describes the way these are built up from a variety of macroscopic physical data and details some points where current models are failing. Section III is devoted to a description of mathematical and numerical methods of describing the defect lattice once the model is specified. It draws attention particularly to the speedy methods now available when access to a computer with a large store is possible. However even when it is not, considerable improvement over many of the methods in use seems possible. Lastly in section IV, lest it should be felt that our critical review has devalued existing calculations too drastically, we comment on the utility of this field of work and, by discussing a particular class of system - rare gas impurities in ionic solids - show how valuable such work may be, not merely qualitatively but also quantitatively.

II. PHYSICAL MODELS OF IONIC CRYSTALS

In this section we review the various classical models of ionic crystals used for calculating defect and other properties. Since the basis for these calculations is generally the quasi-harmonic approximation we begin by briefly repeating the analysis which yields expressions for defect energies, entropies and volumes, all the components of defect free energies, in fact. In the second part of the section we review models of the various non-Coulombic energy terms, e.g., overlap repulsions.

A. The Quasi-Harmonic Approximation

The obvious starting point for the calculation of defect properties in solids in or near thermodvnamic equilibrium is provided by the Born-Oppenheimer and Quasi-Harmonic approximations. Even though the quasi-harmonic approximation has its limitations[3] it provides the only practical general scheme of calculation at the present time. Purely numerical molecular dynamics methods are feasible, but very expensive in computer time and as a result are incapable of satisfying the need for particular defect calculations. As a result these should be used at the present time to answer general questions. The task of developing anharmonic expansions for defect crystals hardly seems worth the effort at present although special methods[4] may provide an economic way to such developments. We shall therefore confine our discussion to calculations based on the quasi-harmonic approximation. In considering optical transitions one, of course, generally also uses the Frank-Condon approximation. However the description of thermal vibrational motions by the quasi-harmonic approximation is basic.

We therefore begin with the familiar expression for the Helmholtz free energy of a crystal as

$$F(T,V) = \Phi(V) + kT \sum_j \ln(2 \sinh \frac{\hbar\omega_j}{2kT}) . \qquad (1)$$

Here the potential energy term and the lattice frequencies, ω_j, are functions of the crystal volume V. In the usual way we have written F as a function of the temperature T and V. The relation of T and V to external pressure, P, (we here consider only uniform hydrostatic stresses) is given by combining (1) with the thermodynamic equation

$$P = -\left(\frac{\partial F}{\partial V}\right)_T \qquad (2)$$

At high temperatures where $\hbar\omega_j << 2kT$ this becomes

$$P = -\left(\frac{d\Phi}{dV}\right) - \frac{kT}{V} \sum_j \frac{d\ln\omega_j}{d\ln V} \qquad (3)$$

or, in the Grüneisen approximation in which all the derivatives $d(\ln\omega_j)/d(\ln V)$ are set equal to $-\gamma$,

$$P = -\frac{d\Phi}{dV} + \frac{3NkT\gamma}{V} \qquad (4)$$

where N is the number of atoms in the volume V. For the low pressures usually encountered, it is convenient for our present purposes to introduce the compressibility $\kappa(P,T)$

$$\kappa(P,T) = -\frac{1}{V}\left(\frac{\partial V}{\partial P}\right)_T \quad (5)$$

and thus to replace P on the l.h.s. of (4) by

$$-\frac{(V(P,T) - V(0,T))}{V(0,T)\ \kappa(0,T)}.$$

For a cubic crystal of lattice parameter r_o, $V(P,T) \propto N\, r_o^3$ so that (4) can be written as

$$-\frac{3(r_o(P,T) - r_o(0,T))}{r_o(0,T)\ \kappa(0,T)} = -\frac{d\Phi}{dV} + \frac{3n\ kT\gamma}{v_m} \quad (6)$$

where $v_m = V/N$ is the volume per lattice cell (containing n atoms).

All the above results are standard. The point for defect theory however is that one can write such expressions for two different configurations of the atoms making up the crystal (e.g. a perfect lattice and a lattice with defects or two different defect configurations) and by comparison obtain expressions for characteristic defect parameters. At high temperatures ($\hbar\omega_j \ll 2kT$) particularly simple results follow. Thus it follows immediately from (1) that the internal energy change is

$$\Delta U = U_2 - U_1 = \Phi_2 - \Phi_1 \equiv \Delta\Phi, \quad (7)$$

while the (non-configurational) entropy change is

$$\Delta S = S_2 - S_1 = -k \sum_j \ln(\omega_j^{(2)}/\omega_j^{(1)}). \quad (8)$$

These can be applied to give the parameters for defect formation, migration, association, etc. In practice we are concerned with defect reactions occurring at constant pressure so that the crystal volumes at which Φ_2, $\omega_j^{(2)}$ etc. are evaluated will differ from those for Φ_1, $\omega_j^{(1)}$. These volume differences are themselves significant quantities and determine the effect of pressure upon the corresponding defect processes. They may be calculated from (6). Consider the change in configuration to take place at zero pressure then $\Phi^{(1)}$ satisfies (6) with the l.h.s. equal to zero. Changing now to $\Phi^{(2)}$ introduces an additional term $-d\Phi^{(2)}/dV + d\Phi^{(1)}/dV$ equivalent to an internal defect pressure. In the absence of a balancing external pressure the lattice parameter of the crystal changes by the amount

$$\frac{\Delta V}{\kappa V} = \frac{3\,\Delta r_o}{r_o\,\kappa} = -\frac{d\Phi^{(2)}}{dV} + \frac{d\Phi^{(1)}}{dV}, \qquad (9)$$

neglecting the small change of κ and γ (of the order of defect fraction). From (9) one may calculate the average change in lattice parameter and thus the volume change per defect.

It will be noted that (7)-(9) do not depend explicitly upon temperature although there is an implicit dependence through the volume dependence of $\Delta\Phi$, ω_j etc. In practice most calculations have been carried out only for a single volume and thus little is known about these implied temperature variations although they are often assumed to be small. A repetition of these calculations to study this point seems now to be overdue, especially in view of the continued difficulty of obtaining perfect fits to transport data (e.g. ionic conductivity and diffusion data) with existing defect models when constant defect energies, entropies etc. are assumed[5]. At this point it is also worth noting that most of these calculations up to now have been directed to the evaluation of defect energies (formation, activation and association) since these are responsible for the strong temperature variation of defect processes through exponential Boltzmann factors in the expressions for defect concentrations, mobilities etc. A much smaller effort has gone into the calculation of entropies. These calculations have often been carried out by using Einstein models to evaluate the changes in frequency ω_j occurring in (8). A more rigorous approach has recently been described[6] but has not so far been applied to models of ionic crystals. Some calculations of defect volumes have recently been made on the basis of Eqn.(9)[7].

B. Potential Models for the Solid

We must now consider the way in which the physical model for this class of solid is built up. The important feature of ionic substances in both solid and liquid states is the fact that for many physical properties they can be regarded as assemblies of ions whose closed-shell electronic structure ensures that their non-electrical interactions are repulsive at short distances and of Van der Waals type farther away[8]. Even when the cations have open shell structures, e.g. transition metal ions, the general pattern is often remarkably close to this classical ionic model - exchange energies and covalency effects are small on the scale of energies we are concerned with. The existence of charges on the ions means, however, that electrical polarisation terms are often large and the correct representation of the dielectric behaviour of these crystals is therefore important. Reasonably satisfactory models for these substances can be constructed by the analysis of four main physical properties (1) cohesion (2) elastic constants (3) dielectric properties and (4) lattice vibrations. In the present paper we can only comment on the situation since there is not space to describe it fully (the reviews by Barr and Lidiard[1] and by Hardy and Flocken[2] supply some of the background to this commentary).

Firstly, the classic approach is that following Born[8]. This analyses cohesion and the compressibility of cubic crystals to derive the non-Coulombic overlap repulsion terms in the energy, generally assumed to be of Born-Mayer form, b exp $(-r/\rho)$. Various basic calculations for pairs of closed-shell atoms show this to be reasonably well justified. The Born approach is however generally elaborated by the inclusion of Van der Waals terms (r^{-6} and r^{-8}) which are still generally included with the magnitudes estimated by Mayer many years ago. Although not very significant for cohesion and elastic constants, these Van der Waals terms are more significant in defect calculations and revision of the Mayer constants is desirable (see e.g. Lynch's revised dipole-dipole constants[9]). A more fundamental difficulty arises in some systems (e.g. the oxides MgO, CaO and UO_2 and the alkaline earth fluorides) when attempts are made to fit anion-anion interactions with a Born-Mayer form; the negative values found are inconsistent with the idea of closed shell repulsions. Of course, these anion-anion interactions are rather small in the perfect lattice since the anions are only second neighbours to one another but they are obviously much more important when we consider anion interstitials, for example, since these involve close contacts.

In the classic Born approach the compressibility is the only elastic constant generally employed in the analysis. But since the models used are central-force models the failure of many of the cubic crystals to satisfy the Cauchy relation $c_{12}=c_{44}$ is an indication that such models can never be perfectly satisfactory. In any event the models cannot be defined to better than the uncertainty introduced by the arbitrary choice from among the three independent elastic constants. In practice, it seems that for many defect problems these uncertainties are not very important. Much more important - at any rate for charged defects - is the consistency of the model with dielectric properties.

In these materials we must analyse correctly both the low- and the high-frequency dielectric constants, ε_o and ε_∞. Analyses of ε_∞ such as those by Tessman et al[10] and by Pirenne and Kartheuser[11] provide sets of characteristic electronic polarisabilities α_i of the ions which are independent of compound. These analyses also give good evidence for the applicability of the Clausius-Mosotti relation

$$\frac{(\varepsilon_\infty - 1)}{(\varepsilon_\infty + 2)} = \frac{4\pi}{3v_m} \alpha_m \tag{10}$$

with the molecular polarisability, α_m, taken as the sum of the characteristic ion polarisabilities in the unit cell. These polarisabilities are used to describe the dipoles induced on the ions by the internal fields generated by the defects, generally in the dipole approximation, by placing equivalent point dipoles, $\underline{m}$, at the nuclear positions and evaluating the resultant interactions by electrostatics. One limitation to this polarisable point-ion model (PPI) is the divergencies or catastrophes inherent in it[12]. Thus the polarisation energy is second order in the moments $\underline{m}$, i.e. $\underline{m}^T \underline{\underline{A}}\, \underline{m}$ in matrix notation, while the only other terms are linear in $\underline{m}$, so that the energy may diverge to $-\infty$ unless the matrix $\underline{\underline{A}}$ is positive definite. A necessary condition for this to be the case is that all ion pairs (i,j) must be farther apart than the critical distance

$$r_{ij}^{crit} = (4\, \alpha_i \alpha_j)^{1/6} \tag{11}$$

otherwise the induced spontaneous dipole-dipole interaction outweighs the self energy of polarisation. Other divergencies are also inherent in the model but this dipole-dipole divergence appears to be the most important because it occurs at a non-zero interionic spacing, e.g. in the alkali halides at between 30 and 50°/o of r_o, the anion-cation nearest neighbour separation in the perfect crystal. Other divergences e.g. the charge-induced dipole divergence ($\sim r^{-4}$) which occurs as $r \to 0$, are thus less important than previously thought[1,2,13]. It is important to note that the dipole-dipole divergence cannot be removed by altering the other terms in the energy, e.g. the overlap repulsion, as long as these remain finite at all interionic separations. The obvious way to remove this weakness is to allow for the reduction in mutual polarisability which evidently takes place as the adjacent ions overlap. A plausible way to represent this has been suggested[13] although not yet much applied in practice. A more fundamental examination might be of value.

The polarisable point-ion model as so far described has another defect, namely that it may not describe the static dielectric constant well. From the interatomic potential ϕ (Born-Mayer plus Van der Waals etc.) one may calculate the displacement polarisability of the ions; for cubic lattices MX the result is

$$\alpha = z^2e^2/p \tag{12}$$

where the force constant

$$p = \frac{2\nu}{3}\left(\phi''(a) + \frac{2}{a}\phi'(a)\right) \tag{13}$$

where ν is the anion-cation co-ordination number and a is the nearest-neighbour anion-cation separation. When we then calculate the static dielectric constant ε_o from the Clausius-Mosotti relation,

$$\frac{\varepsilon_o - 1}{\varepsilon_o + 2} = \frac{4\pi}{3v_m}(\alpha_+ + \alpha_- + 2\alpha), \tag{14}$$

however, we find that it is too large. This is the reason why calculations of the energies of charged defects have often given values considerably too low (by $\sim$ 20°/o or more); for although empirical values of ε_o have generally been used to describe the polarisation in the bulk of the crystal by appeal to limiting continuum results, in the inner region around the defect ("region I") the model is evidently too polarisable, with the result that the (negative) polarisation energy term is overestimated. This has been explicitly verified[7] by progressively enlarging the inner region I and showing that the energy falls systematically as this region of excessive polarisability grows. It now appears that the effect of varying ϕ may be felt most through its influence (via α), on the polarisation energy rather than directly through the overlap energy; this should be kept in mind when comparing different calculations - a 'better' overlap potential may in reality be doing little more than giving an improved description of ε_o.

This failure of the polarisable point ion model to represent ε_o correctly has, of course, been known for a considerable time and its consequences for the lattice dynamics of these crystals have also been well described. The failure is removed in two extensions (i) the 'deformation dipole' model of Szigeti[14], extensively applied in lattice dynamical calculations by Karo and Hardy[15] and (ii) the shell model, extensively employed for the representation of experimental data on lattice vibrations[16]. Both these models allow for the deformation of an ion which occurs when its perfect lattice co-ordination is perturbed (e.g. by the relative motion of anions and cations in an electric field). In the deformation dipole model of an MX compound it is supposed that relative displacement of the anion and cation sublattices by 2ξ leads to formation of an additional moment per ion pair of

$$m_d = (e^* - e)\ \xi\,. \qquad (15)$$

This moment opposes that induced directly upon the ions by the electric field and leads to an apparent displacement polarisability of

$$\alpha = z^2 e^{*2}/p \qquad (16)$$

in place of (12). Values of e*/e are found empirically to lie in the range 0.7 to nearly 1 for the alkali halides. Extension of this model to defect calculations raises questions of the location and representation of the moments m_d in situations of non-uniform distortion and imperfect co-ordination. To some extent they also occur in lattice dynamics where it has generally been found best to assume them to be located entirely on the anions[15], but for defects this still leaves open the question of the proper representation of the model in the vicinity of interstitials and vacancies where the co-ordination is changed. The resulting ambiguities therefore make this model somewhat unattractive for detailed defect studies.

The shell model does not suffer from these ambiguities. It assumes each ion to be composed of a core (charge Xe) and a shell (of charge Ye) coupled together harmonically. Non-Coulombic interactions between adjacent ions are principally shell-shell interactions. Although core-shell and core-core interactions are also often assumed, the simplest shell models with only shell-shell interactions are often quite successful. Indeed when, as in defect calculations, one wishes to go beyond the harmonic approximation to use the simplest shell model has some evident advantages - thus the Born-Mayer form $b \exp(-r/\rho)$ can be used for the shell-shell interaction directly rather than partitioned into several different terms. Since the shell model is a model of the ions and their mutual interactions there are evidently no ambiguities when a defect structure is being considered.

Both the shell and deformation dipole models allow an accurate description of the dielectric constant ε_o which is consistent with the other input data (lattice parameter, compressibility, ε_∞ etc.) and thus are important improvements over the polarisable point ion model. They differ from one another however in that the shell model implies that the non-Coulombic interaction of two ions is changed by their polarisation since it depends (principally) on the relative separation of the shells and not on the positions of the nuclei whereas in the deformation dipole model the non-Coulombic interactions are assumed independent of ionic polarisation. The consequences of this difference for defect calculations have yet to be explored. However it can be seen without much difficulty that the catastrophes inherent in the polarisable point ion model also occur in the deformation dipole model since the $\underline{m}^T \underline{\underline{A}}\, \underline{m}$ term is unchanged, even though additional second order terms in $\underline{\underline{m}}\, \underline{\xi}$ are present. Thus the condition (11) applies here too. By contrast the shell model changes the second order term in $\underline{m}$ from the purely electrostatic and self-energy form $\underline{m}^T \underline{\underline{A}}\, \underline{m}$ by the addition of terms from the non-Coulombic shell-shell interaction, and although these additional terms do not in general prevent such catastrophes they appear to delay their onset and make r_{ij}^{crit} smaller. However the model does not answer the problem of what happens when neighbouring shells strongly overlap. It should nevertheless be mentioned that an elaboration of the shell model - the breathing-shell model[17] which allows for radial deformation of the shell as well as displacement relative to the core - appears a successful innovation in the description of the dynamics of ionic lattices although it has yet to be evaluated for static defect lattices.

III. CALCULATION OF DEFECT ENERGIES

Given the basic specification* of the defect one is interested in, the fundamental problem is to find the minimum energy configuration of the lattice model chosen. For defects bearing a net electrical charge the polarisation of the lattice falls off only slowly, as r^{-2}, and important contributions to the energy of the equilibrium configuration come from regions distant from the defect. For example, the contribution to the polarisation energy of a vacancy defect in an alkali halide coming from that part of the crystal beyond the first 250 ions is typically -½eV. There can be no hope therefore of relaxing just a few ions in the immediate vicinity of the defect and assuming that the rest are undisplaced and unpolarised. Even for defects without net charge (e.g. neutral defects, pairs of oppositely charged defects, etc.) it may be necessary to relax many shells.

---

*The specification of the defect in general will mean not only its description as interstitial, vacancy etc. but, as in the case of colour centres, also its electronic state. The correct calculation of these electronic states introduces methods and problems (e.g. electron correlation) similar to those arising in molecular theory. For recent examples of such quantum mechanical calculations including lattice relaxation see, e.g., Stoneham and Bartram[18] and Norgett[19].

A. General Strategy - Regions I and II

The strategy of all the various approaches to this problem can be described very simply as follows: divide the lattice into two regions, (i) an inner region (I) containing the defect and as many shells of neighbours as can be handled explicitly and (ii) the rest of the crystal (region II) which is assumed to be far enough from the defect that a harmonic or a continuum approximation is valid. Obviously if the harmonic description can be handled accurately the inner region I can be quite small; for example, one such method due to Kanzaki[20] has been applied to a number of vacancy defect problems in each case assuming that it is sufficient to take region I as the vacancy alone - although the error in doing so may be substantial (Tewary[21]).

In this strategy we can evidently write the total energy of the defect lattice as

$$E = E_1(\underline{x}) + E_2(\underline{x},\underline{\xi}) + E_3(\underline{\xi}) \tag{17}$$

where we use the vectors $\underline{x}$ and $\underline{\xi}$ to denote the displacements (both core and shell or equivalently displacements and moments) in regions I and II respectively. We can evidently rearrange terms in $\underline{\xi}$ between E_2 and E_3 so that the harmonic approximation for E_3 takes the purely quadratic form

$$E_3(\xi) = \tfrac{1}{2}\,\underline{\xi}^T\,\underline{\underline{A}}\,\underline{\xi} \tag{18}$$

corresponding to the energy of a distorted region II filled with a perfect undistorted inner region. The term $E_2\,(\underline{x},\underline{\xi})$ then represents the change in interaction between region I and II resulting from the presence of the defect and the distortions. If we minimise E w.r.t. $\underline{\xi}$ we then obtain, as the equilibrium condition for II,

$$\underline{F} = -\frac{\partial E_2(\underline{x},\underline{\xi})}{\partial \underline{\xi}} = \underline{\underline{A}}\,\underline{\xi}, \tag{19}$$

which can be solved formally as long as it is adequate to expand the l.h.s. only as far as the linear terms in $\underline{\xi}$; thus

$$-\frac{\partial E_2}{\partial \underline{\xi}} = \underline{F}^{(o)}(\underline{x}) + \underline{\underline{F}}^{(1)}(\underline{x})\,\underline{\xi} \tag{20}$$

i.e.

$$(\underline{\underline{A}} - \underline{\underline{F}}^{(1)}(\underline{x}))\underline{\xi} = \underline{F}^{(o)}(\underline{x}), \tag{21}$$

or

$$\underline{\xi} \equiv \underline{\underline{G}}\,\underline{F}^{(o)}(\underline{x}), \tag{22}$$

where

$$\underline{\underline{G}} = (\underline{\underline{A}} - \underline{\underline{F}}^{(1)}(\underline{x}))^{-1}, \tag{23}$$

is the perturbed static Green's function.

Methods of explicit solution based on these formal equations have been devised by Kanzaki[20] and by Tewary[21]. In Kanzaki's method equation (21) is Fourier transformed over the crystal lattice. The resulting block-diagonalisation of $\underline{\underline{A}}$ renders the calculation of the transformed displacements

$$\underline{Q}(\underline{q}) = \sum_{l} \underline{\xi}_l\, e^{-i\underline{q}.\underline{R}_l} \tag{24}$$

relatively easy provided that the force $\underline{F}$ is of short range, i.e. has non-zero components only near to the defect. This is not the case for charged defects and the evaluation of the Fourier transform of the force $\underline{F}$ is then far from straightforward. Finally the inversion of the calculated $Q(\underline{q})$ to get the actual displacements $\underline{\xi}$ involves numerical summations over the Brillouin zone. As a result of these substantial numerical problems this method has not been applied widely and only once has it been applied to a charged defect in an ionic lattice[22]. It has however proved convenient for the demonstration of limiting results applicable at large separations[23,24].

The alternative approach due to Tewary[21] calculates the static lattice Green's function directly but only appears to be practicable for localised forces and defects of reasonably high symmetry. It has been applied to vacancies and interstitials in cubic lattices but does not appear to be convenient for charged defects in ionic lattices.

As a result of these difficulties to using accurate harmonic descriptions of region II it has been common to follow the early lead provided by Mott and Littleton[25] and use continuum results as a basis. We first observe that the macroscopic polarisation, $\underline{P}$, at distance $\underline{r}$ from a defect of net charge q is

$$\underline{P} = \frac{1}{4\pi}\left(1 - \frac{1}{\varepsilon_o}\right) \frac{q\underline{r}}{r^3} . \tag{25}$$

The polarisation per unit cell $\underline{P}v_m$ is then divided among the various sources (electronic and displacement) in proportion to the appropriate polarisabilities. In the polarisable point ion model it is divided up directly in proportion to the electronic and displacement polarisabilities (α_+, α_- and α). In the deformation dipole model an additional component is provided by the deformation dipole itself, while in the shell model the partition is slightly different again[26]. This simple description of region II allows the major numerical effort then to go into the evaluation of the equilibrium configuration of region I and into the systematic enlargement of this region. These methods can now be made very general, e.g., they are not limited to simple defects or defects of specified symmetry. Thus as well as stable minimum energy configurations total energy curves corresponding to the displacement of a specified atom or ion through the lattice can also be obtained; this is evidently important in relation to activation energies and metastable configurations.

B. Methods of Minimisation - Region I

When we turn to the methods available for the minimisation (w.r.t. $\underline{x}$) of the energy function $E_1(\underline{x}) + E_2(\underline{x},\underline{\xi})$ there is of course a choice of several well tried approaches and for an up-to-date introduction to the extensive literature on this topic we refer the reader to a recent review by Fletcher[27]. These methods divide into three main groups (i) those which evaluate only the function itself (search procedures) (ii) those which also evaluate the first derivatives of the function (e.g. steepest descents) and (iii) those which evaluate second derivatives as well (variable metric or modified Newton-Raphson methods). We shall comment on the use of these methods in defects calculations.

In the simplest search procedure one seeks the minimum of the energy function by a cycle of linear searches in the variables of the problem, $x_1 \ldots x_n$. This is equivalent to the successive relaxation of equivalent shells of atoms or ions as applied to point defect studies by a number of authors. However the convergence of this method is known to be bad, although in the present application computation times can be reduced ($\sim^n/2$ times) by not evaluating the total energy function at each step but only that part involving the shell of ions being relaxed. Various improvements of this direct search procedure have therefore been developed, of which the most widely used is probably that of Powell[28]. The general idea here is that of a linear transformation to new variables at the end of each cycle which by incorporating the results of that cycle defines new "less strongly coupled" variables. This "conjugate direction" method retains the advantage of not requiring derivatives, but converges satisfactorily. Even so the required number of evaluations of the energy increases as the square of the number of variables, n, and for long-range forces a reasonable practical limit is then reached with $\sim$10-20 variables. These search procedures are convenient for small computers as they do not require large storage.

Methods which utilise derivatives of the energy function, i.e. the forces on the atoms, in general converge faster than search methods, the number of iterations being reduced by a factor $\sim$ n. The calculation of the function derivatives at any point specifies a local optimum direction of search. The choice of a suitable step length in this direction may be made according to several criteria. Thus a linear search may be made to minimise the energy and the various shells of ions may be relaxed independently or collectively - when the method is simply the application of the method of steepest descent. Alternatively, the problem may be related to molecular dynamics. Bullough and co-workers[29] used the calculated forces to solve the Newtonian equations of motion for the atoms in the relaxed region and then quenched out the kinetic energy as it passed through a maximum. This method is claimed to have the advantage of avoiding local minima (i.e. metastable configurations) but it is not clear that in this it is demonstrably better than other gradient methods. In this method, the step length is related to the time step chosen in the iterative solution of the Newtonian equations and this may be chosen to improve convergence. Although no direct comparisons exist between the various methods, the basic similarities suggest that they must have generally similar convergence properties. They have the advantage of requiring relatively little core store.

It is known, however, that the convergence of the steepest descent method is poor and it is also well established in general that it is possible to gain substantially improved convergences within the limitations of this class of methods using the method of Fletcher and Reeves[30]. This does not yet appear to have been applied to defect problems even though it has apparent advantages in circumstances where a large store is not available.

Lastly, we turn to methods derived from the Newton-Raphson method. These incorporate information on the first and second derivatives at each iteration and converge particularly rapidly. The basic equation relating the ith iteration to the (i+1)th is

$$\underline{x}^{(i+1)} = \underline{x}^{(i)} - \underline{\underline{H}}^{(i)}\underline{g}^{(i)} \quad (26)$$

where $\underline{g}$ is the vector of derivatives of the energy and $\underline{\underline{H}}$ the "Hessian" matrix is the inverse of the matrix of second derivatives $\underline{\underline{W}}$. A straightforward application of (26) is however ruled out since the calculation and inversion of the matrix of second derivatives is far too time consuming, as found directly by Sinclair and Pollard[31].

Fortunately, this is not necessary. Thus Sinclair and Pollard[31] evaluated $\underline{\underline{H}}$ at some initial configuration and then used the same value of the inverse during succeeding iterations. Using this method, with a defect system of 150 atoms moving in 2 dimensions, they found that a minimum was obtained after 450 force evaluations while a simpler iterative method required as many as 13,200. However, the use of a constant inverse $\underline{\underline{H}}$ is itself an unnecessary limitation. Methods designed to refine $\underline{\underline{H}}$ during iteration from any well conditioned approximation to the Hessian were initially developed to avoid an initial calculation of the second derivatives[32]. However, in a crystal lattice involving pairwise interactions,the initial evaluation of $\underline{\underline{H}}$ is not difficult since each second derivative involves only one pair of atoms. Thus the following sequence may conveniently be employed for minimisation[33].

(i) The crystal is specified at some convenient initial configuration $\underline{x}^{(0)}$ and first and second derivatives $\underline{g}^{(0)}$ and $\underline{\underline{W}}^{(0)}$ are calculated. $\underline{\underline{H}}^{(0)} = (\underline{\underline{W}}^{(0)})^{-1}$ is also evaluated.

(ii) An improved approximation to the equilibrium configuration is obtained

$$\underline{x}^{(1)} = \underline{x}^{(o)} - \underline{\underline{H}}^{(o)}\underline{g}^{(o)}.$$

(iii) New forces $\underline{g}^{(1)}$ are also calculated and $\underline{\underline{H}}^{(0)}$ is updated by

$$\underline{\underline{H}}^{(1)} = \underline{\underline{H}}^{(0)} + \frac{\left(\underline{x}^{(1)} - \underline{x}^{(0)}\right)\left(\underline{x}^{(1)T} - \underline{x}^{(0)T}\right)}{\left(\underline{g}^{(1)T} - \underline{g}^{(0)T}\right)\left(\underline{x}^{(1)} - \underline{x}^{(0)}\right)} - \frac{\underline{\underline{H}}^{(0)}\left(\underline{g}^{(1)} - \underline{g}^{(0)}\right)\left(\underline{g}^{(1)T} - \underline{g}^{(0)T}\right)\underline{\underline{H}}^{(0)}}{\left(\underline{g}^{(1)T} - \underline{g}^{(0)T}\right)\underline{\underline{H}}^{(0)}\left(\underline{g}^{(1)} - \underline{g}^{(0)}\right)} . \quad (27)$$

whereas before the superscript T denotes the transposed matrix.

(iv) New co-ordinates are obtained then as

$$\underline{x}^{(2)} = \underline{x}^{(1)} - \underline{\underline{H}}^{(1)}\underline{g}^{(1)},$$

and the cycle repeated.

The method has the important advantage that the number of iterations required to converge is only slowly dependent on the number of variables although, of course, since forces as well as energy must be calculated, each iteration takes somewhat longer (typically about twice) than in a search method which evaluates the energy alone. Comparison with search methods is thus very favourable (e.g. 10-20 times faster for 12 variables). The penalty is that considerable storage for the large Hessian matrix is required, so that the method is not suitable for small computers although possibly slow access storage devices would be acceptable. Given that this storage requirement can be met these variable metric methods are speedy and especially suitable for large numbers of variables and for models involving long range forces, e.g. ionic crystals. For short-range forces an advantage will remain, but since W then is a narrow band matrix the gain over force iteration methods, which in effect treat W as a constant diagonal matrix, will be less than for ionic and other polarisable crystals.

Previously, with the slower search methods, the large amount of computing required has meant that it has been possible to study only a limited range of simple symmetric defects. Consequently, it has been usual to regard each particular defect system as an isolated problem and program it accordingly. The application of fast matrix methods means that it is now feasible to consider the calculation of characteristic defect energies and configurations in a wide range of cases by the same program. This new opportunity has led to the development of a comprehensive modular program for ionic crystals, HADES*, which will treat the relaxation of the lattice about an arbitrary defect configuration involving vacancies, interstitials or substitutional atoms. Where appropriate, the defect symmetry is used to minimise the extent of computation but unsymmetric configurations may be studied.

The current versions allow a choice between point polarisable ion and shell model potentials and the program is designed to require only essential data. Thus subsidiary calculations normally necessary, for example, the estimation of Madelung constants and other lattice summations, are carried out as required within the context of the package. The only information required concerns the chosen potential and data specifying the lattice symmetry and the extent of the relaxed regions. Region I is surrounded by an infinite region II described by the Mott-Littleton approximation and the two regions are coupled in a way consistent with the quadratic approximation used in the outer region.

Currently, the lattice-defect interaction is restricted to a particular analytic form since it was designed for defects such as vacancies and closed-shell ions or atoms[34]. However, the modular nature of the program allows the easy alteration of the analytic form of particular interactions. Thus, apart from closed-shell systems (Section IV), we have already used the program in a preliminary study of the migration of the V_k- centre in KCl. In this defect the trapped hole is shared between two Cl^- ions and their interaction is substantially modified from that of two Cl^- ions. Furthermore it is apparent that a further extension of the program to allow the simultaneous determination of defect wave-functions and lattice relaxation is straightforward for those defects where the wave-functions are based on linear combinations of Gaussian atomic orbitals (see e.g. Ref. 19).

For symmetric defects where relatively small numbers of variables, n, may be required the times of computation are very small, e.g., 20 variables requires typically only 1/3 min in total on an IBM 360/75. For the intermediate state of the V_k-centre in a 60° reorientation jump where 228 variables were used the total time required was still less than 5 min. This roughly proportional increase was largely due to the time required for each force evaluation ($\underline{g}$) and associated manipulations - the total number of iterations as expected was insensitive to this increase in n, being only 10.

In concluding this section on numerical methods we might

*Harwell Automatic Defect Evaluation System.

comment that development of efficient methods seems now to be substantially in advance of their application to defect problems.

IV. DEFECTS IN IONIC SOLIDS

It is not difficult to demonstrate the utility of calculations of the properties of defects in ionic solids made by the methods and models described in sections II and III, since there is no doubt that these have often 'hit the nail on the head' in a way which has aided and enlarged the experimental studies. We mention for example: (1) the original Mott-Littleton calculation[25] showing that the dominant defects in alkali halides were Schottky defects; (2) the calculation by Tharmalingam[35] of the activation energies for motion of interstitial ions in alkali halides at a time when the production of interstitial anions in irradiated crystals was being suggested experimentally by low-temperature annealing studies; (3) more recently the confirmation and extension of the models describing the behaviour of rare gas impurities in ionic crystals[34,36]. Other examples of this kind could be quoted. Nevertheless it is also true that many of these calculations have directed attention to the calculation of defect energies already well known experimentally, e.g., the energies of formation and, to a lesser extent, migration of Schottky and Frenkel defects, particularly Schottky defects in the alkali halides. The reason for this is that it was generally felt desirable that the models chosen should evaluate the energy of simple defects correctly before more complex problems were tackled. Unfortunately, the calculated results until relatively recently folded together errors in the model with errors introduced by the method of calculation, e.g., the use of a small region I. The correct inferences to draw from any given comparison of calculation and experiment or from the intercomparison of two calculations were thus always somewhat uncertain (see, e.g., the review by Barr and Lidiard[1]). Recent comparative studies such as that by Faux[7] and the development of fast accurate programs such as the HADES program of Norgett described in the last section, will do much to remove these uncertainties, but until these are more widely applied the full utility of this kind of calculation for the study of defects in ionic crystals will not be realised. However, a good example of the value of calculations of this kind is provided by the studies of the energies of rare gas atoms in ionic crystals and we wish therefore to describe these recent applications in more detail. We conclude by reviewing some of the recent results on Schottky defect energies.

A. Rare Gases in Ionic Crystals

The rare gases by their lack of electron affinity and by their high ionisation potentials form an interesting class of electrically neutral impurity in ionic crystals. By these same characteristics they are highly insoluble, but can be introduced in appreciable quantities as a result of fast neutron induced transmutation reactions when these crystals are exposed in a reactor. There have been many scientific studies of such systems in the last few years, as well as many more technical studies of the behaviour of Kr and Xe in nuclear reactor fuels such as UO_2. There were initially no detailed theoretical predictions of the likely behaviour of these systems but from the experiments a tentative picture of the manner of migration of the gas atoms through the lattice to the surface emerged[37]. It was supposed that the basic mode of migration was as interstitial atoms jumping by thermal activation from one interstitial site to another. The behaviour at elevated temperatures distinguished two temperature regions; from reactor temperature up to several hundred ^{o}C (the low temperature region) the behaviour was supposed to be due to such interstitial migration hindered by trapping into the vacancy defects introduced by the irradiation: the trapping caused the effective activation energy to be high, since the gas had to be released from the trap by thermal activation before it was able to migrate. At higher temperatures beyond this region where now a <u>lower</u> activation energy was observed it was assumed that the radiation induced defects had annealed out and that the interstitials were migrating through essentially perfect lattice. The interpretation, however, posed several questions:

(1) Were the low activation energies ($\sim$ 0.3 eV) observed in the high temperature region in many alkali halides reasonable for the migration of a neutral interstitial atom? It was known that self-interstitial ions and atoms("H-centres") moved with activation energies even lower than these, but in those cases the motion was by an interstitialcy or replacement mechanism in which the interstitial moves to a lattice site and the lattice ion becomes interstitial.

(2) Why was no region of low activation energy seen in KF and RbF?

(3) Although in CaF_2 an analogous division into high and low temperature regions was observed[38] the activation energies were far higher, e.g., $\sim$3 and 6 eV respectively. Was it reasonable to use the same general model and ascribe the much larger activation energies to differences in lattice structure?

Several calculations (the most recent ones using the HADES program described in section III) of the characteristic energies of rare gas atoms in ionic crystals have therefore been made and have not only provided satisfactory answers to the above questions but in so doing have enlarged the above model by demonstrating the importance in some of these systems of trapping by defects other than those introduced by irradiation, e.g., by intrinsic vacancies produced by thermal activation at high temperatures. In so doing these calculations have stimulated the experimental studies and lent precision to their interpretation so that it is now sensible to use rare gas migration experiments as a way of studying the traps. An example of this is provided by the work of Felix and Müller[39] showing how chemical doping affects the gas diffusion rates.

These calculations have initially used polarisable point ion models of the ionic crystals although shell models are currently also being evaluated. In any case with a neutral defect such as a rare gas atom inconsistency with the static dielectric constant is unlikely to be important; the polarisation energy terms are very small.* The least certain aspect of the model was the form of the gas-ion interactions. Here we appealed to the similar electronic structure of rare gas atoms and closed shell ions and estimated the interaction either from corresponding ion-ion interactions or from gas-gas interactions (empirical and calculated). The two sets of results are close to one another and give confidence in our choices therefore. The computations with these models were made by taking region I to contain up to ∿20 shells of ions while region II was either held fixed ($\underline{\xi} = 0$) or described in a Mott-Littleton approximation when dealing with a charged defect, e.g., a gas atom trapped in a vacancy. Some of the results are reproduced in Tables 1-4. These show that:

(1) The absolute energies of the interstitial gas atoms in their stable cube-centre positions are high - 1-3 eV in the alkali halides, higher still in CaF_2 etc. This is consistent with the known low solubilities of these gases.

(2) In the alkali halides the activation energy for migration of the interstitial gas atoms is generally low while in CaF_2, etc. it is high. The difference is due to the F^--F^- contact in the CaF_2 lattice which imprisons the gas atom more tightly at the centre of the F^- cube. These results are thus in good agreement with the experimental results - except that they offer no explanation for the exceptional behaviour of the systems KF:Ar and RbF:Kr.

---

*Preliminary comparison with a shell model of CaF_2 indicates only rather small changes from the polarisable point ion predictions, even for vacancy defects.

Table I

Absolute energies (eV) of gas atom interstitials in alkali halides and their activation energies of motion. The range of values given correspond to two different choices of gas-ion interaction potential. The experimental values are those characterising gas diffusion in the high temperature region except for KF:Ar and RbF:Kr where only one region is observed.

System	Absolute Energy	Activation Energy	Activation Energy (expt)
KF:Ar	1.98 - 2.38	0.19 - 0.26	1.8
KCl:Ar	1.26 - 1.63	0.15 - 0.22	0.38
KBr:Ar	1.08 - 1.44	0.18 - 0.26	0.36
KI:Ar	0.81 - 1.12	0.27 - 0.34	0.30
RbF:Kr	2.71 - 3.01	0.22 - 0.22	1.38
RbCl:Kr	1.47 - 1.77	0.13 - 0.19	0.56
RbBr:Kr	1.22 - 1.51	0.16 - 0.24	0.30
RbI:Kr	0.95 - 1.22	0.22 - 0.30	0.31

Table II

Absolute energies (eV) of gas atom interstitials in the alkaline earth fluorides and their activation energies of motion. The range of values given corresponds to two different choices of gas-ion interaction potential. The experimental values[38,40] are those characterising gas release at high temperatures, but in SrF_2 the correct value for comparison is uncertain.

System	Absolute Energy	Activation Energy	Activation Energy (expt)
CaF_2:Ar	2.96 - 3.61	2.22 - 2.47	3.0
SrF_2:Kr	3.82 - 3.99	2.20 - 2.39	(1.2 - 3.7)
BaF_2:Xe	3.62 - 4.56	2.56 - 2.96	3.5 - 3.7

(3) The explanation for this exceptional behaviour is to be found in the trapping of the gas atoms into thermally produced vacancy defects, in the same way as high activation energies are found at low temperatures in the other systems as a result of trapping into radiation-induced defects. From a simple application of Maxwell-Boltzmann statistics it is easy to show that the criterion for significant trapping into thermally produced defects is

$$B - h_f > 0$$

where B is the energy gained when the interstitial gas atom drops into the vacancy and h_f is the effective energy of formation (i.e. the vacancy fraction is set $\sim \exp(-h_f/kT)$). For Schottky vacancies h_f is one half the Schottky energy whichever vacancy is being considered. Tables 3 and 4 show the quantity $B - h_f$ for several systems and it will be seen how trapping into vacancy pairs is particularly important in KF:Ar and RbF:Kr. This trapping is

Table III

Values (in eV) of the binding energy, B, of gas atoms into various vacancy defects in the alkali halides minus the corresponding (effective) vacancy formation energy, h. The small differences between the two models have here been averaged out in view of corresponding uncertainties in the values of h. When B - h is negative trapping is unimportant at high temperatures; otherwise B - h must be added to the interstitial migration energy to obtain the effective activation energy, Q, for gas diffusion.

System	B - h			Q
	Cation Vacancy	Anion Vacancy	Vacancy Pair	
KF:Ar	0.5	-0.8	1.0	1.2
KCl:Ar	0.2	0.1	0.2	0.4
KBr:Ar	-0.2	-0.1	-0.1	0.2
KI:Ar	-0.15	0.0	-0.2	0.3
RbF:Kr	1.5	-0.7	1.9	2.1
RbCl:Kr	0.5	0.3	0.5	0.7
RbBr:Kr	0.4	0.3	0.2	0.6
RbI:Kr	0.1	0.2	0.0	0.5

Table IV

Values (in eV) of the binding energy, B, of gas atoms into vacancy defects in the alkaline earth fluorides minus the corresponding (effective) vacancy formation energy, h. The range of values corresponds to the same two models as in Table II, but these small model differences in Q have been averaged out.

System	B - h		Q
	Anion Vacancy	Cation Vacancy	
CaF_2:Ar	0.0 - 0.2	-0.1 - 0.3	2.5
SrF_2:Kr	0.2 - 0.2	0.7 - 0.8	3.0
BaF_2:Xe	-0.2 - 0.0	0.3 - 1.2	3.5

the reason why KF and RbF appeared anomalous. The results also show that trapping is responsible for a part of the measured Q in other cases too. In CaF_2 where the dominant thermal defects are anion Frenkel defects there is little effective trapping into F^- vacancies in the high temperature region. The much larger Ca^{2+} vacancies - a minority defect in this system - offer deeper traps and even though they are fewer in number it is seen that they can have the greater effect. Indeed by their consideration we obtain not only a correct qualitative description of these systems but one that is surprisingly quantitative too. The calculations are now being extended also to the caesium halides, in particular CsCl where interesting changes in the relative rates of diffusion of different gases are observed as one passes through the α-β transition.

B. Energies of Schottky Defects

As we have previously mentioned many previous calculations have been directed to the evaluation of the formation energy of Schottky defects and it is appropriate therefore to conclude by reference to the recent results of Faux comparing a polarisable point ion model with a shell model for varying region I.

Firstly in Table 5 we show the variation of Schottky formation energy for a polarisable point model of NaCl (Tosi-Fumi overlap potential) as the size of region I is enlarged. Region II was described in a Mott-Littleton approximation supplemented by a self-consistent description of elastic distortion based on (correction of)

Table V

Values (in eV) of the predicted formation energy, h_S, of Schottky defects in NaCl for a polarisable point ion model and various sizes of region I.

Shells of ions included in region I	h_S	h_S (expt)
(100)	1.56	
(100), (110)	1.65	
(100), (110), (111)	1.64	
(100), (110), (111), (200)	1.39	2.18 - 2.50
(100), (110), (200)	1.39	
(100), (110), (200), (300)	1.19	
(100),(110),(200),(300),(400)	1.08	

the limiting results obtained by Hardy and Lidiard for the elastic strength of a vacancy in an ionic crystal[23]. It will be seen that the predicted Schottky energy falls quite rapidly as region I is enlarged and, as we have previously commented, analysis shows that this is due to the failure of the model to predict ε_o correctly. Indeed this must be the reason why many previous calculations (all made with small region I) have also predicted values which were too low.

Secondly in Table VI we show the corresponding results for a simple 5-parameter shell model fitted to r_o, c_{11}, c_{12}, ε_∞ and ε_o. It will be seen that in this case the predicted energies are very stable against variations in the size of region I - and in better agreement with experiment. The lesson seems rather plain; only by use of models which give an accurate description of the dielectric as well as the elastic behaviour can one hope to make reliable predictions of the energies of charged defects in ionic crystals. Although other refinements and elaborations of the ionic model are required the above condition appears the minimum base on which to build.

Table VI

Values (in eV) of the predicted formation energy, h_S, of Schottky defects in NaCl for a 5-parameter shell model and various sizes of region I.

Shells of ions included in region I	h_S	h_S (expt)
(100)	2.23	
(100), (110)	2.25	
(100), (110), (111)	2.26	2.18 - 2.50
(100), (110), (200)	2.27	
(100), (110), (200), (300)	2.24	
(100),(110),(200),(300),(400)	2.24	

V. CONCLUSION

The general conclusions which we draw from this survey are (1) that certain improvements are necessary in the basic physical models used, and (2) that by using efficient minimisation methods much greater advantage could be taken of existing computers especially where these have core stores large enough to apply variable metric methods. The quasi-harmonic approximation is the basis of all these calculations at the present time and while this may be a source of some error it is difficult to accept or present evidence that it is seriously misleading. Nevertheless the study of the temperature dependences implicit in the quasi-harmonic approximation could well be valuable.

When we turn to specific details of these ionic models our examination has shown the importance of correctly describing di-electric properties accurately. Shell models have evident advantages in this and other respects. Overlap potentials in the alkali halides appear to be well enough described by the Born-Mayer form and are well parameterised by the work of Fumi and Tosi[41]. However, in the oxides some basic questions about oxygen-oxygen interactions and even about the oxygen ion itself remain. Shell models can evidently be enlarged beyond their usual harmonic form by assuming a Born-Mayer interaction when two ions undergo a large relative displacement. Nevertheless in all models there remains the difficulty of describing the polarisability of ions when their electron clouds begin to overlap appreciably; the plausible suggestion of Quigley and Das[13] should be evaluated and applied more widely. Lastly, although there presently are no strong indications of their importance it would seem very desirable to study the role of 3-body interactions in these defect calculations. These interactions describe the departures from the Cauchy relation[42,43].

On the mathematical side there is clearly much to be gained from the application of variable metric methods when a large computer is available. Even when this is not the case there is probably considerable gain over the frequently used search methods to be obtained from applying force iteration methods[30] and, as the work of Faux on shell models indicates, there may be many problems for which is is not necessary to take a large region I in order to achieve accurate and useful results. Clusters of defects and defects with low symmetry will remain the province of those with access to a large computer, but even then only when coupled with a fast efficient minimisation procedure.

REFERENCES

1. L. W. Barr and A. B. Lidiard, Physical Chemistry - an Advanced Treatise, Vol. 10, (Academic Press, New York and London, 1970) p.151.

2. J. R. Hardy and J. W. Flocken, CRC Critical Reviews in Solid State Sciences, 1, 605 (1970).

3. See e.g., C. P. Flynn, Z. Naturforsch. 26a, 99 (1971) for ionic crystals and M. L. Klein, G. K. Horton and J. L. Feldman, Phys. Rev., 184, 968 (1969) for rare gas solids.

4. A. R. Allnatt and L. A. Rowley, J. Phys. Chem. Solids, 30, 2187 (1969).

5. See e.g., A. R. Allnatt and P. Pantelis, Solid State Commun., 6, 309 (1968); R. G. Fuller, M. H. Reilly, C. L. Marquardt and J. C. Wells, Phys. Rev. Letters, 20, 662 (1968); A. R. Allnatt, P. Pantelis and S. J. Sime, J.Phys.C. 4, 1778 (1971).

6. J. Mahanty and M. Sachdev, J. Phys. C., 3, 773 (1970).

7. I. D. Faux and A. B. Lidiard, Z. Naturforsch., 26a, 62 (1971); I. D. Faux, Ph.D. thesis, University of London (1971).

8. See e.g., M. P. Tosi, Solid State Physics, 16, 1 (1964).

9. D. W. Lynch, J. Phys. Chem. Solids, 28, 1941, (1966).

10. J. R. Tessman, A. H. Kahn and W. Shockley, Phys. Rev. 92, 890 (1953).

11. J. Pirenne and E. Kartheuser, Physica 30, 2005 (1964).

12. I. D. Faux, J. Phys. C., 4, L 211, (1971); also Ref. 7.

13. R. J. Quigley and T. P. Das, Solid State Commun., 5, 487 (1967).

14. B. Szigeti, Proc. Roy. Soc. A 204, 51 (1950).

15. J. R. Hardy, Phil. Mag., 7, 315 (1962); A. M. Karo and J. R. Hardy, Phys. Rev. 129, 2024 (1963) and J. Chem. Phys., 48, 3795 (1968).

16. See e.g., A.D.B. Woods, B. N. Brockhouse, R. A. Cowley and W. Cochran, Phys. Rev., 131, 1025, (1963) (NaI and KBr) ; G. Peckham, Proc. Phys. Soc., 90, 657 (1967) (MgO) ; J. S. Melvin, J. D. Pirie and T. Smith, Phys. Rev., 175, 1082 (1968) (Na halides).

17. U. Schröder, Solid State Commun., 4, 347 (1966); U. Nüsslein and U. Schröder, Phys. Stat. Solidi, 21, 309 (1967); J. S. Melvin et al. Ref. 16.

18. A. M. Stoneham and R. H. Bartram, Phys. Rev., B2, 3403, (1970).

19. M. J. Norgett, J. Phys. C., 4, 1289 (1971).

20. H. Kanzaki, J. Phys. Chem. Solids, 2, 24 and 37 (1957).

21. V. K. Tewary, AERE Report TP 388 (1969).

22. A. M. Karo and J. R. Hardy, Phys. Rev. B3, 3418 (1971).

23. J. R. Hardy and A. B. Lidiard, Phil. Mag., 15, 825 (1967); for corrections see Ref. 7.

24. J. R. Hardy and R. Bullough, Phil. Mag., 15, 1237 (1967) and 16, 405 (1967); R. Bullough and J. R. Hardy, Phil. Mag., 17, 833 (1968).

25. N. F. Mott and M. J. Littleton, Trans. Faraday Soc., 34, 485 (1938).

26. I. D. Faux, Ref. 7.

27. R. Fletcher, Computer Aided Engineering (Univ. of Waterloo Press, 1970) p.123.

28. M.J.D. Powell, Comput.J., 7, 155 (1964).

29. R. Bullough, these proceedings.

30. R. Fletcher and C. M. Reeves, Comput.J., 7, 149 (1964).

31. J. E. Sinclair and H. F. Pollard, Phys. Letts. 32A, 93 (1970).

32. R. Fletcher and M.J.D. Powell, Comput.J., 6, 163 (1963).

33. R. Fletcher, Comput.J., 13, 317 (1970) and M. J. Norgett and R. Fletcher, J. Phys. C., 3,L190 (1970).

34. M. J. Norgett, J. Phys. C., 4, 298 and 1284 (1971).

35. K. Tharmalingam, J. Phys. Chem. Solids, 24, 1380 (1963) and 25, 255 (1964).

36. M. J. Norgett and A. B. Lidiard, Phil. Mag., 18, 1193 (1968); also Radiation Damage in Reactor Materials, Vol.1 (IAEA, Vienna 1969) p.61.

37. For references to key experimental papers up to 1968 see Ref.36. See also F. Felix and M. Müller, Phys. Stat. Sol. (b) 46, 265 (1971) for later work.

38. F. Felix and S.Y.T. Lagerwall, Phys. Stat. Sol. (a) 4, 73 (1971).

39. Ref. 37.

40. P. Schmeling, Phys. Stat. Sol. 20, 127 (1967).

41. F. G. Fumi and M. P. Tosi, J. Phys. Chem. Solids, 25, 31 and 45 (1964).

42. P. O. Löwdin, Adv. Phys., 5, 1 (1956).

43. E. Lombardi, L. Jansen and R. Ritter, Phys. Rev., 185, 1150 (1969).

COMPUTER SIMULATION OF POINT AND LINE DEFECTS IN IRON AND COPPER

R. Bullough

Theoretical Physics Division, A.E.R.E, Harwell, Didcot,

Berkshire, ENGLAND

1. INTRODUCTION

Computational techniques are playing an ever-increasing role in the calculations of both the electronic and structural properties of defects in solids. Although all such calculations should ideally be done within a fundamental quantum mechanical framework, the complex nature of the atomic configuration associated with most defects has precluded this and many defect properties cannot yet be discussed with a fundamentally satisfactory model. The situation is particularly acute when a principle objective of the calculation is the determination of the atomic configuration in the neighborhood of the defect or of a property that depends directly on the configuration. To deal with such problems a procedure which we shall refer to as computer simulation has been devised and applied to various defect solids. The present paper will outline the essential features of such procedures and give some justification for the approach. In fact, we shall see that the main justification follows by looking at some specific examples where computer simulation has provided explanations of phenomena that could not otherwise be explained.

The computer simulation can be of either a static[1-6] or dynamic[7,8] defect solid and in either case consists essentially of setting up a large assembly of discrete interacting atoms and allowing some interatomic potential to prevail between the atoms. This potential, which for reasons of computational feasibility must be as simple as possible, should at the very least hold the assembly in its perfect crystal equilibrium configuration and correctly define the (linear) response of the system from the equilibrium configuration. The defect is then introduced into the crystallite and the atoms are allowed to relax under the potential into the defect configuration. By following this configuration and its energy variation as a function of defect position, various properties of the defect can be easily deduced. It is clear that the most difficult part of such a calculation to justify is the construction of an interatomic potential that is relatively simple and is at the same time appropriate for both the perfect and defect body. We shall see in Section 2 that for metals such potentials can be best constructed by forcing an empirical fit to as much physical information as possible.

The actual relaxation procedure is briefly described in Section 3 and then two specific examples are given of point and line defects and their interactions in iron[3] (Section 4) and copper[4,5] (Section 5). Both these are strictly static defect calculations but do serve to illustrate and justify the procedure. Finally in Section 5 we discuss very briefly the relevance of computer simulation to the overall study of defects in crystalline solids.

2. THE POTENTIAL CONSTRUCTION

It is well known that a strictly pairwise interatomic potential is not valid for metals.[9,10] Nevertheless such potentials have been frequently used in defect calculations, the hope being that by suitably parametrizing such potentials the volume-dependent electronic contributions can be somehow subsumed into the pairwise interaction. It is clear at the outset that such a procedure is untenable if the pairwise interaction is solely responsible for the equilibrium of the lattice. For example, such a central force description requires the elastic constants to satisfy the Cauchy relations and since in general these relations are not obeyed, an equilibrium pair potential model cannot be even consistent with the linear elastic model. It therefore cannot be an improvement on the elastic model and its use must be regarded as of dubious value. For this reason we believe that equilibrium potentials such as the Morse potential should cease to be used in defect calculations. A comprehensive survey and critique of the interatomic potentials used in defect calculations has recently been given by Hardy and Flocken[11] and we will not therefore attempt to review the situation; it will suffice to outline the potentials for α-iron and copper constructed by Johnson[2] and Englert, Tompa and Bullough,[4] respectively and extensively used for defect calculations on these metals. Both these potentials are pairwise but implicitly contain the volume-dependent term in the lattice energy by not requiring the lattice to be in equilibrium under the pair potential lone.

Johnson has published two useful potentials (I and II) for α-iron.[2] By the method of long waves with an axially symmetric force constant model, the elastic constants C_{ij} for a bcc lattice can be easily related to the first and second derivatives of the pair potential V(r) at the first and second neighbor equilibrium sites:[12]

$$C_{11} = \frac{2}{3r_2}\left[\frac{2V'}{r_1}(r_1) + V''(r_1) + 3V''(r_2)\right]$$

$$C_{12} = \frac{2}{3r_2}\left[-\frac{4V'}{r_1}(r_1) + V''(r_1) - \frac{3V'}{r_2}(r_2)\right] \quad (1)$$

$$C_{44} = \frac{2}{3r_2}\left[\frac{2V'}{r_1}(r_1) + V''(r_1) + \frac{3V'}{r_2}(r_2)\right] \quad .$$

To construct his potential I, Johnson[2] fitted the right-hand side of (1) to the short-range elastic moduli (the long-range electronic contributions were removed by the method due to Fuchs using a free-electron model). The potential was represented by three splines such that V(r) was set to zero with zero alope and curvature midway between second and third neighbors and was matched at short range to the radiation damage potential of Erginsoy, Vineyard and Englert.[13] Later Johnson[2] considered that in view of the semi-empirical nature of this potential it was questionable whether or not the electronic contribution should be subtracted from the experimental elastic constants and he therefore constructed potential II by fitting the right-hand side of (1) to the total elastic constants. The short-range elastic moduli used in the construction of potential I satisfy the Cauchy relations and thus potential I is a central force equilibrium potential and the iron lattice is in equilibrium, without external constraint, at the observed lattice spacing. The volume dependence has thus been included by direct modification of the elastic constants. In contrast, potential II will not hold the iron lattice at its equilibrium spacing; the total elastic moduli violate the Cauchy relations, and an external constraint must be applied to maintain the lattice in equilibrium at the correct lattice spacing. This implicit external pressure now represents the volume-dependent part of the total lattice energy. In practice, for α-iron, there is very little difference between the two potentials; they can both be written in the form

$$V(r) = A(r-r_0)^3 + B\,r + C \quad , \quad (2)$$

where A, B and C are given in Table I for the various ranges of r. In fact, of course, there is appreciable cohesive energy which is not accounted for by potential II and it is tacitly assumed that this energy plays only a minor role in determining the properties of defects; thus it is not yet clear whether the long-range oscillations, which can describe such additional forces, will be significantly distorted in the neighborhood of a defect.

TABLE I. The Coefficients in Johnson's α-Iron Potentials[2] (Eq. 2)

Potential	Range	A	B	C	R_0
I	<2.40	-2.195976	2.704060	-7.436448	3.097910
	2.40-3.00	-0.639230	0.477871	-1.581570	3.115829
	3.00-3.44	-1.115035	0.466892	-1.547967	3.066403
II	<2.40	-4.719041	-0.395690	0.809049	2.569932
	2.40-3.00	-0.886887	0.437641	-1.450695	3.083269
	3.00-3.44	-1.057432	0.434665	-1.441617	3.069839

Most of the point defect calculations reported by Johnson were performed using potential I and, as we shall described in Section 3, our point defect-dislocation[3] calculations also made use of potential I. Potential II has been used by Gehlen[14] to discuss the self energy of a dislocation; it gives strain energy results that are somewhat easier to correlate with the long-range elastic results because it involves the total elastic moduli and the correlation is direct. A particular advantage of potential I is that is somewhat fortuitously yields very sensible absolute vacancy and interstitial formation energies;[2] although in fairness to Johnson, he is very careful to emphasize that these potentials can only be of value in investigating relative energies such as interaction or migration energies of defects and the "good" formation energies are simply a bonus and not to be taken too seriously. It should be noted that these potentials were not fitted to the phonon dispersion data although, again rather fortuitously, both potentials yield phonon dispersion curves in close agreement with observation.[15] In addition, no stacking fault energy information was included and therefore, conceivably, the bcc lattice could be unstable under inhomogeneous deformation with these potentials. This possibility is particularly important when the potential is being used to study the core configuration of dislocations (as in the copper work described in Section 4) since unstable dissociation can occur if any of the possible stacking faults have negative energy. In fact, to deliberately safeguard against this possibility Bullough and Perrin[16] have constructed a longer-ranged potential for α-iron which has defined stacking fault energies and is carefully fitted to the dispersion data for iron. This potential is constructed from a set of quintic polynomials and has been used to discuss the stability of cavities in α-iron. It has not yet been used to study point or line defects and will therefore not be discussed further here.

The copper potential was originally constructed by Englert and Tompa[17] and subsequently used by Englert, Tompa and Bullough[4] and Perrin, Englert and Bullough[5] to discuss various point and line defects in copper. It consists of ten splines and its analytic form is given in Table II. In essence this potential is analogous to

TABLE II. The Coefficients of the Pair Potential for Copper, Given by a Spline Function Consisting of 10 Cubic Equations $V(r) = A_K(r-r_K)^3 + B_K(r-r_K)^2 + C_K(r-r_K) + D_K$ each valid for $r_K \leqslant r \leqslant r_{K+1}$. V is in eV and r_K in Å.

K	r_K(Å)	A_K	B_K	C_K	D_K
1	1.	-667.9458	1081.8838	-628.5649	138.1100
2	1.5	-49.0449	79.9655	-47.6408	10.8050
3	2.0	-3.2382	6.3981	-4.4591	0.8453
4	2.551	-0.258148	1.045285	-0.357854	-0.210930
5	3.061199	-2.221407	0.650164	0.507164	-0.155699
6	3.341810	1.507669	-1.219882	0.347295	-0.011272
7	3.607658	-0.080144	-0.017445	0.018353	0.023168
8	4.209149	2.186182	-0.162063	-0.089620	0.010455
9	4.311190	-1.575972	0.507171	-0.054405	0.001945
10	4.418461	0.0	0.0	0.0	0.0

Johnson's potential II in that the derivatives of the potential are fitted to the total elastic constants via the fcc relations equivalent to (1):[12]

$$
\begin{aligned}
C_{11} &= \frac{\sqrt{2}}{r_1}\left[\frac{V'}{r_1}(r_1) + V''(r_1) + 2V''(r_2)\right] \\
C_{12} &= \frac{\sqrt{2}}{r_1}\left[-5\frac{V'}{r_1}(r_1) + V''(r_1) + \frac{2V'}{r_2}(r_2)\right] \qquad (3) \\
C_{44} &= \frac{\sqrt{2}}{r_1}\left[3\frac{V'}{r_1}(r_1) + V''(r_1) - \frac{2V'}{r_2}(r_2)\right] .
\end{aligned}
$$

However, it was also carefully fitted to other physical data: (a) to the force constant data[18] which provided the best fit to the phonon dispersion data of Sinha;[19] (b) to the best value of vacancy formation energy (1.09 eV); (c) to the experimentally observed intrinsic stacking fault energy (70 ergs/cm^2); and (d) to the Born-Mayer radiation damage potential determined by Gibson et al.[1] for interatomic distances less than the first neighbor separation. Finally, it was truncated at the third neighbor with zero slope and curvature and together with the volume-dependent term it was constrained to hold the copper lattice in equilibrium at the correct lattice parameter (3.608 Å). A particular feature of this potential is its oscillatory form between the second and third neighbor. The presence of such an oscillation is consistent with the long-range potentials deduced by second-order perturbation theory with a pseudopotential;[10] it has appeared in the present empirical construction as a direct result of the constraint imposed by the fit to the stacking fault energy ((c) above).

3. THE RELAXATION PROCEDURE

Defect simulation involves setting up a large parallelepiped of discrete interacting atoms. The atoms are first arranged to form the desired perfect lattice by allowing the pairwise potential to prevail between the atoms with appropriate constraints on the surface of the atomic assembly. The defect is then introduced into the assembly and suitable boundary conditions are imposed. The atomic configuration associated with the defect is then found by integrating the classical equations of motion for the complete set of interacting atoms[1,2,3]

$$x_\ell^{ijk}(t) = m^{-1} \sum_{\substack{i',j',k' \\ \neq i,j,k}} F_\ell(\underline{x}^{ijk}(t) - \underline{x}^{i'j'k'}(t)) \quad , \qquad (4)$$

where x_ℓ^{ijk} (ℓ=1,2,3) are the Cartesian components of the position vector $x^{ijk}(t)$ of an atom at time t relative to a fixed laboratory system of axes; the triplet of integers ijk is a convenient labelling system to identify a particular atom, and F_ℓ are the components of the net force on the (i,j,k)th atom from the (i'j'k')th atom: thus in obvious notation,

$$F_\ell(\underline{r}-\underline{r}') = \partial V(|\underline{r}-\underline{r}'|)/\partial x_\ell \quad . \qquad (5)$$

Several different relaxation procedures have been used to solve such large sets of differential equations. These range from static atom by atom relaxation[2,20] with or without symmetrization to fully dynamic relaxation[3] in which the finite difference form of (4) are solved and the configuration is periodically "quenched" as the total kinetic energy passes through a maximum. This dynamic procedure has been used in the two examples described in the following sections;[3,5] it ensures very rapid convergence to the absolute minimum in total potential energy and has the distinct advantage that metastable configurations can usually be avoided and the true stable configuration found.

4. INTERSTITIAL AGGREGATION IN α-IRON[3]

When certain bcc metals, including α-iron, are irradiated with neutrons, the displaced interstitial atoms are observed to aggregate in the form of planar circular platelets or edge dislocation loops oriented on {111} planes.[21,22] To understand why the aggregation occurs with this particular morphology involves a knowledge of the mode of nucleation of the aggregate; thus we wish to know: what is the form of the interstitial aggregate nuclei and at what stage in its growth and by what process does it become the observed dislocation loop?

It is clear that these questions involving the motion and interactions of several defects can only be answered by using a discrete atom model of the kind we have been discussing. Thus an atomic parallelepiped containing almost 6000 atoms interacting under the Johnson potential I (Table I) was set up[3] and the interstitial aggregation process was simulated by using the dynamic relaxation procedure. To avoid any spurious effects of possible boundary constraints on the defect configuration the equilibrium configuration was first obtained with the surface atoms held rigid in their perfect crystal positions and then forces were imposed on these atoms to simulate the surrounding infinite crystal. The whole assembly, including the surface atoms, was then allowed to further relax and the forces on the latter were raised in direct proportion to their subsequent displacements. In this way the internal atomic configuration was not prejudiced.

The procedure was to successively insert interstitials near the center of the assembly, relaxing completely between each addition, and allow the aggregation to proceed. We found, in agreement with Johnson's original work on the single interstitial,[2] that the isolated interstitial adopts a split dumbbell configuration along [110] axis. When subsequent interstitials were put in the vicinity of this interstitial they were found to aggregate with all their axes parallel to the same [110] axes. At this stage we were able to identify the nucleation plane as the (110) plane and it was apparent that the "two sides" of the nucleus were beginning to shear over to try to remove the high-energy stacking fault across the nucleus. The transition to the observed morphology actually occurred when the nucleus had incorporated sixteen interstitials. At this stage the (110) fault completely shears over and the rhombus-shaped aggregate became a glissile dislocation with a [111] Burgers vector. This loop then lowers its self energy by slipping on its {110} faced glide prism into a pure edge orientation such that it lies on the observed (111) plane.[3,23] Thus the simulation study provided a consistent and detailed explanation of the origin of the interstitial {111} dislocation loops observed in irradiated α-iron.

5. DISLOCATIONS AND POINT DEFECTS IN COPPER[4,5]

The direct simulation method is particularly useful for the study of dislocations in crystalline solids since the complex topological features of such defects preclude the use of the somewhat simpler lattice statics method[18,24,25] that have proved so valuable in the study of point defects in metals and ionic solids. To obtain the atomic configuration associated with a straight dislocation in copper it is again necessary to set up a parallelepiped of discrete atoms such that the atoms form a perfect face centered cubic lattice with the copper lattice spacing subject to the interatomic potential given in Table II. The faces of the parallelepiped were appropriate

{110}, {111} and {112} crystallographic planes and the dislocation was arranged to lie through the center of the assembly and orthogonal to the two {112} faces. In the direction of the dislocation line, specifically the $[11\bar{2}]$ direction, the assembly was only six lattice planes thick and periodic boundary conditions were imposed across these two $(11\bar{2})$ faces; the dislocation was thus automatically long and straight. In order to accommodate the long-range strain field of the dislocation, the assembly was made as extensive as possible in the two directions orthogonal to the dislocation line. About 900 atoms were actually relaxed and the required dislocation configuration was imposed on the initially perfect lattice by giving the boundary layer of atoms the appropriate anisotropic elastic displacements.

Several important dislocations and their interactions with point defects have been studied. It will suffice to describe briefly here some of the results for the dissociated pure edge dislocation. This is one of the common glissile dislocations in an fcc metal like copper[26] and clearly an understanding of its properties will provide useful insight into the overall deformation properties of copper. It lies along the $[11\bar{2}]$ direction with a total Burgers vector $\underline{b} = \frac{a}{2}[\bar{1}10]$ and can dissociate into a pair of Shockley partial dislocations by the reaction:[26]

$$\frac{a}{2}[\bar{1}10] \quad = \quad \frac{a}{6}[\bar{1}2\bar{1}] + \frac{a}{6}[\bar{2}11] \quad . \tag{6}$$

The equilibrium separation of these partials arises from a balance between the elastic repulsion between the partials and the attraction from the stacking fault separating them. In the present atomic model this separation was obtained by a self-consistent method which involved repeatedly recalculating the boundary displacements as the partials move towards their equilibrium separation. It is essential, because of the unavoidable restrictions on the size of the assembly, that the boundary setting should be exactly consistent with the configuration near the center of the dislocation. With an initial guessed separation of 8b the iterative process converged to give a a final partial separation of 9.2b.

An important result of this study of the glissile edge dislocation in copper was the observation that the individual partials increased their "Peierls" widths from an elastic width of about 2b to just over 5b. A result which is in striking agreement with the known ductility of copper; the critical shear stress to move a dislocation is an exponentially decreasing function of the "Peierls" width.[26] Also a careful study of the tensile strains in the slip plane of the dislocation shows that, in contrast to the elastic model, the strains have an oscillatory form; a result which is in beautiful agreement with Parson's recent high-resolution electron microscopy observations of dislocation cores in such metals.[27]

Finally, the model has been used to calculate the interaction energy between intrinsic point defects (vacancies and interstitials) and dislocations. To do this it was necessary to extend the parallelepiped in the dislocation direction and then drop the periodic boundary conditions. The boundary atoms of a smaller three-dimensional subassembly were held in their previous relaxed dislocation positions and the point defect was placed near the center of the subassembly; the internal atoms were then relaxed and the interaction energy deduced. The position of the point defect relative to the dislocation was then varied by simply changing the location of this atomic subassembly relative to the dislocation itself. These calculations have shown that interstitials have a much larger interaction with dislocations than do vacancies; a not unexpected result but nonetheless gratifying since, for example, such a preferential interstitial attraction was a fundamental hypothesis of our theoretical explanation of void growth in metals irradiated to a high dose with fast neutrons.[28] Of particular interest to dislocation theory was the observation that at very short range the vacancies are bound to the dislocation by the second-order inhomogeneity interaction, whereas at distances only one neighbor separation from the slip plane the first-order size effect appears to dominate and the vacancies are repelled from the dilated regions and attracted to the compressed regions.[29] Furthermore, it was found that the point defects interacted strongly with the entire faulted region and not just with the partials themselves. The interstitial interaction was dominated completely by the size effect interaction with strong attraction to the dilated regions.[29]

6. DISCUSSION

The two examples of computer simulation described in Sections 4 and 5 were chosen to illustrate the kind of enhancement of our understanding of defect properties that the procedure can give us. It is important to realize that the simulation procedure is particularly valuable for the exposure and clarification of complex geometrical features of defects and their interactions. It cannot give us really quantitative information about defects in α-iron or copper but it does, for example, indicate the geometrical form of the single interstitials and the morphology of the interstitial aggregate in α-iron and in copper it does indicate the basic reason for the ductiligy of copper and many other interesting features of dislocations in such a metal. In the future, it is clear that computer simulation will be best directed to clarifying such essentially structural problems. Such problems are particular y prevalent in the field of dislocation theory. For example, what is the structure of a jog; how exactly do dislocations intersect at short range; the mobility of a tilt boundary depends on the widths of the individual dislocations in the boundary--how does this width vary with orientation across the boundary; how do glissile martensitic interfaces actually transform one phase into another; etc., etc. All these problems and many others including dynamic properties of defects can be investigated with the aid of computer simulation methods and, of course, with the inevitable development of our ability to construct reliable and more physically meaningful interatomic potentials, the quantitative content of the results of such simulations must increase.

REFERENCES

1. Gibson, J. B., Goland, A. N., Milgram, M. and Vineyard, G. H., Phys. Rev. 120, 1229 (1960).

2. Johnson, R. A., Phys. Rev. 134, A1329 (1964); and Phys. Rev. 145, 423 (1966).

3. Bullough, R. and Perrin, R. C., Proc. Roy. Soc. A305, 541 (1968).

4. Englert, A., Tompa, H. and Bullough, R., Fundamental Aspects of Dislocation Theory. N.B.S. Special Pub. 317, 1, 273 (1970) (ed. by J. A. Simmons, R. deWit and R. Bullough).

5. Perrin, R. C., Englert, A. and Bullough, R., Interatomic Potentials and the Simulation of Lattice Defects. Battelle Colloquium at Harrison Hot Springs, B.C. (1971) (ed. by J. Beeler and P. C. Gehlen) (to be published).

6. Cotterill, R.M.J. and Doyama, M., Phys. Rev. 145, 465 (1966); and Calculation of the Properties of Vacancies and Interstitials. N.B.S. Misc. Pub. 287, 47 (1966); and Phys. Letters 25A, 35 (1967).

7. Tsai, D. H., Bullough, R. and Perrin, R. C., J. Phys. C. Solid St. Phys. 3, 2022 (197).

8. Cotterill, R.M.J. and Pedersen, L. B., Interatomic Potentials and the Simulation of Lattice Defects. Battelle Colloquium at Harrison Hot Springs, B.C. (1971) (ed. by J. Beeler and P. C. Gehlen) (to be published).

9. Ashcroft, N. W., Fundamental Aspects of Dislocation Theory. N.B.S. Spec. Pub. 317, 1, 179 (1970) (ed. by J. A. Simmons, R. deWit and R. Bullough).

10. Harrison, W., Interatomic Potentials and the Simulation of Lattice Defects. Battelle Colloquium at Harrison Hot Springs, B.C. (1971) (ed. by J. Beeler and P. C. Gehlen) (to be published).

11. Hardy, J. R. and Flocken, J. W., Crit. Reviews in Solid State Sciences 1, 605 (1970).

12. Squires, G. L. Arkiv. für Fysik, Bd25 (3) 21 (1963).

13. Erginsoy, C., Vineyard, G. H. and Englert, A., Phys. Rev. 133, A595 (1964).

14. Gehlen, P. C., J. Appl. Phys. 41, 5165 (1970).

15. Bullough, R., unpublished work.

16. Bullough, R. and Perrin, R. C., Radiation Damage in Reactor Materials. I.A.E.A., Vienna, II, 233 (1969).

17. Englert, A. and Tompa, H., Tech. Rep. ERA, Union Carbide Ref. 26/69 (1969).

18. Bullough, R. and Hardy, J. R., Phil. Mag. 17, 833 (1968).

19. Sinha, S. K., Phys. Rev. 143, 422 (1966).

20. Girafalco, L. A. and Weizer, V. G., J. Phys. Chem. Solids 12, 260 (1960).

21. Eyre, B. L. and Bartlett, A. F., Phil. Mag. 12, 261 (1965).

22. Masters, B. C., C.E.G.B. Rep.(B), N145 (1964).

23. Eyre, B. L. and Bullough, R., Phil. Mag. 12, 31 (1965).

24. Flocken, J. W. and Hardy, J. R., Phys. Rev. 175, 919 (1968); and Phys. Rev. 177, 1054 (1969).

25. Kanzaki, H., J. Phys. Chem. Solids 2, 24 (1957).

26. Cottrell, A. H., Dislocations and Plastic Flow in Crystals. (Oxford Clarendon Press, 1953).

27. Parsons, J. R., Interatomic Potentials and the Simulation of Lattice Defects. Battelle Colloquium at Harrison Hot Springs, B.C. (1971) (ed. by J. Beeler and P. C. Gehlen) (to be published).

28. Bullough, R. and Perrin, R. C., Proc. of the Albany Conference on Radiation-Induced Voids in Metals (Albany, 1971) (ed. by J. W. Corbett).

29. Bullough, R. and Newman, R. C., Rep. on Progress in Physics 33, 101 (1970).

DISLOCATIONS IN ANISOTROPIC MEDIA

Jens Lothe

Institute of Physics, University of Oslo, Oslo

INTRODUCTION

The role of computer calculations in the problems of solid state physics is the unifying topic at this conference. In dislocation theory there are in particular two types of problems that require computer calculations. The problems of the dislocation core is one important class of problems. Extensive relaxation calculations are necessary to determine from given interatomic potentials the atomic structure in and near the core, the core contributions to the dislocation energy, and the lattice resistance to dislocation motion. Some work on these problems has been done, and the paper of Bullough and Perrin [1] is a good example. The problems of the elastic strains and stresses outside the core is the other important class of problems, and at least for anisotropic media, extensive computer calculations are again needed in the general case. These are the problems we shall consider in this article. Of course, the elastic stresses will to some extent depend on the details of the core structure. However, the details of the core make themselves felt only in terms which decay much faster with distance from the core than the main terms. In this asymptotic sense, the elasticity problem is a well defined separate problem.

In the case of general anisotropy one meets the problem of solving for the roots in a sextic equation, and computer calculations are necessary. However, theoretical work is nevertheless very important for a well organized presentation to the computer of the remainder of the problem so that the computer is used efficiently. One wants a formulation which is as explicit as possible, say a solution which would be explicit if the roots and the eigenvectors of the secular equation were known. In this paper the word explicit will have this meaning. The standard textbook theory for dislocations in anisotropic media is not explicit and does not provide a basis for economical computations. Only recently were more explicit theories developed (Willis [2], Malén and Lothe [3]). After some introductory theory, an account of these recent developments will be given. The close relation with surface wave theory and Greens function theory will also be pointed out.

2. TEXTBOOK THEORY

More detailed expositions can be found in the books of Nabarro [4] and Hirth and Lothe [5]. Here only the most necessary basic theory will be given in the briefest possible way.

Let C be a closed curve in an elastic medium (fig. 1). The curve is given a sense $\underline{\xi}$. Let S be any surface bounded by C. The sign convention for the normal $\underline{n}$ of S is such that $\underline{\xi}$ defines a right hand rotation about $\underline{n}$. Now make a cut over the surface S, give the surface on the negative side of the cut a constant displacement $\underline{b}$ relative to the surface on the positive side, remove or add material after the operation so that the two surfaces are just in contact,

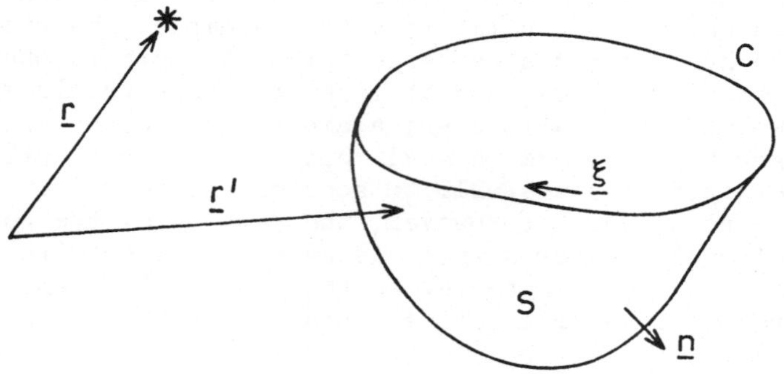

Fig. 1

A dislocation loop C formed by operations on the surface S.

glue together, and then leave the system to relax. This operation produces a dislocation with Burgers vector $\underline{b}$ along C. In a crystal $\underline{b}$ must be a translational vector for lattice register over S to be restored so that only a line defect C results.

More complicated dislocation structures involving dislocation branching can readily be constructed in a similar fashion. At an n-fold node the law of conservation of Burgers vector

$$\underline{b}_1 + \underline{b}_2 + \dots \underline{b}_n = 0 \tag{2-1}$$

must hold.

The straight edge dislocation and the straight screw dislocation are well known special cases (fig. 2a and 2b). In general, the straight dislocation is of mixed character, with an edge component b_e and a screw component b_s, as illustrated in fig. 2c.

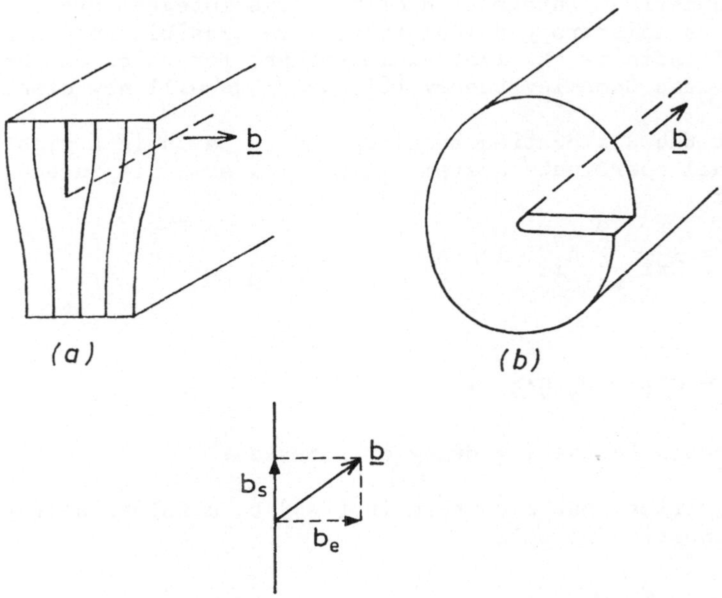

Fig. 2

Edge dislocation, screw dislocation, and dislocation of mixed character respectively.

The dislocations give rise to internal stresses. In terms of the Greens functions of elasticity, G_{km}, Kröner [6] showed that the displacements due to an arbitrary configuration such as in fig. 1 is

$$u_m(\underline{r}) = - \int_S d\, S_j b_i \, c_{ijk\ell} \frac{\partial G_{km}}{\partial x'_\ell} \tag{2-2}$$

Here $c_{ijk\ell}$ are the elastic coefficients. The stresses are obtainable from (2-2) by

$$\sigma_{ij} = c_{ijk\ell} \frac{\partial u_k}{\partial x_\ell} \tag{2-3}$$

However, the formal solution (2-2) and (2-3) is not a very convenient basis for actual calculations. After insertion of (2-2) into (2-3), the surface integral can be transformed to a line integral along the dislocation (Mura [7]), and this is a considerable simplification. However, convenient methods for tabulating the Greens function derivatives in the integrand would be needed for making numerical integration of the line integral practical. For many years anisotropy calculations were feasible only for simple cases of infinite straight dislocations, for which one has the Eshelby-Read Shockley theory [8], which we will now discuss.

Let the dislocation be along the $\underline{\tau}$ axis in a right-handed orthogonal coordinate system $\underline{m}$, $\underline{n}$, $\underline{\tau}$. Assume displacements of the form

$$u_i = \frac{1}{2\pi i} \sum_{\alpha=1}^{6} A_{i\alpha} D_\alpha \log z_\alpha \tag{2-4}$$

where

$$z_\alpha = \underline{m}\cdot\underline{x} + p_\alpha \, \underline{n}\cdot\underline{x} \quad , \tag{2-5}$$

consistent with the 1/r decay of stresses.

Requiring that each term in (2-4) be a solution of the equilibrium condition in bulk,

$$c_{ijk\ell} \frac{\partial^2 u_k}{\partial x_j \partial x_\ell} = 0 \quad , \tag{2-6}$$

one obtains

$$(mm)\underline{A}_\alpha + p_\alpha((nm) + (mn))\, \underline{A}_\alpha + p_\alpha^2\, (nn)\, \underline{A}_\alpha = 0 \tag{2-7}$$

where a matrix (ab) is defined from the vectors $\underline{a}$ and $\underline{b}$ as

$$(ab)_{jk} = a_i c_{ijk\ell} b_\ell \tag{2-8}$$

For (2-7) to be satisfied non-trivially, the condition

$$||(mm) + p_\alpha((nm) + (mn)) + p_\alpha^2 (nn)|| = 0 \tag{2-9}$$

must be fulfilled, which gives six roots p_α occuring in pairs of complex conjugates. The associated eigenvectors are $\underline{A}_\alpha$. Let $p_{-\alpha}$ mean p_α^*. Then we can choose $\underline{A}_{-\alpha} = \underline{A}_\alpha^*$.

However, having solved the secular equation for roots and eigenvectors, it still remains to determine the coefficients D_α. The requirement of real u_k can be stated as

$$D_{-\alpha} = - D_\alpha^* \quad , \tag{2-10}$$

but six more conditions are needed for full determination of D_α. These are the boundary conditions imposed by the dislocation.

Firstly, the discontinuity in the displacement (2-4) must just be the Burgers vector $\underline{b}$,

$$\sum_\alpha \pm \underline{A}_\alpha \underline{D}_\alpha = \underline{b} \tag{2-11}$$

The plus and minus sign is for positive and negative imaginary part in p_α respectively.

Secondly, the displacement (2-4) should not be associated with external forces along the $\underline{\tau}$ - axis. Requiring that the stresses associated with (2-4) transmit no net force through a fictitious cylindrical surface surrounding the dislocation, one derives the condition

$$\sum_\alpha \pm \underline{L}_\alpha D_\alpha = 0 \tag{2-12}$$

where

$$\underline{L}_\alpha = -((nm) + p_\alpha (nn)) \underline{A}_\alpha \tag{2-13}$$

(2-11) and (2-12) provide six conditions, and D_α is now also completely determined.

With p_α, $\underline{A}_\alpha$ and D_α now in principle determined, all quantities of interest can easily be found. For example, from (2-3) and (2-4) the stresses in the plane $\underline{n}\cdot\underline{x}=0$ are found to be

$$\sigma_{ij} = \frac{1}{2\pi i} \sum c_{ijk\ell} A_{k\alpha} D_\alpha (m_\ell + p_\alpha n_\ell) \frac{1}{r} \tag{2-14}$$

If one now considers the dislocation formed by operations at a cut at the surface $\underline{n}\,\underline{x} = 0$, the relative displacement $\underline{b}$ will have to be created against restoring forces $\sigma_{ij}n_j$, and the work done per unit area in building up the dislocation is

$$\tfrac{1}{2}b_i\sigma_{ij}n_j = \frac{1}{4\pi i}\sum_\alpha (\underline{b}\cdot\underline{L}_\alpha)D_\alpha \cdot \frac{1}{r} \,, \tag{2-15}$$

according to (2-13) and (2-14). The dislocation energy factor is defined as

$$E = \frac{1}{4\pi i}\sum_\alpha (\underline{b}\cdot\underline{L}_\alpha)D_\alpha \tag{2-16}$$

The work done per unit length in the strip between r_1 and r_2 is thus

$$W/L = E\int_{r_1}^{r_2} \frac{dr}{r} = E\log\left(\frac{r_2}{r_1}\right) \tag{2-17}$$

and this is then the elastic dislocation energy per unit length between two coaxial cylinders of radii r_1 and r_2.

The above constitutes the Eshelby-Read-Shockley theory for straight dislocations. This theory is not an explicit theory of the form one wishes, because the D_α are determined implicitly by the conditions (2-11) and (2-12). Thus, even as regards calculating just straight dislocation data by computers for a general dislocation direction, the theory was not in a good form. The above scheme found extensive application only for dislocations in high symmetry directions.

3. CURVED DISLOCATIONS IN TERMS OF STRAIGHT ONES

One notices that in the preceding section, curved dislocations are described by Greens functions and straight dislocations by a separate theory that is simple in special cases. However, a more uniform description would be desirable. One could use Greens functions throughout, characterizing the elastic properties of the medium by its respons to point forces. Or one could use a Fourier representation throughout. Finally, one can imagine that one characterizes the elastic properties of the medium by its response to straight dislocations and construct a general theory on that basis. This in itself of course does not solve the problem of more explicit solutions. However, from the standpoint of dislocation theory, it is the most informative representation. Also, as we shall see in the next section, an improved theory for straight dislocations is possible which then also immediately makes the curved dislocation case more explicit.

The first steps towards a straight dislocation representation for curved dislocations was made by the present author with some simple considerations on dislocation bends [9]. Brown [10] then generalized this work very ingeniously and in an important way and arrived at an expression for the stresses from a planar loop in the plane of the loop in terms of straight dislocation parameters. We shall give Browns result without proof.

Consider a planar loop of Burgers vector $\underline{b}$. Also consider a straight dislocation in the same plane and of orientation Θ, with the same Burgers vector $\underline{b}$. The stress from the straight dislocation at a point M in the plane a distance d from the dislocation is

$$\sigma_{ij} = \Sigma_{ij}(\Theta)/d \tag{3-1}$$

The stress factor $\Sigma_{ij}(\Theta)$ is a function of Burgers vector and orientation. The anisotropy is contained in $\Sigma_{ij}(\Theta)$. Browns theorem now says that in terms of straight dislocation stress-factors the stress from the planar loop at a point P in the plane is

$$\sigma_{ij} = \tfrac{1}{2}\oint \frac{(\Sigma_{ij}(\Theta) + \Sigma_{ij}''(\Theta))\sin(\Theta-\alpha)}{r^2}\, d\ell \tag{3-2}$$

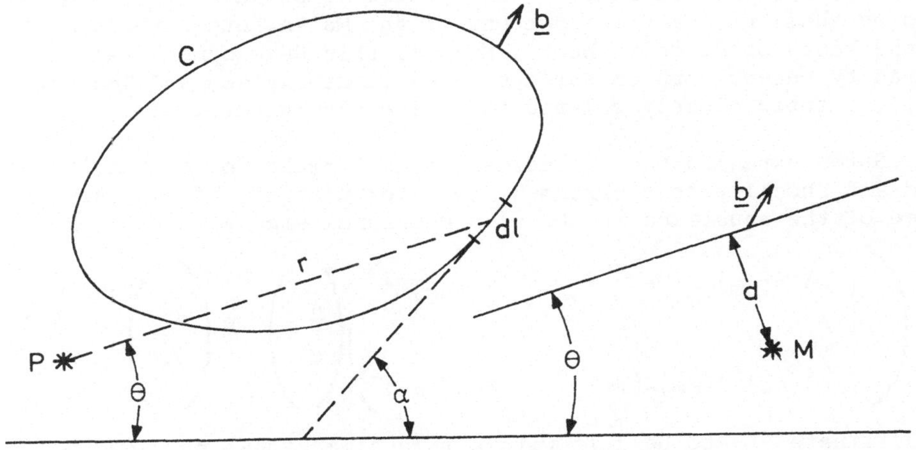

Fig. 3

Coordinates for Browns formula.

It is noted that $\sum_{ij}(\Theta)$ refers to a straight dislocation parallel with the vector between P and $d\ell$ and not to a dislocation parallel with $d\underline{\ell}$. The derivatives in the integrand refer to the Θ-dependence.

Browns development was for the planar case. Indenbom and Orlov [11] have given an elegant analysis that generalizes Browns result to the threedimensional case. Just as in the planar case, the integrands are entirely determined by straight dislocation stress-factors and their angular derivatives. Browns formula suffices here as an illustration of what has been accomplished. The Brown-Indenbom-Orlov theory in the dislocation representation corresponds to Muras [7] theory in the Greens function representation.

4. EXPLICIT SOLUTIONS

The Brown-Indenbom-Orlov theory was developed with the aim of making anisotropic dislocation calculations more tractable. However, as observed by Willis [2], it was still an implicit theory as far as the possibility of actual calculations was concerned. Willis made a fresh start on the problem with Fourier transform methods and did indeed achieve solutions explicit in the sense of this paper. However, at about the same time (Malén and Lothe [3]), it became apparent that Stroh [12] already in 1962 had extended the Eshelby-Read-Shockley theory for straight dislocations to an explicit theory, and that Stroh's theory could be extended to give explicitly also the angular derivatives needed in the Brown-Indenbom-Orlov theory. The two schemes can be said to be largely equivalent. Here we shall give a brief account of the Malén-Lothe development on the basis of Stroh's theory. Much of this development was inspired by recent work on surface waves (Ingebrigtsen and Tonning [13]), a topic closely related to dislocation theory.

Stroh expanded the threedimensional formulation of Eshelby, Read and Shockley to a sixdimensional formulation. If one uses three of the equations in the sixdimensional eigenvalueproblem

$$-\begin{pmatrix} (nn)^{-1}(nm) & (nn)^{-1} \\ (mn)(nn)^{-1}(nm)-(mm) & (mn)(nn)^{-1} \end{pmatrix} \begin{pmatrix} \underline{A} \\ \underline{L} \end{pmatrix} = p \begin{pmatrix} \underline{A} \\ \underline{L} \end{pmatrix} \qquad (4\text{-}1)$$

to eliminate $\underline{L}$ to be left with a threedimensional problem in $\underline{A}$ only, one returns identically to the Eshelby-Read-Shockley secular equation (2-7),

$$(mm)\underline{A} + p((nm) + (mn))\underline{A} + p^2(nn)\underline{A} = 0 \,. \qquad (4\text{-}2)$$

Also, three of the equations in (4-1) are just the same as the definition of $\underline{L}$ in (2-13).

Thus, the $\underline{A}_\alpha$ and $\underline{L}_\alpha$ introduced in section 2 are the components of the sixdimensional eigenvectors

$$\xi_\alpha = \begin{pmatrix} \underline{A}_\alpha \\ \underline{L}_\alpha \end{pmatrix} \qquad (4\text{-}3)$$

of the sixdimensional eigenvalueproblem

$$N\xi_\alpha = p_\alpha \xi_\alpha \quad , \qquad (4\text{-}4)$$

where N is brief for the sixdimensional matrix on the left hand side of (4-1). The $\underline{A}_\alpha$ and $\underline{L}_\alpha$ are the eigensolution displacement and force components respectively.

The matrix N is not symmetric, and therefore the different ξ_α are not simply orthogonal. However, enough symmetry is present to lead to useful orthogonality relations. Because of the general symmetries in $c_{ijk\ell}$, the transpose of (ab) is (ba). Thus, the nondiagonal block elements in N are symmetric, and one diagonal block element is the transpose of the other diagonal block element, and this is enough to assure the orthogonality relation

$$\underline{A}_\alpha \cdot \underline{L}_\beta + \underline{A}_\beta \cdot \underline{L}_\alpha = 0 \quad , \alpha \neq \beta \qquad (4\text{-}5)$$

It is convenient to normalize so that

$$\underline{A}_\alpha \cdot \underline{L}_\beta + \underline{A}_\beta \cdot \underline{L}_\alpha = \delta_{\alpha\beta} \qquad (4\text{-}6)$$

Applying this orthogonality relation to the boundary conditions (2-11) and (2-12),

$$\sum_\alpha \pm \underline{A}_\alpha D_\alpha = \underline{b} \qquad (4\text{-}7)$$

$$\sum_\alpha \pm \underline{L}_\alpha D_\alpha = 0 \qquad (4\text{-}8)$$

the explicit solution

$$D_\alpha = \pm \underline{b} \cdot \underline{L}_\alpha \qquad (4\text{-}9)$$

is readily obtained. With an explicit D_α, the stress- and energy factor expressions also become explicit. For the energy factor E, (2-16), one now simply has

$$E = \frac{1}{4\pi i} \sum_\alpha \pm (\underline{b} \cdot \underline{L}_\alpha)^2 \qquad (4\text{-}10)$$

The above results are all contained in Stroh's paper of 1962. It was not till very recently that it was recognized that Stroh actually had an explicit solution to the Eshelby-Read-Shockley problem.

In connection with the Brown-Indenbom-Orlov theory, explicit expressions for the angular derivatives of the straight dislocation parameters are also needed. These can be obtained by applying standard perturbation theory to Stroh's eigenvalue problem (Malén and Lothe, [3]).

The unperturbed problem is

$$N^o \xi_\alpha^o = p_\alpha^o \xi_\alpha^o \tag{4-11}$$

The orthogonality relations (4-6) can be written as

$$\xi_\alpha^o T \xi_\beta^o = \delta_{\alpha\beta} \quad , \tag{4-12}$$

where in block notation

$$T = \begin{pmatrix} 0 & 1 \\ 1 & 0 \end{pmatrix} \tag{4-13}$$

The operator N depends on the direction $\underline{\tau}$ of the dislocation line. If the orientation is changed by a small rotation $\delta\Theta$ about the $\underline{n}$ vector, then

$$N = N^o + \frac{\partial N}{\partial \Theta} \delta\Theta \tag{4-14}$$

and in the new direction we have the eigenvalue problem

$$N\xi_\alpha = p_\alpha \xi_\alpha \tag{4-15}$$

The solution of (4-15) in terms of the eigenvectors and eigenvalues of (4-11) just gives the angular derivatives, and by standard perturbation theory

$$\partial p_\beta / \partial\Theta = (\partial N/\partial\Theta)_{\beta\beta} \tag{4-16}$$

and

$$\partial \xi_\beta / \partial\Theta = \sum_{\substack{\alpha=1 \\ \alpha\neq\beta}}^{6} \frac{(\partial N/\partial\Theta)_{\alpha\beta}}{p_\beta - p_\alpha} \xi_\alpha \tag{4-17}$$

The matrix elements

$$(\partial N/\partial\Theta)_{\alpha\beta} = \xi_\alpha T (\partial N/\partial\Theta) \xi_\beta \tag{4-18}$$

are readily explicitly calculated to be

$$(\partial N/\partial\Theta)_{\alpha\beta} = \underline{A}_\alpha \left[-((\tau m)+(m\tau)) - p_\alpha (n\tau) - p_\beta (\tau n) \right] \underline{A}_\beta \quad , \tag{4-19}$$

and the components of (4-16) and (4-17) thus explicitely give angular derivatives for p_α, $\underline{A}_\alpha$ and $\underline{L}_\alpha$. Using these in the expressions for the stressfactors and the energy factors, explicit expressions for the angular derivatives of the latter immediately follow. Higher order angular derivatives are also easily obtained. As an example of the type of formulae obtained, I give the expression for $\partial E/\partial\theta$,

$$\partial E/\partial\theta = \sum_{\alpha=1}^{6} \pm \sum_{\substack{\beta=1\\ \beta\neq\alpha}}^{6} \frac{1}{\pi i(p_\alpha - p_\beta)} b_i L_{i\alpha} \{\underline{A}_\beta(-(\tau m) - p_\alpha(\tau n))\underline{A}_\alpha\} L_{s\beta} b_s \quad (4\text{-}20)$$

The above completes the program of constructing an explicit theory on the basis of the Brown-Indenbom-Orlov representation.

5. RELATION TO OTHER PROBLEMS

The surface wave problem is closely related to the dislocation problem, as already noted by Stroh [12]. Consider surface waves at the surface with normal $\underline{n}$, travelling in the $\underline{m}$ direction. The half-infinite solid occupies the space $\underline{n}\cdot\underline{x}>0$. Then similar to equation (2-4) for the dislocation, we can assume a solution of the type

$$u_i = \sum_{\alpha=1}^{3} A_{i\alpha} D_\alpha \quad \exp\,[ik(z_\alpha - vt)] \quad (5\text{-}1)$$

Here we sum only over the three roots with positive imaginary part, since we only want waves which decay as we move into the solid. The equilibrium condition in bulk is now

$$c_{ijk\ell} \frac{\partial^2 u_k}{\partial x_j \partial x_\ell} = \rho \frac{\partial^2 u_k}{\partial t^2} \quad (5\text{-}2)$$

(5-1) and (5-2) lead to a secular equation identical to (2-7), except that

$$(mm) - \rho v^2 I \quad ,$$

where I is the unit matrix, replaces (mm). If we consider moving dislocations, moving in the $\underline{m}$ direction with velocity v, this would also be the change in the dislocation theory. Thus, the vectors $\underline{A}_\alpha$ and the eigenvalues p_α are the same in the Rayleigh wave and the moving dislocation problem. Expanding to a sixdimensional formulation, the vectors $\underline{L}_\alpha$ are also the same and related to $\underline{A}_\alpha$ by (2-13). However, the condition on D_α is different in the surface wave theory and can only be satisfied for a certain v, the Rayleigh wave velocity. The condition is that the surface is force free

$$\sum_1^3 \underline{L}_\alpha D_\alpha = 0 \quad , \tag{5-3}$$

and there is no condition on the amplitudes. (5-3) is the requirement that the three $\underline{L}_\alpha$ are linearly dependent, i.e.

$$||L_{i\alpha}|| = 0 \quad , \tag{5-4}$$

or

$$||\sum_{\alpha=1}^3 L_{i\alpha} L_{j\alpha}|| = ||L_{i\alpha}||^2 = 0 \quad . \tag{5-5}$$

The $\underline{L}_\alpha$ are functions of v, and (5-5) is satisfied only for the Rayleigh velocity. One notices the similarity between the components in (5-5) and the energy factor for straight dislocations, (4-10). For moving dislocations, (4-10) would in fact be the Lagrange factor, and one can show that the Lagrange factor vanishes for dislocations with a Burgers vector contained in the plane of elliptic polarization of the surface wave and moving with the Rayleigh wave velocity [14].

A complete derivation of the sixdimensional theory for Greens functions is too lengthy to be included here. We shall just give the result, as derived by Malén [15]. Consider the Greens function $G_{ij}(\underline{r})$. Construct the dislocation coordinate system $\underline{m}, \underline{n}, \underline{\tau}$ of section 2, with the $\underline{\tau}$ axis parallell with $\underline{r}$. Then, in terms of the eigenvectors $\underline{A}_\alpha$ for these coordinates,

$$G_{ij}(\underline{r}) = -\frac{i}{4\pi r} \sum_{\alpha=1}^6 \pm A_{i\alpha} A_{j\alpha} \quad , \tag{5-6}$$

when the vectors are normalized according to (4-6).

Thus, dislocation calculations will quite directly give Greens functions as a side benefit.

A particularly important aspect of this theory for Greens functions is that it is suited for derivation of explicit expressions for Greens function derivatives, by the very same technique as we used for obtaining angular derivatives for straight dislocation parameters. With explicit expressions for Greens function derivatives, the Mura [7] line integral for curved dislocations becomes an equally convenient basis for computation as the Brown-Indenbom-Orlov integrals.

For a more extensive discussion of the interrelation between the various problems and the various ways of treating them, I refer to Malén [15].

6. CLOSING COMMENTS ON THEORY

The foregoing theory works well when applied directly to the anisotropic solid. However, it is not well suited for studying anisotropy as a perturbation from isotropy. Weird degenerate behavior is encountered in the limit of isotropy. The matrix N becomes nonsemisimple with just two roots $p = \pm i$, and it does not possess a complete set of eigenvectors. As one approaches isotropy, the components $A_{i\alpha}$ and $L_{i\alpha}$ increase without limit.

However, it is possible to study the approach to isotropy by means of certain sum rules, which can readily be deduced from the general theory. Two of these are

$$(nn)^{-1} = \sum_\beta p_\beta \underline{A}_\beta \underline{A}_\beta \tag{6-1}$$

and

$$(mm)^{-1} = \sum_\beta \frac{1}{p_\beta} \underline{A}_\beta \underline{A}_\beta \quad , \tag{6-2}$$

where on the right hand side we have a dyadic notation for matrices.

With a small anisotropy present,

$$p_\beta = \pm i + \delta p_\beta \tag{6-3}$$

To first order in δp_β, (6-1) and (6-2) become

$$(nn)^{-1} = - i \sum \pm \underline{A}_\beta \underline{A}_\beta - \sum \delta p_\beta \underline{A}_\beta \underline{A}_\beta \tag{6-4}$$

and

$$(mm)^{-1} = - i \sum \pm \underline{A}_\beta \underline{A}_\beta + \sum \delta p_\beta \underline{A}_\beta \underline{A}_\beta \tag{6-5}$$

which can be solved to give

$$\sum \pm \underline{A}_\beta \underline{A}_\beta = \tfrac{1}{2} i ((nn)^{-1} + (mm)^{-1}) \tag{6-6}$$

and

$$\sum \delta p_\beta \underline{A}_\beta \underline{A}_\beta = \tfrac{1}{2} (mm)^{-1} - (nn)^{-1}) \tag{6-7}$$

The right hand sides of (6-6) and (6-7) are well behaved. One can show that in the limit of isotropy

$$(nn)^{-1} + (mm)^{-1} \rightarrow \frac{1}{\mu} \{2 \underline{\tau}\, \underline{\tau} + \frac{\lambda+3\mu}{\lambda+2\mu} \underline{n}\, \underline{n} + \frac{\lambda+3\mu}{\lambda+2\mu} \underline{m}\, \underline{m} \} \tag{6-8}$$

Here μ is the shear modulus, and λ is Lamés constant.

Thus, from (5-6) and (6-6), the isotropic Greens function must be

$$\tilde{G}(\underline{r}) = \frac{1}{8\pi\mu r} \{2 \underline{\tau}\,\underline{\tau} + \frac{\lambda+3\mu}{\lambda+2\mu} \underline{n}\,\underline{n} + \frac{\lambda+3\mu}{\lambda+2\mu} \underline{m}\,\underline{m} \} , \qquad (6\text{-}9)$$

which indeed checks with isotropic theory.

Further work on the degeneracy problem is in progress (Nishioka and Lothe, [16]).

Barnett and Swanger [17] have very recently demonstrated that it is possible to reformulate the theories explicit in the roots of the sextic and the associated eigenvectors to a theory explicit in certain integrals, which, if performed formally by the residue theorem, would again involve the roots of the sextic. Barnett and Swangers method is numerical integration of these integrals instead of numerical solution for the roots of the sextic. Although one may say that both ways are equally explicit (or implicit), it may be that Barnett and Swangers formulation is a significant further step in the direction of more convenient numerical methods in the anisotropy problem. Barnett and Swangers methods are applicable both to dislocations and Greens function derivatives. Also, isotropy does not present a special problem in the Barnett-Swanger formulation.

7. APPLICATIONS

The anisotropy problem for dislocations is now in such a form that computer calculations for quite complex dislocation structures have become feasible. One of the simpler problems but an important one is the stability of straight dislocations with respect to zig-zagging (Head [18]). The new methods should make numerical investigations on this problem much simpler, and work on this problem is being done at the present (Pettersson and Malén [19], Barnett and Swanger [17]). Another, and related problem, is elastic jog-jog interactions. Work on this problem for the case of dislocations in ionic crystals is nearly completed in Oslo (Brækhus and Lothe [20]). Pettersson [21] is presently extending the theories for dissociated nodes, which are important for stacking fault energy determinations, to include anisotropy effects. This is a complex problem, and would hardly be feasible without the recent progress in theoretical formulation. An equally important problem would be the stacking fault tetrahedron. In conclusion, the progress in formulation has already led to significant applications, and many more important applications remain to be done.

8. CONCLUSION

Stroh, Willis and Malén and Lothe have developed explicit theories for dislocations in anisotropic media, which make anisotropy computer calculations easier. These theories are explicit in the sense that the basic problem is solving for the roots in the sextic equation and that the procedure from there on is simple and explicit. Barnett and Swangers method of replacing the problem of solving for the roots in the sextic equation with the problem of performing certain numerical integrals, may be a further significant contribution towards efficient procedures. Anisotropy computer calculations for quite complex dislocation structures are now feasible, and many important applications can be expected. The problems of surface waves and Greens functions in anisotropic media are analogous with the dislocation problem. Dislocations, surface waves, and Greens functions can be treated in a unified manner, and unified computer program for all three types of problems would seem to be the best way to organize work in this field.

REFERENCES

[1] R. Bullough and R.C. Perrin, in "Dislocation Dynamics", McGraw-Hill, New York, 1968, p.175.

[2] J.R. Willis, Phil.Mag. 21, 931 (1970).

[3] K. Malén and J. Lothe, phys.stat.sol. 39, 289 (1970).

[4] F.R.N. Nabarro, "Theory of Crystal Dislocations", Oxford University Press, 1967.

[5] J.P. Hirth and J. Lothe, "Theory of Dislocations", McGraw-Hill, New York, 1968.

[6] E. Kröner, Ergeb.angew.Math. 5 (1958).

[7] T. Mura, Phil.Mag. 3, 625 (1963).

[8] J.D. Eshelby, W.T. Read and W. Shockley, Acta Met. 1, 251(1953).

[9] J. Lothe, Phil.Mag. 15, 353 (1967).

[10] L.M. Brown, Phil.Mag. 15, 363 (1967).

[11] V.L. Indenbom and S.S. Orlov, Soviet Phys. (Cryst) 12, 849 (1967/68).

[12] A.N. Stroh, J. Math.Phys. 41, 77 (1962).

[13] K.A. Ingebrigtsen and A. Tonning, Phys.Rev. 184, 942 (1969).

[14] K. Malén and J. Lothe, phys.stat.sol. (b) 43, K139 (1971).

[15] K. Malén, phys.stat.sol. (b) 44, 661 (1971).

[16] K. Nishioka and J. Lothe, unpublished work.

[17] D.M. Barnett and L.A. Swanger, to be published.

[18] A.K. Head, phys.stat.sol. 19, 185 (1967).

[19] B. Pettersson and K. Malén, Report AE-426 (1971), AB Atomenergi, Studsvik, Nyköping.

[20] J. Brækhus and J. Lothe, unpublished work.

[21] B. Pettersson, to be published as AE-report, AB Atomenergi, Studsvik, Nyköping.

LIST OF CONTRIBUTORS

Balkanski, M., University of Paris 293, 367

Baym, G., University of Illinois 267

Berko, S., Brandeis University 59

Biermann, L., Max-Planck Institute for Physics and Astrophysics, Munich 257

Bilz, H., Technical University of Munich 309

Börner, G., Max-Planck Institute for Physics and Astrophysics, Munich 261

Bross, H., University of Munich 143

Bullough, R., Atomic Energy Research Establishment, Harwell 413

Calais, J.-L., Uppsala University 253

Cardona, M., Brown University; presently at Max-Planck Institute for Solid State Research, Stuttgart 7

Christensen, N.E., Technical University of Denmark, Lyngby . 155

Cowley, R.A., University of Edinburgh 299

Dalton, N.W., IBM Research, San Jose; presently at Atomic Energy Research Establishment, Harwell 81, 113, 183

Eastman, D.E., IBM Research, Yorktown Heights 23

Haensel, R., University of Hamburg and German Electron-Synchrotron (DESY), Hamburg 43

Hedin, L., University of Lund 233

Henderson, D., IBM Research, San Jose; formerly at University of Waterloo 175

Herman, F., IBM Research, San Jose . . . . v, 245

Horner, H., Institute for Solid State Research, Jülich . . 351

Lidiard, A.B., Atomic Energy Research Establishment, Harwell 363, 385

Lothe, J., University of Oslo 425

Löwdin, P.-O., Uppsala University 191

Lundqvist, B.I., Chalmers University of Technology 219

Lundqvist, S., Chalmers University of Technology 219

Koehler, T.R., IBM Research, San Jose 289, 339

Madelung, O., University of Marburg 3

March, N.H., University of Sheffield 205

Norgett, M.J., Atomic Energy Research Establishment, Harwell 385

Ortenburger, I.B., IBM Research, San Jose 179

Pick, R.M., Center for Nuclear Studies, Saclay and University of Paris 325

Rössler, U., University of Marburg 161

Rudge, W.E., IBM Research, San Jose 179

Schneider, T., IBM Research, Zurich 99

Schreiber, D.E., IBM Research, San Jose 183

Schwartz, K., IBM Research, San Jose; presently at University of Vienna 245

Schweitzer, P., IBM Germany, Sindelfingen v

Sonntag, B., University of Hamburg and German Electron-Synchrotron (DESY), Hamburg 43

Sperber, G., Uppsala University 253

Stoddart, J.C., University of Sheffield 205

Stöhr, H., University of Munich 143

Stoll, E., IBM Research, Zurich 99

Treusch, J., University of Dortmund 85

Weger, M., Hebrew University, Jerusalem 59

Williams, A.R., IBM Research, Yorktown Heights 23

Yu, P.Y., Brown University; presently at University of California, Berkeley 7

SUBJECT INDEX

A

Adiabatic approximation 310
Alkali halides 46, 50, 55
Alternant molecular orbital (AMO) method 201, 253
Amorphous semiconductors 96, 175, 181
 atomic arrangements 175, 179
 germanium and silicon 175
 radial distribution function 175, 179
 random tetrahedral network 176, 179
 selenium 95
Anharmonicity 302, 339ff, 351ff
 computational aspects of 339ff
 cubic anharmonic correction 341, 343, 344, 357
 effect on defects 279, 381
 effect on neutron scattering 301, 350
 in quantum crystals 351
 leading anharmonic corrections 341ff, 357
 multiple scattering effects 305
 normal mode coupling 302
 theory of 351
 two-phonon bound state 298, 306
APW (see band calculations)
Astrophysics 257ff
 black dwarf 257
 black hole 258, 267
 dust particles 258
 interstellar gas 258
 solid state physics, influence on 257, 258, 261, 272, 283
 stellar evolution 258, 267, 275
 supernova event 259, 262

B

Band calculations
 APW 34, 35, 63, 68ff
 complex 96
 empirical 85ff
 KKR 161ff
 modified APW 143ff
 modified OPW 113ff
 pseudopotential 99ff, 180
 relativistic APW 155ff
 relativistic OPW 113ff
Born-Oppenheimer approximation 294, 386
Born-von Karman theory 296
Brillouin theorem 197
Brown-Indebom-Orlov theory of curved dislocations 432

C

Cauchy relation 390, 410
CdTe-HgTe 170
Cellular method 114
Charge density
 electron correlation, effect on 205ff
 valence electron contour map 94, 187
Chemical potential 100, 220
Clausius-Mosotti relation 390
Cohesive energy 105, 192, 221, 253

Compton
potential 216, 217
scattering 206, 300
Computer simulation of defects 386, 413
energy minimization methods 397ff, 410
relaxation procedure 418
Copper
3d states 23, 123, 127, 131, 151
4s states 135, 151
dislocations in 419
halides 168
MAPW band calculation 147
optical absorption 27
photoemission 27
point and line defects in 413
Correlation effects 191ff
charge density, influence on 205ff
correlation hole 208
coulomb hole 222
energy density 212
in atoms 193
momentum density, influence on 205ff
perturbation aspects 194
self energy 222
symmetry aspects 199
Critical points 8, 161, 171
Cuprous halides 168

D

Defect vibrations 369, 370
Defects (see point defects, dislocations)
Deformable bond model 396
Deformable dipole model 392, 397
Density matrix 191, 206, 235
Dirac equation 136
Dislocations 363
complex structures 438
core of 425
curved 430ff
elastic stresses and strain outside core 425
in anisotropic media 425ff
in copper 419
straight 427
theory, explicit 425, 432
theory for almost isotropic solid 437
theory, standard 426
Disordered materials (see amorphous semiconductors)
Dynamical matrix 311, 341, 354
in microscopic theory 295, 327

E

Eshelby-Read-Shockley theory of straight dislocations 428
Europium chalcogenides 36
Exchange effects
density gradient expansion 210, 239, 243, 247
exchange hole 208
inhomogeneity correction 247
local exchange-correlation potential 219ff, 233ff
screened exchange 222
statistical exchange approximations 199, 245ff
$X\alpha$ method 199, 245
$X\alpha\beta$ method 199, 247
Excitons 50ff, 172

F

Fluctuation-dissipation theorem 355
Force constants
from experiment 293, 311
Frenkel defects 407

G

Gallium arsenide 7
Germanium 7, 175, 334
3d states 131
polytypes 179
Gilat-Raubenheimer method 25, 27, 179
Gold 31, 155
Green's function
defect 396
elasticity 428ff, 436

matrix 369
phonon 301ff, 353, 355

H

HADES program 402, 404
Hard core problem 352
Hohenberg-Kohn-Sham scheme 219, 222, 230, 239, 335
Homopolar crystals
bond charges in 319
overlap charges in 317
phonons in 319ff
Huang theory 296

I

Inelastic scattering 299ff
Infrared spectroscopy 368
Ionic solids
defects in 385ff, 402
dielectric properties of 390
infrared absorption 379
localized modes in 378
phonons in 311ff
potential for 390, 409
rare gases in 403
Inverse dielectric function 327, 328
Iron, point and line defects in 413ff

K

KKR (see band calculations)

L

Lattice dynamics, anharmonic 339ff, 351ff
Linear response theory 233, 238
Lithium 99ff, 127, 147, 253
band structure 110
cohesive energy 105
effective pair potential 107
phase transition 109, 111
phonon dispersion 107
total energy 109
Localized modes 363, 364, 367
in ionic crystals 378, 379
observation of 367
Raman scattering from 379ff

M

Mercury chalcogenides 169
Microscopic theory of phonons 325ff
in insulators 330, 336
in metals 330, 333
in semiconductors 334
in transition metals 336
Model Hamiltonian 114
Model potentials
Heine-Abarenkov 88, 99
Heine-Animalu 90
Modified APW band calculations 143ff
Modified OPW method 113ff
Molybdenum 155, 158
Momentum density
correlation effects, influence on 205ff

N

N-representability problem 235
Neutron scattering
determination of eigenvectors by 300
from harmonic crystals 299
inelastic 299ff
interference effects in 303
(see also anharmonicity)
Neutron stars 267ff
degenerate neutron, proton, and electron fluids 269
densities of 268, 278
equation of state of 258, 270, 272, 274
ground state of 270, 278
hadronic core of 269
masses of 268, 276
models of 274
neutron drip regime 271
nuclear phase transitions 271

properties in enormous magnetic fields 273
radii of 268, 277
solid crust 258, 269
stability of 274
superfluidity and superconductivity in 279ff
temperatures of 269
transition from nuclear to fluid phase 272
Nickel 23, 127
Niobium 127

O

Optical properties 5, 26
amorphous selenium 95
copper 27
due to defects 370
europium sulfide 38
gold 31, 155
polytypes of germanium 179
spectral calculation 26
tellurium 93
(see also photoemission, spatial dispersion, and synchrotron radiation)
OPW (see band calculations)
Overlap and effective charges 317

P

Pair defects 366, 372
Palladium hydride 34
Photoemission 5, 23ff, 222, 231
chemisorbed CO on nickel 38ff
copper d bands 27ff
energy distribution 26, 32, 34
europium sulfide 36ff
gold 31ff, 155
momentum conservation 34
palladium hydride 34ff
relation of theory to experiment 24
silver 158
ultraviolet spectroscopy 23, 38
zinc oxide 169
zonal integration 27
Point defects 363, 367, 385
energies, calculation of 395
impurity modes 368
in copper 419
interatomic potentials for 366, 414ff
interstitial aggregate in alpha-iron 418
in ionic solids 385, 402
normal modes of crystals with defects 369
optical absorption by 370
pairing of 368, 372
rare gases in ionic crystals 404ff
Schottky defects 402, 407
Polarizable point ion model 391, 397, 401, 404, 407
Polytypes
germanium 179
silicon carbide 90
Positron annihilation 5, 59ff
angular correlation of emitted gamma rays 59, 65, 68
β-W system, properties of 61ff
computer simulation of experiments 74
corrections to V_3Si APW band calculation 71
coupled chain model for β-W system 63, 70
density of states of V_3Si 70ff
drawbacks of method 60
electronic wave-function information 60
electron momentum distribution 59ff
experimental measurements on V_3Si 64ff
Fermi surface determination 60, 64, 67, 73
linear chain model for β-W system 61
principles of method 59
simplified treatment of V_3Si band structure 68, 72
successful applications of method 61

Potassium halides 167
Potentials
exchange-correlation 223
for iron and copper 414
ionic crystals, models for 389
Kohn-Sham 220, 236
momentum dependent 225
non-local 222
Slater 247
spin dependent 229
Pseudopotential method 85
Austin-Heine-Sham 86
cancellation theorem 86
form factors 87, 88
non-local 91
Phillips-Kleinman 86, 99, 111
phonon theory 333
pseudo-wave function 86
relativistic corrections 91
Pulsars 258, 261ff
elasticity of crust, influence on 264, 284
electrical and magnetic properties of 262
observational evidence for 261ff
plastic flow in crust 285
radiation from 263
rotational dynamics 259, 261, 266, 281
speed-up 263, 266, 284
starquakes 258, 264, 265, 284
timing 269, 282

Q

Quasi-harmonic approximation 309, 340, 345, 355, 386, 410

R

Raman scattering 379
RAPW (see band calculations)
Rare gas solids 46, 50, 166, 351
Relativistic APW (see band calculations)
Relativistic OPW (see band calculations)
Resonant electronic polarization 320
Rigid ion model 311

S

Selenium 95
Self-consistent harmonic approximation 345, 347
Self-consistent phonon theory 345ff, 351
computational aspects of 346
application of 347
Shell model 295, 311, 392, 397, 401, 404, 407, 409
breathing 314, 336
extended 313
Silicon 7, 175, 334, 371
Silver 155
Spatial dispersion 5, 7ff
birefringence below fundamental absorption edge 11, 17
dielectric constant tensor 10
E_o, $E_o + \Delta_o$ region 8, 9, 12
E_1, $E_1 + \Delta_1$ region 8, 10, 15
E_2 region 8, 10, 16
electro-optic tensor 9
experimental results for Ge, GaAs, and Si 17ff
microscopic theory 11
model band structure 8
Penn model 10, 16
perturbation effects 8, 12
phenomenological theory 10
piezobirefringence 9
pressure dependence of optical constants 9
Raman tensor 9
temperature dependence of optical constants 9
Spin-orbit coupling 91
Superfluidity and superconductivity in neutron stars 279ff
cooling rates, influence on 283
flux diffusion time 282
neutron superfluidity 264, 280
proton superconductivity 282

quantized vortices 281
Tkachenko modes 281
Surface waves 435
Susceptibility
magnetic 235, 241
pauli 240
self-consistent field theory 328
spin 234, 244
Symmetry 199, 201
dilemma 200
group 200
Synchrotron radiation 5, 43ff
alkali halides 46, 50, 55
autoionization 52, 53
configuration interaction 52, 54
continuum absorption 48ff
coulomb interaction (electron-hole) 50ff
density of states 46, 48
exchange interaction 52
exciton effects 50ff
inner shell excitation 45, 52
interband transitions 46, 52
lanthanides 50, 52, 53
lifetime broadening 45
metals 50, 55
multiple excitations 54
multiplet splitting 52
optical absorption spectrum 46, 48
semiconductors 50
simultaneous electron-plasmon excitations 55
soft X-ray radiation sources 43, 45
solid rare gases 46, 50, 54
synchrotron and storage ring facilities 43, 45
transition metals 53
vacuum ultraviolet radiation sources 43, 45

T

Tellurium 93
Thallous halides 167
Tungsten (β-W) 61

V

V_3Si 59ff
Vanadium 155, 158
Vibrational modes 367ff
Virial theorem 192, 198

W

Wave operator 195, 196, 198
White dwarfs 267ff
densities of 278
equation of state of 258, 279
ground state of 278
heat capacity of 283
masses of 267, 276
models of 274
nuclear phase transitions 276
radii of 267, 277
solid core of 270
stability of 276
temperature of 279
Wigner's formula 193, 194
Wigner-Seitz sphere 104, 242, 253

X

X-ray scattering
inelastic 299ff
interference effects in 303

Z

Zinc oxide 127